国外优秀数学著作
原　版　系　列

Finsler Geometry of Hadrons and Lyra Geometry(Cosmological Aspects)
—Finsler Geometry of Hadron Structure and Lyra Geometry (Topological Defects)

强子的芬斯勒几何和吕拉几何（宇宙学方面）
——强子结构的芬斯勒几何和吕拉几何（拓扑缺陷）

（英文）

[印] 萨提亚·桑卡尔·德（Satya Sankar De）
[印] 法鲁克·拉哈曼（Farook Rahaman）
著

HITP
哈爾濱工業大學出版社
HARBIN INSTITUTE OF TECHNOLOGY PRESS

黑版贸审字 08 - 2019 - 180 号

图书在版编目(CIP)数据

强子的芬斯勒几何和吕拉几何:宇宙学方面:强子结构的芬斯勒几何和吕拉几何:拓扑缺陷 = Finsler Geometry of Hadrons and Lyra Geometry(Cosmological Aspects):Finsler Geometry of Hadron Structure and Lyra Geometry (Topological Defects):英文/(印)萨提亚·桑卡尔·德(Satya Sankar De),(印)法鲁克·拉哈曼(Farook Rahaman)著.—哈尔滨:哈尔滨工业大学出版社,2023.3
ISBN 978-7-5767-0200-2

Ⅰ.①强… Ⅱ.①萨… ②法… Ⅲ.①芬斯拉空间 - 研究 - 英文②拓扑空间 - 研究 - 英文 Ⅳ.①O186.14 ②O189.11

中国版本图书馆 CIP 数据核字(2022)第 118940 号

QIANGZI DE FENSILE JIHE HE LÜLA JIHE(YUZHOUXUE FANGMIAN):QIANGZI JIEGOU DE FENSILE JIHE HE LÜLA JIHE(TUOPU QUEXIAN)

策划编辑 刘培杰 杜莹雪
责任编辑 刘立娟
封面设计 孙茵艾
出版发行 哈尔滨工业大学出版社
社　　址 哈尔滨市南岗区复华四道街 10 号 邮编 150006
传　　真 0451 - 86414749
网　　址 http://hitpress.hit.edu.cn
印　　刷 黑龙江艺德印刷有限责任公司
开　　本 880 mm×1 230 mm 1/32 印张 6.625 字数 258 千字
版　　次 2023 年 3 月第 1 版 2023 年 3 月第 1 次印刷
书　　号 ISBN 978-7-5767-0200-2
定　　价 38.00 元

PREFACE

In the last several years a few books and monographs on Finsler space, a generalization of Riemannian space have come out for expounding the geometrical aspects of the space as well as the applications of it in different areas of theoretical physics and biological systems. The purpose of the present book is not to overcrowd them. On the other hand, here we want to introduce a Finsler space of the extended structure of the sub-atomic particles called hadrons for the development of internal symmetry of them. This new line of application of Finsler geometry together with the quantization of this space-time in microlocal domain , in fact, can generate the quantum field equations. Subsequent discussions have been made on the quantum fields and states and their interactions. As cosmological consequences, the non-singular origin of the universe with the early evolution and the resolution of problems encountered by the standard cosmology have been described in detail. In fact, present application shows the anisotropic character of the Finslerian geometry of the microlocal space-time. Here, we also present an explanatory account of Lyra geometry which is a modification of Riemannian geometry with the introduction of gauge function into the structure-less manifold. Such an account is not available in any other book form. In the framework of Lyra geometry that has a close resemblance of Weyl geometry , we have discussed some physical features of topological defects like domain wall, cosmic strings, monopoles, textures arising out from phase transitions in the early universe near the putative big-bang. We would like to express our sincere gratitude to our wives Mrs. Krishna De and Pakizah Yasmin for their patience and support during the gestation period of the manuscript. It is a pleasure to thank Dr. M Kalam, Dr. A Bhattacharya and Dr. A Ghosh for their technical assistance in preparation the book.

CONTENTS

Prologue

As early as in 1957, the measurement of the electromagnetic form factor of nucleon by Hofstadter et al. may be regarded as an experimental support for the concept of extended structure for the elementary particle although theoretically it was an older idea. In fact, this concept can be traced back in the electron theory of Lorentz. On the contrary, the elementary particles as the field quanta which are essentially point like entities came as the outcome of relativistic quantum mechanics in the framework of local field theory that suffers from the well known divergence difficulties. With the underlying motivation of resolving these divergence difficulties and also for a unified description of elementary particles, H. Yukawa, around 1950, came up with his theory of bilocal field which opened the possibility of intrinsic extension of the elementary particles. Subsequently, in the mid fifties M.A. Markov made an early attempt in applying bilocal model to the hadrons. Later this simple bilocal field theory model was generalized to multilocal theories. The motivation behind all these theories is to understand multifariousness of subatomic particles (whose number went on increasing in the last fifty years) as the ultimate manifestation of their extension in space and time, the macroscopic spacetime.
The composite character of hadrons can also be regarded as the manifestation of hadronic matter extension in the microlocal spacetime. With the earlier studied of such composite picture of extended hadron structure by Fermi-Yang and Sakat it was in1964. Gellmann arrived at the quark model. So far, no free quarks are observed directly and consequently confinement of quarks in the structure of hadron has to be assumed. Thus, we can only conclude that these quarks are not the 'ordinary' particles. Moreover, not only their fractional charges but also their flavour quantum numbers such as isospin, strangeness, charm etc. have to be assigned for the generation of the quantum numbers of hadrons. Even in Weinberg-Salam unified theory of weak and electromagnetic interactions such a phenomenolgy of assigning individual hypercharges to leptons and quarks has to be adopted. Thus, in a sense the constituent model which is a form of extended structure of hadronds remains incomplete unless one can find the origin of the internal quantum numbers of the constituents. On the other hand, if one considers the extended hadron structure in spacetime then there remains the problem of specification of the microlocal sapcetime in which the hadronic matter whether it is composed of constituents or not is really extended. It is , indeed, the fact that the microscopic spacetime may be different from the macroscopic one (the laboratory or even the large scale spacetime of the Universe) incited Yukawa et al. to introduce a theory of elementary domain in 1968. Thus the important questions one has to face, namely,
(a) one the hadrons extended in microscopic spacetime different from the usual Minkowskian macroscopic spacetime?
(b) can we have a geometric origin of the internal uantum numbers of the constituents of the extended hadrons?
(c) can we formulate a theory of interaction for hadrons on the basis of their extended structure in microlocal spacetime?

An effort has been made in the present monograph to accomplish reasonable solutions to these posers with n underlying principle of keeping the theory ' as simple as possible' - a principle that worked per excellence in any fundamental achievements, such as, Einstein's theory of general relativity, Dirac equation etc. In order to comprehend the microlocal sapcetime of extended hadrons one has to remember some imporatnt points. One of them is that the spacetime distance less or equal to the Planck length is not meaningful concept. In fact, recently, Adler and Santiago (Mod.Phys.Lett.A 14 (1999), 1371-1381) have modified the uncertainty principle by considering the gravitational interaction of the photon and the particle.

From this modified gravitational uncertainty principle it follows that there is an absolute minimum uncertainty in the position of any particle and it is of the order of Planck length. This is also a standard result of super string theory. In Y. Jack Ng and H. VanDam (Mod.Phys.Lett.A 9 (1994), 335-340 and references therein) the intrinsic limitation to quantum measurements of spacetime distances has been

obtained. The intrinsic uncertainty of a distance is shown there to be proportional to the one third power of the distance itself. This uncertainty in spacetime measurements does also imply an intrinsic uncertainty of the spacetime metric and yields a quantum decoherence for perticle heavier than Planck mass. Thus because of the intrinsic uncertainty of spacetime metric, it suffices to give a particle heavier than Planck mass a classical treatment. Also the spacetime can be defined only as averages over local regions and can not have any meaning locally. This also indicates that one should treat the spacetime as quantized below a fundamental length scale.
The microspace of extended hadron structure, which is different from macrospace is regarded here as an anisotropic spacetime. In fact, there are some indirect indications that in our epoch the spacetime, on the average, has a weak relic local anisotropy and consequently it should be described by Finsler geometry instead of Rimannian one. A strong local anisotropy of the spacetime might have occurred at an early stage in the evolution of the Universe as result of high temperature phase transitions in its geometric structure, caused by breaking of higher gauge symmetries and due to the appearance of massive elementary particles. Because of the expansion of the Universe this anisotropy might have decreased to the present weak level of it. The breaking of the discrete spacetime symmetries in weak interactions, an anisotropy in the relic background radiation of the Universe and the absence of the effect of cutoff in the spectrum of primary ultra-high energy cosmic protons are all indirect indications of the existence of a local anisotropy in spacetime.

In chapter one, the microlocal spacetime has been specified as a Finsler space which is a generalization of Riemannian space. Historically, such a generalization was suggested by G.F.B Riemann himself in the year 1854, although this geometry was fully generated from the works of Paul Finsler only in the early period of twentieth century. The microlocal spacetime considered here is a special Finsler space that is anisotropic. In this connection, we may point out the approach of Busemann to the mathematical theory of Finsler space. According to this a flat anisotropic space can be understood as a linear normalized vector space in which the norm of a vector is determined not exclusively by its Euclidean length but also by its orientation with respect to some preferred direction. In the present case, the macroscopic spaces, that is, the usual Minkowski spacetime and the background spacetime of the Universe (the Robertson Walker spacetime) appear as the associted Riemannian spaceof the Finslerian microscopic space. Of course, as an alternative procedure in transiting to the macrospaces, an ' averaging ' of the metric tensor of the microscopic space which manifests below a characteristic length scale (a fundamental length scale) is suggested here. In this way one can recover the metric tensors of the macroscopic spaces of physical interest.

Derivation of the field equations for the microspaces have been made in the second chapter. Firstly, from an equivalent property of a classical field along the neighbouring points on the auto parallel curve of the Finslerian microlocal space, the equation for it has been derived. This classical field equation is form invariant under general coordinate transformations. Then the quantum field equation (or wave equation) has been found by quantizing the spacetime in two steps in succession. These steps are the quantization of the coordinate differentials and the differentials of the directional variables of the Finsler space. This procedure gives rise to the Finsler space field equation for the field which is a bispinor in this case. We have also obtained photo-field equations for this space. These equations have similarity with the Bargmann-Wigner equations for the photon field in the usual Minkowski spacetime. The fields of the Finsler space are separable as the direct products of two spinors, one depending on the coordinates (which are also the coordinates of the associate Riemann space representing macrospaces) and the other on the directional variables. The coordinate dependent spinors are found to satisfy the usual field equation (for the free field as well as the interacting cases) of the Minkowski spacetime. On the other hand, the directional variable dependent spinorial parts of the field satisfy a new equation. These spinorial parts carry additional quantum numbers for the particles represented by these fields.

It is possible to find the Dirac equation for the field of the macrospace in the possible case of separation of the bispinor field. An interesting fact is that an additional mass term appears as a constant in the process of separation. This can be regarded as manifestation of the anisotropic Finslerian nature of microscopic spacetime. The Dirac equation for the curved spacetime (FRW spacetime) so obtained differes from the usual one only in respect of the additional mass term therein. As the consequence, the particle mass consists of two parts - the usual mass term (the 'inherent' mass) and a time dependent additional mass. In fact, this cosmological time or epoch dependent part of the particle-mass is found to be proportional to the Hubble function. Of course, for some species of elementary particles the epoch dependent part of the mass may be zero. In particular, for the constituents of hadrons such situation is assumed to take place resulting the generation of an additional quantum number. The field equation for the directional variable dependent spinors has been solved for the homogeneous solutions for them. A relation between the parameter of the space and the additional mass term has been obtained there. This relation specifies the mass of the particle completely. It is found that the epoch dependent part of the mass is relevant only in the very early epochs of the Universe. It has practically no contribution to the present mass of the particle. The present mass corresponds to the inherent mass to an extremely high degree of accuracy. For electromagnetic interaction the field equations have been found to be associated Riemannian spaces of Finsler space. Apart from this, the solutions of the field equation for the 'directional variable' spinor field have been obtaIt is an interesting fact that for non zero additional mass the solution is singular at ${\nu^0}^2 - \overrightarrow{\nu}^2 = 0$, (ν^μ being the directional variables), although it is finite for all other values of the directional variables. On the other hand, the solution for the case of no additional mass remains finite for all $\widetilde{\nu} = (\nu^0, \nu^1, \nu^2, \nu^3)$ even on ${\nu^0}^2 - \overrightarrow{\nu}^2 = 0$. Thus there is an intrinsic connection between the additional mass and the directional variable for which ${\nu^0}^2 - \overrightarrow{\nu}^2 = 0$. This situation is comparable to the fact that for a massive particle its mass becomes infinite if its velocity approaches to that of light. Although the analogy is not exact, one may guess that the directional variables may represent the four velocity of " internal fluctuations".

In the subsequent chapter not only the internal symmetry group of hadrons has been accomplished but also discussions have been made on hadron fields and states. In fact, for constituents particles (leptons) of the hadrons, one can have additional quantum numbers for them in the form of internal helicities. This is a manifestation of the anisotropic Finslerian character of the microlocal spacetime of extended hadrons. Then an important suggestion made by Budini in 1979 regarding the geometric study of hadrons is followed. We have considered the assumption that the internal symmetry algebra can be generated from the conformal reflection group. Presently, the " bispinor doublets" both in the semispinor basis and in Dirac basis are constructed from the bispinor fields of the Finsler space. These doublets which represent the conformal spinors in Budini's formalism may be regarded as the constituents of a hadron such that they are in the internal spin half angular momentum state. Also, the semispinors (bispinors in this case) are eigenstates of an internal helicity operator with opposite eigenvalues, apart from the fact that they have particle - antiparticle relation. The bispinor doublet formed by the semispinors with opposite eigenvalues of the internal helicity operator can be treated as in Budini's approach to generate the internal symmetry algebra. Moreover, the fixed values of third components of the internal spin angular momenta for particle and antiparticle states give rise to another quantum number representing the algebra of U_1. These SU_2 and U_1 algebras indicate isospin and hypercharge respectively and consequently a Lie group structure $SU_3 \longrightarrow SU_2 \times U_1$ can be achieved. An important fact is that the existence of continuous groups (rotation Lie groups) which may operate in a space other than $M^{4,2}$, orthogonal to $M^{3,1}$ was necessitated by Budini et al (1979) and postulated as an ad-hoc gauge group. This makes the internal symmetry algebra an internal symmetry group. For the present case, the existence of such a continuous group is natural and not a postulate. Consequently, the SU_3 Lie group formalism for the internal symmetry of hadrons is accomplished.
The baryonic multiplet corresponding to the internal symmetry group SU_3 representing baryon with spin half is shown to arise from the mesonic SU_3 multiplet with spin zero by introducing a constituent spinor having a specific internal helicity. In fact, the two opposite values of the third component of the internal spin angular momentum for this spinorial constituent correspond to baryon and antibaryon.

In the mesonic state, the two constituents of it are in opposite internal helicity states and therefore no signature of internal spin half angular momentum (or anistropy of internal spacetime) is carried outside the configuration. It is discussed here how the quantum numbers such as baryonic charge number, strageness, hypercharge can arise from the internal quantum numbers of the constituents. Also, the mass difference of the number of different isomultiplets of the SU_3 multiplet as well as that between the highest massive meson with hypercharge and the lowest massive baryon are obtained here in good agreement with the experimental values.
Next task is to construct hadron fields from those of the constituents in the anisotropic microdomain. This procedure is elaborate and general in nature. The pseudo scalar meson field is constructed from the fields of its two leptons constituents as a specific example. Also, the macrospace fields are constructed from these 'microscopic space' fields. Interestingly, the resulting field of pseudo scalar meson has a similarity with the field obtained by Haag (1958), Zimmermann (1958) and Nishijima (1958, 1961, 1964). The hadron sates are also been specified here. It is shown that the hadron fields and states carry the quantum numbers such as baryon number, strangeness, hypercharge etc. The composite fields satisfy the asymptotic conditions necessary for S-matrix theory and they can be used in the reduction formulae which express the S-matrix element in terms of time ordered products of field operators.
From a phenomenological point of view the dynamics of strong interaction of hadrons is described here. In fact, these interactions are regarded as the interactions among the constituents. Specially, these reactions are dependent on the $\pi\pi^-$ interaction where the interacting pion are in the structure of the incident hadrons. The two constituents (leptons) having particle - antiparticle relation in the structure of hadron form a pseudo scalar meson (π^-, π^+, or π^0). The hadrons are regarded as the clusters of pions together with a single spinor constutient (muon or neutrino). The amplitude for the two body hadron reaction (non hypercharge exchange) depends on a term arising out due to the rearrangement of constituents apart from the contribution of $\pi\pi^-$ interaction derived from Lanrangian field theory. The axioms of S-matrix theory are shown to satisfy in these reactions. The amplitude does not violet the Froissart bound at the high energies. The hypercharge exchange reactions are dominated by the direct interactions such as Knockout or stripping which are very familiar reactions in Nuclear Physics. The amplitudes of these reactions are obtained from the one-particle-exchange model with a modification by the Distorted Wave Born Approximation (DWBA). The DWBA can account for the dip phenomenon where it has been observed experimentally as well as unitarity. As the exchanged particle has spin zero, the unitaryity limit is not violated at sufficiently high energy. Here also, the postulates of S-matrix elements are satisfied. For both the cases the dynamics is thus generated from the structural aspects of hadrons. The field theory in a modified form (as it shown that the perturbation expansion remains valid) and the S-matrix approach for strong interaction become united in the general premises of composite character of hadrons are the extended hadron structure in the microlocal spacetime.

In the next part of the monograph, the cosmological consequences of the epoch dependence of particle mass have been discussed. In chapter five, the early Universe is treated as an open thermodynamic system which allows irreversible is based upon the modified thermodynamical energy conservation law of Prigogine (1961) for an open system. It modifies the Bianchi identities for homogeneous and isotropic Universe. For a perfect fluid model in a spatially flat Universe the conservation law (Bianchi identities) holds good with a phenomenological pressure instead of the true thermodynamic pressure. Particle creation occurs as long as the true thermodynamic pressure remains greater than the phenomenological pressure and the irrreversibility of this process is implied by the second law of thermodynamics. The governing equations that decide the ' Cosmology ' of the creation are in the very early Universe are derived here. The trivial solution of these equations in the usual (exponential) inflation. On the other hand, the equation of state for this creation era is completely specified from these governing equations for the case of non trivial solution which is matter dominated FRW Universe. The equation of state has been found to be the maximally stiff equation of state s given by Zel'dovich in 1962. The observable parameter such as the present value of the specific entropy per baryon can be calculated in this framework. Next we turn to an important topic of quantum creation of matter. The quantum creation of matter in the perturbed 'anisotropic' Minkowski spacetime is discussed here.

It is interesting to find that the created very massive matter constituents (mass about fifty times of Planck mass) due to quantum effect can make the Minkowski spacetime unstable, as discussed by Nardone (1989), and ushers in to an expansion phase of the Universe. A very brief period of this anisotropic perturbation (compared to the Planck time) signifies the 'origin' of the expanding Universe without an initial singularity. Also, the quantum creation of particles after this Planck order time, that is, for expanding phase which is a matter dominated FRW Universe has been considered. The creation phenomenon continues up to a transition epoch $\alpha = 10^{-23}$ sec. (after a cosmological zero time) when the Universe enters in to its usual expansion phase of radiation-dominated FRW era. Both the energy density and the Universe temperature at the transition epoch are calculated. The present value of the photon to baryon ratio is also found in agreement with the observation.
Grib (1989) has considered the quantum effect of vacuum polarization in the early FRW spacetime. This gives rise to an effective change of the gravitational constant. Such change of gravitational constant due to quantum creation of very massive particles is considered here. Consequently, the Planck time is effectively changed and the gravity may remain classical in the epochs of creation of these very massive particles. Here, an experimental determination for the particles with such Planck order masses in the early Universe, as suggested by Parker (1989) through detection and the study of the nature of non thermal cosmic gravitational wave background is discussed.

Apart from considering particle creation due to bulk viscosity in the early Universe regarded as both the thermodynamically closed systems, discussions have been made on the modified general relativity (MGR) of Rastall (1972), Al-Rawaf and Taha (1996). Although MGR can not be derivable from a variational principle, a prototype of it with a variable gravitational 'constant' can be obtained from this principle. The MGR formalism contains an adjustable parameter η in the range $0 < \eta \leq 1$ apart from the usual Newton's constant. The standard general relativity (GR) can be achieved with the value of the parameter $\eta = 1$. An important fact about MGR is that it can be moulded in to a model with a variable cosmological 'constant' and consequently, instead of matter energy momentum conservation, a sum of tensors corresponding to this usual energy momentum and a 'vacuum energy momentum' is conserved as in the 'decaying vacuum' cosmologies. We discuss the observational scenario in respect of MGR and apply it in the early Universe taken as an adiabatic perfect fluid with spatially flat FRW metric. Irreversible creation of particles is allowed here apart from the matter creation due to the non conservation of usual energy momentum in MGR. In fact, for the early Universe taken as a thermodynamically open system, we have the modified conservation law and the field equations. These equations are incorporated with the relation for the epoch dependence of particle masses and one obtains the governing equations for the matter dominated 'creation era' of the early Universe. The interesting fact is that both the equation of state and expansion scalar can be determined from these equations. The solution for the expansion scalar corresponds to a 'mild inflation' in the epoch time $t \leq \alpha$. The scalar factor for this period is found to be proportional to the cosmological time. The mild inflation switches off automatically at the epoch time α in contrast to the fact that the usual exponential inflation is required to be stopped by 'some' phase transition. The creation phenomenon also stops around this epoch time α. The production of specific entropy per baryon is an effect of bulk viscosity in the early Universe in the framework of MGR. The produced specific entropy per particle is found to be in good agreement with the observational data for the values of the non-Newtonian parameter η in the range $.75 \leq \eta \leq 1$. It is interesting to note that the flatness problem of the standard cosmology can be resolved in the present formalism.

Other topics like causal homogenization condition, the cosmological constant problem are also discussed. The hypersurface $t = \alpha$ is shown to be homogenized causally. There remains no horizon problem because the Universe went through the mild inflationary phase after its generation from a perturbed Minkowski spacetime. In the present consideration of the early period of our Universe, the cosmological constant problem can be resolved in the context of 'changing gravity approach' where the determinant of the metric is not dynamical but the action remains stationary only with respect to variations in the metric maintaining this determinant fixed.

In 1951, Lyra proposed a modification of Riemannian geometry by introducing a gauge function into the structure less manifold that bears a close resemblances to Weyl's geometry. In general relativity Einstein succeeded in geometrising gravitation by identifying the metric tensor with gravitational potentials. Lyra's geometry is more in keeping with the spirit of Einstein's principle of geometrization since both the scalar and tensor fields have more or less intrinsic geometrical significance. In the chapter seven, we give some outlines of the modified Riemannian geometry Proposed by Lyra.

The origin of the structure in the Universe is one of the greatest cosmological mysteries even today. It is believed that the early Universe had undergone a number of phase transitions as it is cooled down from it hot initial state. Phase transitions in the early Universe can give rise to various forms of topological defects. A defect is a discontinuity in the vacuum and depending on the topology of the vacua , the defects could be domain walls , cosmic string , monopoles and textures. In the final chapter, we have discussed some physical features of topological defects within the framework of Lyra geometry.

CHAPTER ONE:

FINSLER SPACE OF HADRONIC MATTER EXTENSION

1.1: METRIC FUNCTION: POSITIVE n^{th} ROOT OF A nth ORDER FORM

Riemann in his famous lecture Über die Hypothesen welche der Geometric zu Grunde liegen (1854) suggested that the positive fourth root of a fourth order differential form might serve as a metric function. The common property with Riemannian quardratic form is that they are both positive, homogeneous of first degree in the differentials and also convex in the latter. Thus, the notion of distance between two neighbouring points x^i and $x^i + dx^i$ can be generalized as it is given by a fundamental function $ds = F(x^i, dx^i)$ satisfying these three conditions. This programme was, in fact, carried out by Paul Finsler (1918) and subsequently developed by Cartan (1934), Berwald (1941), Rund (1957), Matsumoto (1986), Asanov (1985) and many others. The theory of Finsler spaces finds its applications in various branches of Physics and Biology (Antonelli et al., 1993).
We first introduce the Finlser pace with the fundamental function $F(x^i, dx^i)$ as the nth root of a nth order form in the directional arguments or variables $y^i = \frac{dx^i}{dt}$, which are tangents to the points x^i. That is, we write

$$F(x^i, y^i) = [g_{\mu_1\mu_2....\mu_n}(x_k)y^{\mu_1}y^{\mu_2}.....y^{\mu_n}]^{\frac{1}{n}} \tag{1}$$

The metric tensor for a Finsler space can be defined as (Rund, 1957 , Asanov , 1985)

$$g_{ij}(x^k, y^k) = \frac{1}{2}\frac{\partial^2 F^2(x^k, y^k)}{\partial y^i \partial y^j} \tag{2}$$

For the fundamental function (1) the metric tensor can be calculated to be

$$g_{ij}(x^k, y^k) = \widehat{g}_{ij}(x^k, y^k) + \omega_{ij}(x^k, y^k) \tag{3}$$

where

$$\widehat{g}_{ij}(x^k, y^k) = \left[\frac{g_{ij\mu_3\mu_4....\mu_n}(x_k)y^{\mu_3}y^{\mu_4}.....y^{\mu_n}}{F^{n-2}(x^k, y^k)}\right] \tag{4}$$

and

$$\omega_{ij}(x^k, y^k) = (n-2)\left[\frac{g_{ij\mu_3\mu_4....\mu_n}(x_k)y^{\mu_3}y^{\mu_4}.....y^{\mu_n}}{F^{n-2}(x^k, y^k)} - \frac{g_{i\mu_2\mu_3....\mu_n}(x_k)y^{\mu_2}y^{\mu_3}.....y^{\mu_n}g_{\nu_1 j\nu_3....\nu_n}(x_k)y^{\nu_1}y^{\nu_3}.....y^{\nu_n}}{F^{2(n-1)}(x^k, y^k)}\right] \tag{5}$$

It is easy to see that

$$\omega_{ij}(x^k, y^k)y^i y^j = 0 \tag{6}$$

and consequently

$$F^2(x^k, y^k) = g_{ij}(x^k, y^k)y^i y^j = \widehat{g}_{ij}(x^k, y^k)y^i y^j \tag{7}$$

It is to be noted that $\omega_{ij}(x^k, y^k) = 0$ for $n = 2$ and the components of the metric tensor $g = (g_{ij})$ are given by $g_{ij}(x^k, y^k) = \widehat{g}_{ij}(x^k, y^k) = g_{ij}(x^k)$. Thus the case $n = 2$ with the components of the metric tensor independent of the directional arguments y^k corresponds to the Riemannian space which is a special case of Finsler space. For $n > 2$, $g_{ij}(x^k, y^k)$ are dependent of both the position x^k and the directional arguments y^k.

It is obvious that the fundamental function defined in (1) is a positively homogeneous function of degree one (for brevity, (1)p-homogeneous) in the directional variables y^k. The components of the metric tensor derived from this fundamental function, which are given in (3) are (0)p-homogeneous functions in the directional arguments. Thus, if the metric function is defined as the positive n^{th} root of a nth order form in the differentials or directional variables then the corresponding metric tensor becomes dependent of the directional arguments. This property of metric tensor and (1)p-homogeneity and (0)p-homogeneity in the directional arguments for fundamental function and metric tensor respectively are the common properties for specification of any Finsler space. In fact, there is a proposition (Antonelli et al. 1993) that for a segment of a curve given by

$$x^i = x^i(t), a \leq t \leq b \tag{8}$$

the length

$$s = \int_a^b F\left(x^i(t), \frac{dx^i}{dt}\right) dt \tag{9}$$

is independent of the choice of the parameter if and only if $F\left(x^i(t), \frac{dx^i}{dt}\right)$ or $F\left(x^i(t), y^i\right)$ is (1)p-homogeneous in y.

The case $n = 4$ in (1), which Riemann suggested for the metric function will be the case of special interest in future discussions as the space of hadronic matter extension and will be characterized to be a Finsler space with such a fundamental metric function. The metric function for this case is of the type

$$F\left(x^i(t), y^i\right) = \left[g_{\mu\nu\rho\sigma}(x^k) y^\mu y^\nu y^\rho y^\sigma\right]^{\frac{1}{4}} \tag{10}$$

The components of the metric tensor are given by

$$f_{\mu\nu}\left(x^i(t), y^i\right) = g_{\mu\nu}\left(x^i(t), y^i\right) + \omega_{\mu\nu}\left(x^i(t), y^i\right) \tag{11}$$

where

$$g_{\mu\nu}\left(x^i(t), y^i\right) = \frac{1}{F2} g_{\mu\nu\rho\sigma}(x^k) y^\rho y^\sigma \tag{12}$$

and

$$\omega_{\mu\nu}\left(x^i(t), y^i\right) = \frac{2}{F^2}\left[g_{\mu\nu\rho\sigma}(x^k) y^\rho y^\sigma - \frac{1}{F^4} g_{\mu\nu'\rho\sigma}(x^k) y^{\nu'} y^\rho y^\sigma g_{\mu'\nu\rho'\sigma'}(x^k) y^{\mu'} y^{\rho'} y^{\sigma'}\right] \tag{13}$$

Here, $g_{\mu\nu\rho\sigma}(x^k)$ must be symmetric with respect to the first two indices in order that $\left[f_{\mu\nu}\left(x^i(t), y^i\right)\right] = f\left(x^i(t), y^i\right)$ is symmetric. The tensor field $g = g_{\mu\nu\rho\sigma}(x^k)$ for the space of hadronic matter extension will be specified later.

1.2: FINSLER SPACES

Now, the generalized homogeneous metric geometry or Finsler space can be defined. A Finsler space F_n is an n-dimensional smooth manifold M_n endowed with a metric function $F\left(x^i, dx^i\right)$, the fundamental function of the coordinates x^i and the differential, where,

$$ds = F\left(x^i, dx^i\right) \tag{14}$$

$$(i = 1, 2, ...n)$$

The differential (dx^i) does no longer represent a general increment in the coordinate space, but is a direction tangent to the curve $x^i = x^i(t)$. For this curve in the manifold, the finite distance between any two of its points $x^i(t_1)$, $x^i(t_2)$ is

$$s = \int_{t_1}^{t_2} F\left(x^i(t), \frac{dx^i}{dt}\right) dt \tag{15}$$

In order that this distance be independent of the choice of the parameter, the fundamental function F of Finsler space must be (1)p-homogeneous in dx^i (Carathéodory, 1935, Antonelli et al., 1993) i.e. ,

$$F\left(x^i, \lambda dx^i\right) = \lambda F\left(x^i, dx^i\right), \lambda > 0 \tag{16}$$

The quantities $\frac{dx^i}{dt}$ are the components of a tangent vector to the manifold, namely, the one given by the curve $x^i(t)$. The dependence of the metric function or of any geometric object derived from it, on these quantities is referred to as the directional dependence.
Now, from the fundamental function, the Finsler metric tensor can be defined as

$$g_{ij}\left(x^k(t), \dot{x}^k\right) = \frac{1}{2}\frac{\partial^2 F^2\left(x^k(t), \dot{x}^i\right)}{\partial \dot{x}^i \partial \dot{x}^j} \tag{17}$$

where $\dot{x}^k = \frac{dx^k(t)}{dt}$.
This metric tensor $g = (g_{ij})$ transforms as a covariant tensor of rank two under a coordinate transformation and is symmetric under interchange of indices.
Under the coordinate transformation

$$\overline{x}^i = \overline{x}^i(x^j) \tag{18}$$

the directional arguments or variables $\dot{x}^i$ and the differentials are transformed , respectively, as

$$\dot{\overline{x}}^i = \frac{\partial \overline{x}^i}{\partial x^j}\dot{x}^j \tag{19}$$

$$d\overline{x}^i = \frac{\partial \overline{x}^i}{\partial x^j}dx^j \tag{20}$$

The tensor $g = (g_{ij})$ is (0)p-homogeneous in the directional arguments and the Finsler metric function F may be written in terms of g_{ij} as

$$F^2(x^k, \dot{x}^k) = g_{ij}(x^k, \dot{x}^k)\dot{x}^i\dot{x}^j \tag{21}$$

or, in terms of the differentials

$$ds = F^2(x^k, dx^k) = g_{ij}(x^k, dx^k)dx^i dx^j \tag{22}$$

It should be noted that if g_{ij} does not depend on dx^k the space is Riemannian.
Also the Finsler metric function can be defined by

$$F^2(x^k, \dot{x}^k) = \widehat{g}_{\mu\nu}(x^k, \dot{x}^k)\dot{x}^\mu\dot{x}^\nu \tag{23}$$

The tensor $\widehat{g} = (\widehat{g}_{\mu\nu})$ is not in general the Finsler metric tensor, but simply a homogeneous tensor of degree zero in $\dot{x}^k$, which might be used to define the metric function. The relation between g and $\widehat{g}$ is

$$g_{\mu\nu} = \widehat{g}_{\mu\nu} + \omega_{\mu\nu} \tag{24}$$

where

$$\omega_{\mu\nu} = \frac{\partial \widehat{g}_{\mu\alpha}}{\partial \dot{x}^\nu}\dot{x}^\alpha + \frac{\partial \widehat{g}_{\alpha\nu}}{\partial \dot{x}^\mu}\dot{x}^\alpha + \frac{1}{2}\frac{\partial^2 \widehat{g}_{\alpha\beta}}{\partial \dot{x}^\mu \partial \dot{x}^\nu}\dot{x}^\alpha\dot{x}^\beta \tag{25}$$

Obviously, for Riemanninan space $g = \widehat{g}$. It is also easy to see that

$$\omega_{\mu\nu}(x^k, \dot{x}^k)\dot{x}^\mu\dot{x}^\nu = 0 \tag{26}$$

Writing $\frac{dx^i(t)}{dt} = y^i$ for the directional argument as

$$y_i = g_{ij}(x^k, y^k)y^j = \frac{1}{2}\frac{\partial F^2(x^k, y^k)}{\partial y^i} \tag{27}$$

It is generally assumed for the Finsler spaces F_n (denoted $F_n = (M_n, F(x^k, y^k)$) that the fundamental function $F(x^k, y^k)$ is positive valued. We may relax this condition by the assumption that F is non negative for all y^k. also, we can assume the symmetry property of f, i.e., $F(x^k, y^k) = F(x^k, -y^k)$. This fundamental function $F(x^k, y^k)$ is interpreted as the length of the vector y^μ at the position x^μ. Thus the condition of homogeneity of F in the directional arguments y^k given in (16) means that for any collinear vectors , y_1^μ and y_2^μ, the ratios of their components are the same. Similarly, for any vector, v^i the length of it relative to y^μ can be defined as $g_{ij}(x^k, y^k)v^i v^j$. Some of the important Finsler spaces are as follows: A Finsler space $F_n = (M_n, F)$ is called (α, β)-metric if the fundamental function F is a (1)p-homogeneous function of two arguments $\alpha(x^k, y^k)$ and $\beta(x^k, y^k)$, where $\alpha(x^k, y^k)$ is a Riemannian fundamental function given by $\alpha(x^k, y^k) = \left(\alpha_{ij}(x^k)y^i y^j\right)^{\frac{1}{2}}$ and $\beta(x^k, y^k) = \left(b_i(x^k)y^i\right)^{\frac{1}{2}}$ is a differential 1-form. The space $R_n = (M_n, \alpha)$ is called the associated Riemannian space and the covariant vector field b_i is the associated vector field. In particular, the (α, β)-metric $F = \frac{\alpha^2}{\beta}$ is called a Kropina metric (Kropina, 1959) and a metric of the form $F = \alpha^{m+1}\beta^{-m}$ ($m \neq 0$, -1) a generalized m-Kropina metric. The Finsler space $F_n = (M_n, F = \alpha + \beta$ with a metric $F = \alpha + \beta$ (Randers metric) is called a Randers space.

1.3: GEODESIC OF THE FINSLER SPACES

The extremal of the length integral (15) given by $\delta \int ds = 0$ is the geodesic of the Finsler space. This variation produces the Euler-Lagrange equation which, in turn, gives rise to the following equations for the geodesic.

$$\frac{d^2x^i}{ds^2} + 2G^i\left(x^k, \frac{dx^k}{ds}\right) = 0 \tag{28}$$

or, equivalently

$$\frac{dy_i}{ds} + 2G^i\left(x^k, y^k\right) = 0, y^k = \frac{dx^k}{ds} \tag{29}$$

where s is the arc length and

$$2G^i = 2g_{ij}(x^k, y^k)G^i(x^k, y^k) = \frac{1}{2}\left[\frac{\partial^2 F^2(x^k, y^k)}{\partial y^j \partial x^i} y^i - \frac{\partial F^2(x^k, y^k)}{\partial x^j}\right] \tag{30}$$

The Finsler-Christoffel connection can be introduced as

$$\gamma_{ijk}(x^k, y^k) = g_{ih}\gamma^h_{jk} = \frac{1}{2}\left[\frac{\partial g_{ji}}{\partial x^k} + \frac{\partial g_{ik}}{\partial x^j} - \frac{\partial g_{jk}}{\partial x^i}\right] \tag{31}$$

Also,

$$\gamma^i_{jk} = g^{ih}\gamma_{jhk} \tag{32}$$

Here, g^{ih} is such that

$$g_{ij}(x^k, y^k)g^{ih}(x^k, y^k) = \delta^h_j \tag{33}$$

With these Christoffel symbols of second kind γ^i_{jk} the geodesic in a Finsler space can be given in the following form:

$$\frac{d^2x^i}{ds^2} + \gamma^i_{jk}\frac{dx^j}{ds}\frac{dx^k}{ds} = 0 \tag{34}$$

or, equivalently,

$$\frac{d^2x^i}{ds^2} + \gamma^i_{jk}y^j y^k = 0 \tag{35}$$

Covariant derivative in Finsler spaces may be defined in different ways which depend on the explicit expression chosen as the connection. In fact, different choices can be found in the literature (Cartan,

1934, Berwald, 1947, Rund, 1959) but presently we shall confine ourselves to the δ-derivative of a vector X^i defined along a curve $x^i(t)$, which is supposed to be a function of the position only, is given by

$$\frac{\delta X^i}{\delta t} = \frac{dX^i}{dt} + \Gamma^i_{jk} y^j X^k \tag{36}$$

where the connection Γ^i_{jk} is defined by

$$\Gamma^i_{jk} = \gamma^i_{jk} - C^i_{jl}\gamma^l_{pk} y^p \tag{37}$$

Here, the C-tensor of Cartan or Cartan torsion tensor is given by

$$C_{ijk}(x^l, y^l) = g_{jh}C^h_{ik} = \frac{1}{2}\frac{\partial g_{ij}}{\partial y^k} \tag{38}$$

It is evident that C_{ijk} are symmetric in i, j and k and also that

$$C_{ijk}(x^l, y^l) y^k = 0 \tag{39}$$

One can regard each tangent space as to be an n-dimensional Riemannian space with the Riemannian metric $g_{ij}(x^k, y^k)dy^i dy^j$ with fixed x^k. Thus, $C^i_{jk} = g^{ih}C_{jhk}$ are Christoffel symbols constructed from the Riemannian metric. Of course, this is a very special type of Riemannian space, since, in general the Christoffel symbols are not symmetric in all indices.

Now, if all the C_{jhk} are identically zero then, the δ-derivative becomes the usual Riemannian covariant derivative. It is to be noted that the connection Γ^i_{jk} are, in general, symmetric in their lower indices. The partial δ-derivative of a vector $X^i(x^k)$ is given by

$$X^i_{;j} = \frac{\partial X^i}{dx^j} + \Gamma^{*i}_{jk} X^h \tag{40}$$

where

$$\Gamma^{*i}_{hk} = g^{ik}\Gamma^*_{hkj} \tag{41}$$

$$\Gamma^*_{hkj} = \gamma_{hkj} - \left(C_{jki}\Gamma^i_{hl} + C_{khi}\Gamma^i_{jl} - C_{hji}\Gamma^i_{kl}\right) y^l \tag{42}$$

Now, it is easy to see that

$$\frac{\delta X^i}{\delta t} = X^i_{;j} y^j \tag{43}$$

and also that Γ^{*h}_{ij} are symmetric in their lower indices.

Now, for the curve $x^i = x^i(s)$ where s is the Finslerian arc length, the vector $y^i = \frac{dx^i}{ds}$ (which has unit Finslerian length since $F(x^k, y^k) = 1$) tangent to the curve is transported along it and the covariant δ-derivative of it takes the following form:

$$\frac{\delta y^i}{\delta s} = \frac{dy^i}{ds} + \Gamma^i_{jk} y^j y^k = \frac{dy^i}{ds} + \gamma^i_{jk} y^j y^k \tag{44}$$

The curve $x^i = x^i(s)$ is called autoparallel if $\frac{\delta y^i}{\delta s} = 0$. Thus the geodesic of a Finsler space are autoparallel curves.

1.4: INDICATRIX

For the tangent space M^x_n the n-1 dimensional hypersurface of it given by the st

$$I^x = \{y^i \in M^x_n \mid F(x^k, y^k) = 1\} \tag{45}$$

is known as the indicatrix at the point x^i. It can, thus, be parametrized in the form $y^i = y^i(u^\alpha)$, $\alpha = 1, 2, ...n-1$. If M_n^x is regarded as the Riemannian space, then the indicatrix I^x has the following induced Riemannin metric

$$g_{\alpha\beta}(u^k) = g_{ij}(x^k, y^k(u^l))\frac{\partial y^i}{\partial u^\alpha}\frac{\partial y^j}{\partial u^\beta} \tag{46}$$

Now, let us write

$$l^i = \frac{y^i}{F(x^k, y^k)} \tag{47}$$

It is obvious that $F(x^k, y^k) = 1$. Thus l^k has unit Finslerian length. By using (27) and on differentiation of (47) by u^α we can find

$$l^i\left(\frac{\partial y^i}{\partial u^\alpha}\right) = 0 \tag{48}$$

Thus, l_i can be regarded as the components of the covariant normal vector of I^x. Also we can write $g_{\alpha\beta}(u^k)$ in the following form

$$g_{\alpha\beta}(u^k) = h_{ij}(x^k, y^k(u^l))\frac{\partial y^i}{\partial u^\alpha}\frac{\partial y^j}{\partial u^\beta} \tag{49}$$

with

$$h_{ij}(x^k, y^k) = g_{ij} - l_i l_j = F\frac{\partial^2 F}{\partial y^i \partial y^j} \tag{50}$$

$h = (h_{ij})$ is called the angular metric tensor and the Riemannian metric $g_{\alpha\beta}(u^\alpha)$ may be regarded as the induced one from the tensor h. Also, it is easy to see that

$$h_{ij}(x^k, y^k)y^j = 0 \tag{51}$$

Thus, we have (Antonelli et al., 1993) the proposition:
The angular metric h constitutes the matrix (h_{ij}) of rank $n-1$. A system of algebraic equations $h_{ij}v^j = 0$ for v^j implies that the solutions v^j must be proportional to y^j. The set

$$I^x(y_0^k) = \{y^k \mid g_{ij}(x^k, y_0^k)y^i y^j = 1\}, y_0^k \in I^x \tag{52}$$

is called the osculating indicatrix at the point y_0^k. In fact, $F^2(x^k, l^k)$ can be expanded around a fixed $l^k = l_0^k$ to obtain

$$F^(x^k, l^k) = F^2(x^k, l_0^k) + \left(\frac{\partial F^2}{\partial l^i}\right)_{l^k = l_0^k}(l^i - l_0^i) + \frac{1}{2}\left(\frac{\partial^2 F^2}{\partial l^i \partial l^j}\right)_{l^k = l_0^k}(l^i - l_0^i)(l^j - l_0^j) + ...$$

Now, since ,

$$\left(\frac{\partial F^2}{\partial l^i}\right)_{l^k = l_0^k} l_0^i = 2F^2(x^k, l_0^k)$$

and

$$\left(\frac{\partial^2 F^2}{\partial l^i \partial l^j}\right)_{l^k = l_0^k} l_0^j = \left(\frac{\partial F^2}{\partial l^i}\right)_{l^k = l_0^k}$$

we have

$$F^2(x^k, y^k) = g_{ij}(x^k, l_0^k)y^i y^j + ...$$

Thus the indicatrix I^x osculated with the quardratic hypersurfsce given by the osculating indicatrix $I^x(y_0^k)$ at each point y_0^k. It should be noted that I^x coincides with the osculating with the osculating indicatrix for each point y_0^k if the space in Riemannin.

1.5: CURVATURE TENSOR

Out of the several ways of defining curvature tensor in Finsler geometry the one which appears in the equation of geodesic deviation is presented in the following:

Let us suppose that two geodesics pass through a given point P in the Finsler space F_n and that they have tangent vectors at P with a very small difference between them. If $\xi = (\xi(s))$ represents the small vector joining the two points of the geodesic, which have the same geodesic distance s from the point P, then the second covariant δ-derivative of $\xi(s)$, that is, $\frac{\delta^2\xi(s)}{\delta s^2}$ can be regarded as an invariant measure of the curvature of F_n in the neighbourhood of the point. In fact, in Riemannian geometry the expression of this covariant second derivative is given by the geodesic deviation equation and it represents the curvature. For a Finsler space this equation takes the following form:

$$\frac{\delta^2\xi^i(s)}{\delta s^2} + K^i_{jhk}y^j y^h y^k = 0 \tag{53}$$

where $y^i = \frac{dx^i}{ds}$ are the tangent vectors along the geodesic, and the tensor given by K^i_{jhk} represents one of the several curvature tensors that might be defined in a Finsler space. The expression for K^i_{jhk} is given as

$$K^i_{jhk} = \left(\frac{\partial\Gamma^{\star i}_{jh}}{\partial x^k} - \frac{\partial\Gamma^{\star i}_{jh}}{\partial y^l}\frac{\partial G^l}{\partial y^k}\right) - \left(\frac{\partial\Gamma^{\star i}_{jk}}{\partial x^h} - \frac{\partial\Gamma^{\star i}_{jk}}{\partial y^l}\frac{\partial G^l}{\partial y^h}\right) + \Gamma^{\star i}_{mk}\Gamma^{\star m}_{jh} - \Gamma^{\star i}_{mh}\Gamma^{\star m}_{jk} \tag{54}$$

where $\Gamma^{\star i}_{jh}$ are given in (41) and (42). Here, G^l are also given in (30) and in fact, are found to be as

$$G^l = \frac{1}{2}\gamma^l_{jk}(x^i, y^i)y^j y^k \tag{55}$$

This definition of curvature tensor for the Finsler space has the correspondence with that for the Riemannian space in the sense that the tensor (K^i_{jhk}) reduces to the Riemannian curvature tensor for the Riemannian space.

1.6: MINKOWSKI SPACETIME AS A FINSLER SPACE

Blokhintsev (1978) pointed out that the ardinary Minkowski flat spacetime or Lorentz space can be regarded as a particular homogeneous Finsler space. In fact, the distance element of this spacetime can be written as $ds = N_\mu ds^\mu$ where N_μ is a zero order form in the differential $\widetilde{dx} = (dx^\mu)$. This form is different for space like, time like directions and light cone. Specially, there are three possible values $\widetilde{N^2} = \mp 1$ and 0 for these three directions respectively. Thus, the element ds depends on the directional argument $\widetilde{dx}$ in the sense that the space like, time like and light like directions are distinguished.
In fact, the metric for this Finslerian Minkowski spacetime can be defined with the fundamental function $F(x^i, y^i)$ given by

$$F^2(x^i, y^i) = e\left(\frac{ds}{dt}\right)^2 = e\eta_{ij}y^i y^j \tag{56}$$

with the directional variables $y^i = \frac{dx^i}{dt}$ and

$$ds^2 = \eta_{ij}dx^i dx^j \tag{57}$$

Here, $\eta_{ij} = diag(1, -1, -1, -1)$ and the indicator e is +1 or −1 according as y^i and consequently dx^i represent a time like or space like direction in the Minkowski spacetime. The fundamental function for this Finsler space is non negative (not a positive definite one, as we have pointed out earlier that such a relaxation will be allowed for the present consideration). This also defines a non negative $(interval)^2$ as

$$eds^2 = (d\widehat{s})^2 \tag{58}$$

and consequently, one obtains a real metric $d\widehat{s}$. Also, we can write $d\widehat{s}$ as

$$d\widehat{s} = L_\mu dx^\mu = L^\mu dx_\mu \tag{59}$$

Evidently,

$$L_\mu = e\frac{dx_\mu}{d\widehat{s}} \tag{60}$$

where

$$dx_\mu = \eta_{\mu\nu}dx^\nu \tag{61}$$

Also,

$$L^\mu = e\frac{dx^\mu}{d\widehat{s}} \tag{62}$$

Thus,

$$L_\mu L^\mu = \widetilde{L}^2 = e^2\frac{dx_\mu}{d\widehat{s}}\frac{dx^\mu}{d\widehat{s}} = e \tag{63}$$

There fore, we have,

$\widetilde{L}^2 = 1$ for time like directions.
$= -1$ for space like directions.
For light like direction one can take $L_\mu = Ke\frac{dx_\mu}{dt}$ and $L^\mu = Ke\frac{dx^\mu}{dt}$ where t is a parameter defining the curve $x^\mu(t)$ whose tangent vector or the direction (dx^μ) is being considered here and K is (-)p-homogeneous form in the differential dx^μ. Also, K is supposed to be positive definite for all directions (for example, $K^{-2} = \delta_{\mu\nu}dx^\mu dx^\nu$). Then, we have

$$d\widehat{s} = L_\mu dx^\mu = Ke\frac{dx_\mu}{dt}dx^\mu = 0 \tag{64}$$

and

$$\widetilde{L}^2 = L_\mu L^\mu = K^2e^2\frac{dx_\mu}{dt}\frac{dx^\mu}{dt} = 0 \tag{65}$$

Here, the value of the indicator e can be taken to be 1. Now, the line element of this Finsler space can be written as

$$d\widehat{s}^2 = F^2(\widetilde{x}, d\widetilde{x}) = g_{\mu\nu}(d\widetilde{x})dx^\mu dx^\nu \tag{66}$$

where
$g_{\mu\nu}(d\widetilde{x}) = diag(1,-1,-1,-1)$ for time like directions.
$= diag(-1,1,1,1)$ for space like directions.
Alternatively, we can write

$$F^2(\widetilde{x},\widetilde{y}) = \widetilde{y}^2 \in (\widetilde{y}^2) \tag{67}$$

where

$$\widetilde{y}^2 = \eta_{ij}y^iy^j \tag{68}$$

$$\eta_{ij} = diag(1,-1,-1,-1)$$

and

$\in (\widetilde{y}^2) = 1$ for $\widetilde{y}^2 \geq 0$
$= -1$ for $\widetilde{y}^2 < 0$

1.7: FINSLERIAN MICROLOCAL SPACETIME

The above simplest type of Finsler space can be generalized by taking the line element to be of the form:

$$ds^2 = F^2(\widetilde{x}, d\widetilde{x}) = \widehat{g}_{\mu\nu}(d\widetilde{x})dx^\mu dx^\nu \tag{69}$$

where

$$\widehat{g}_{\mu\nu}(d\widetilde{x}) \equiv \widehat{g}_{\mu\nu}(\widetilde{x}, d\widetilde{x}) = \widehat{g}_{\mu\nu}(\widetilde{x},\widetilde{\nu}) = g(\widetilde{x})f(\widetilde{\nu})\eta_{\mu\nu} \tag{70}$$

with

$$\nu^i = \frac{dx^i}{dt}$$

$$\eta_{ij} = diag(1, -1, -1, -1)$$

Here,

$$f(\widetilde{\nu}) = \in (\widetilde{\nu}^2)\widehat{f}(\widetilde{\nu}) \tag{71}$$

where $\widehat{f}(\widetilde{\nu})$ is (0)p-homogeneous in ν^i. Also, the function g($\widetilde{x}$) can be any of the physically relevant functions of the coordinates of the underlying manifold (the base space). For example, one can choose

$$g(\widetilde{x}) = exp(-b_k x^k), (b_k x^k)^n or (1 + b_k x^k)^n \tag{72}$$

In order to describe the anisotropy of the hadronic matter extension, which is supposed to be manifested in the order of a length scale given by a fundamental length l the microlocal spacetime has recently been characterized as to be Finslerian with $\hat{f}(\widetilde{x}) = 1$ (De, 1997 and reference therein). Thus, this Finsler sapce of the microlocal spacetime of the hadronic matter extension is given by the fundamental function

$$F^2(\widetilde{x}, \widetilde{\nu}) = \hat{g}_{ij}(\widetilde{x}, \widetilde{\nu})\nu^i\nu^j$$

where

$$\hat{g}_{ij}(\widetilde{x}, \widetilde{\nu}) = \eta_{ij} g(\widetilde{x}) \in (\widetilde{\nu}^2) \tag{73}$$

Here, $\hat{g}$ and also in (69) is not in general the Finslerian metric tensor, but simply a homogeneous tensor of degree zero in $\widetilde{\nu}$ that is used for the purpose of defining F. The Finsler metric tensor is given by (2). This Finsler space is, in fact, of the type described in section 1.2. The metric function in this case is the positive forth root of a forth order differential form. This can be evident if we write the tensor field $g = (g_{\mu\gamma\rho\sigma}(x^k))$ as

$$g_{\mu\gamma\rho\sigma}(x^k) = [g(x_k)]^2 \eta_{\mu\gamma}\eta_{\rho\sigma} \tag{74}$$

Consequently,

$$F^4(\widetilde{x}, \widetilde{\nu}) = [g(x_k)]^2 \eta_{\mu\gamma}\eta_{\rho\sigma}\nu^\mu\nu^\gamma\nu^\rho\nu^\sigma = [g(\widetilde{x})\widetilde{\nu}^2]^2 \tag{75}$$

or,

$$ds^4 = [g(\widetilde{x})\eta_{\mu\gamma}dx^\mu dx^\gamma]^2 \tag{76}$$

Two possibilities are obvious and they are

1. The metric is given by

$$ds^2 = g(\widetilde{x})\eta_{\mu\gamma}dx^\mu dx^\gamma \tag{77}$$

This corresponds to a Riemannian metric with the metric tensor

$$g_{\mu\gamma}(\widetilde{x}) = g(\widetilde{x})\eta_{\mu\gamma} \tag{78}$$

2. The metric

$$ds^2 = g(\widetilde{x}) \in (\widetilde{\nu}^2)\eta_{\mu\gamma}dx^\mu dx^\gamma \tag{79}$$

Since

$$[\in (\widetilde{\nu}^2)]^2 = [\in (d\widetilde{x}^2)]^2 = 1$$

for all $\widetilde{\nu}^2$, where $d\widetilde{x}^2 = \eta_{\mu\gamma}dx^\mu dx^\gamma$.

This corresponds to the Finsler space given by the fundamental function (73), the microlocal space we are considering. It is interesting to note that if we insist on the condition of non negativeness of ds^2 then the second case which gives rise to a Finsler geometry is the only possibility from (75). This Finsler space may be termed as Finslerian Riamann spacetime and the space described in section 1.6 as the Finslerian Minkowski spacetime.

1.8: THE MACROSPACE

If the spacetime is the order of a length scale given by a fundamental length l is a Finsler space as described above then the macrospaces like Minkowski flat spacetime and the spacetime that describes the large scale structure of the Universe should come out of it in the length scale greater or far greater than l. For this purpose, let us recall from section 1.2 a kind of Finsler space $F_n = (M_n, F)$ which is (α, β) metric. Here, α is a Riemannian fundamental function and β is a differential 1-form. The present Finsler space (73) that describes the microlocal spacetime of hadronic matter extension is, in fact, a very simple type of Finsler space and also a simple type of (α, β) metric Finsler space. The fundamental function F of this Finsler space can be written as

$$F(\widetilde{x}, \widetilde{\nu}) = \alpha\sqrt{\theta(\alpha^2)} = \alpha\beta^0\sqrt{\theta(\alpha^2)} \tag{80}$$

if, $g(\widetilde{x}) > 0$.
Here

$$\alpha(\widetilde{x}, \widetilde{\nu}) = [g(\widetilde{x})\eta_{ij}\nu^i\nu^j]^{\frac{1}{2}} \tag{81}$$

and β is an arbitrary differential 1-form. Here,
$\theta(z) = 1$, if $z \geq 0$
$= -1$ if $z < 0$.
Obviously, the function F is (1)p-homogeneous in α and β. Thus the Finsler space is given in (73) is (α, β) metric. The associated space $R_n = (M_n, \alpha)$ is Riemannian and may be regarded as as the macrospace. This space is conformal to the Minkowski spacetime. For $g(\widetilde{x}) = 1$, it is Minkowskian itself describing the spacetime of common experience. The other physically relevant spacetimes such as background spacetime of the Universe can be obtained if the function $g(\widetilde{x})$ is suitably chosen. In fact, the Robertson-Walker spacetime will comeout from a function $g(\widetilde{x})$ depending only on time and through a pure time transformation that will be discussed later.
A good physical comprehension of the notion of macrospaces coming out from the Finslerian microspace can be made if one regards the space at larger length scales (than that of the fundamental length l) are only the 'averaged' spaces. This 'averaging' is done on the metric tensor $g_{\mu\gamma}(\widetilde{x}, \widetilde{\nu})$ of the Finsler space and in fact, the line element of the space, the length or norm of vectors and other geometric objects are all dependent on this metric tensor. Physically, one regard the effect of directional dependence is smoothen out at the larger length scales and thus the averaging procedure should be the integration of the metric tensor over the tangent space ($\widetilde{\nu} - space$) with weight functions or probability density function (pdf) of some specific types. In an article, (De, 1989) this has been done to obtain macrospaces but with complex weight functions. Recently, (De, 1997) the real weight functions have been used for this procedure.
First we write the Finsler metric tensor for the sapce we are considering, as given in (73). It is given as

$$g_{ij}(\widetilde{x}, \widetilde{\nu}) = \hat{g}_{ij}(\widetilde{x}, \widetilde{\nu}) + 4g(\widetilde{x})\delta(\widetilde{\nu}^2)\eta_{\alpha i}\eta_{\beta j}\nu^\alpha\nu^\beta \tag{82}$$

where $\delta(\widetilde{\nu}^2)$ is the Dirac δ-function. Obviously, the metric tensor $g = (g_{ij})$ is of the form (24) and the relation (26) is satisfied since

$$\widetilde{\nu}^2\delta(\widetilde{\nu}^2) = 0 \tag{83}$$

Now, the metric tensor of the macrospace can be found on integration over the tangent space with the weight function $\phi(\widetilde{\nu})$. Writing $g = (g_{ij}(\widetilde{x}))$ for the metric tensor of the macrospace, we have

$$g_{ij}(\widetilde{x}) = \int g_{ij}(\widetilde{x}, \widetilde{\nu})\phi(\widetilde{\nu})d^4\nu \tag{84}$$

We take a real weight function given as

$$\phi(\widetilde{\nu}) = \frac{1}{8\pi^2}\widetilde{\nu}^2 \in (\widetilde{\nu}^2)exp\left(-\frac{1}{2}\delta_{ij}\nu^i\nu^j\right) \tag{85}$$

Then we have

$$g_{ij}(\widetilde{x}) = \int g_{ij}(\widetilde{x}, \widetilde{\nu})\phi(\widetilde{\nu})d^4\nu = \widehat{g}_{ij}(\widetilde{x}, \widetilde{\nu})\phi(\widetilde{\nu})d^4\nu = g(\widetilde{x})\eta_{ij} \tag{86}$$

(Here, the relation (83) has been used) If, for example, $g(\widetilde{x}) = exp(-b_k(l)x^k)$ where the parameter $b_k(l)$ are functions of the fundamental length l such that $b_k(l) \to 0$ as $l \to 0$; then $g(\widetilde{x}) \to 1$ as $l \to 0$. Thus it is evident from (86) that $(g_{ij}(\widetilde{x}))$ becomes the metric tensor (η_{ij}) of the Minkowski flat spacetime.
The Robertson-Walker background metric of the Universe can be deduced from (86) if one chooses

$$g(\widetilde{x}) = (1 + b_0 x^0)^n or (b_0 x^0)^n \tag{87}$$

with $x_0 = ct$.
Actually, the metric tensor thus obtained corresponds to a spacetime conformal to the Minkowski sapcetime but a pure time transformation can usher into the Robertson-Walker sapcetime. To see it explicitly, let us write the line element of the macrospace for the case

$$g(\widetilde{x}) = (b_0 x^0)^n = \Omega^2(x^0)(say) \tag{88}$$

It is given by

$$ds^2 = \Omega^2(x^0)\eta_{ij}dx^i dx^j \tag{89}$$

We make a pure time transformation

$$\Omega^2(x^0)dx^0 = cdT \tag{90}$$

where T is known as cosmological proper time or Robertson-Walker time. By integration we can obtain a relation between t and T. Thus, we can set

$$\Omega^2(x^0)dx^0 = R(T) \tag{91}$$

and the conformal time t is given by

$$x^0 = ct = c\int \frac{dT}{R(T)} \tag{92}$$

The line element then becomes

$$ds^2 = c^2 dT^2 - R^2(T)[(dx^1)^2 + (dx^2)^2 + (dx^3)^2] \tag{93}$$

which represents a Robertson-Walker metric with flat spatial cross section. R(T) is the scale factor of the Robertson-Walker Universe. For $n = 4$ in the expression for Ω^2 in (88), $R(T) \propto T^{\frac{2}{3}}$ and represents the Robertson-Walker matter dominated Universe. Thus, the large scale spacetime structures are also contained into the basic microscale Finslerian structure of the spacetime. In fact, the underlying manifold of the Finsler space (the $\widetilde{x}$-space) represents the macrospaces which is also regarded as the associated (Riemannian) space of the microspace, the (α, β) metric. Finsler space.
It is to be noted that we have taken η_{ij} to be of signature +2. On the contrary, if η_{ij} has signature -2 in (73) then we should have the weight function or pdf as

$$\phi(\widetilde{\nu}) = \frac{1}{24\pi^2}\widetilde{\nu}^2(\nu_0)^4 \in (\widetilde{\nu}^2) exp\left(-\frac{1}{2}\delta_{ij}\nu^i\nu^j\right) \tag{94}$$

to arrive the same result.

1.9: GEODESIC OF THE MICROSPACE

In section 1.3, we have written the equations of geodesic of the Finsler space in terms of Christoffel symbols of second kind (the Finsler Christoffel connection). Also, the equation of an autoparallel curve (given by $\frac{\delta\nu^i}{\delta s} = 0$) can be expressed in terms of the connection coefficients Γ^i_{jk}. It is easy to show that

$$\Gamma^i_{jk}\nu^j\nu^k = \gamma^i_{jk}\nu^j\nu^k \tag{95}$$

This also expresses the fact that the geodesics of a Finsler space are autoparallel curves. γ^i_{jk} are given as

$$\gamma^i_{jk} = g^{ih}\gamma_{jhk} = \frac{1}{2}g^{ih}\left[\frac{\partial g_{jh}}{\partial x^k} + \frac{\partial g_{hk}}{\partial x^j} - \frac{\partial g_{jk}}{\partial x^h}\right] \tag{96}$$

For the Finslerian microspace we are considering g_{ij} are given in (83) and $g^{ij}(\widetilde{x}, \widetilde{\nu})$ can be found to be

$$g^{ij}(\widetilde{x}, \widetilde{\nu}) = \widehat{g}^{ij}(\widetilde{x}, \widetilde{\nu}) - \frac{4\delta(\widetilde{\nu}^2)\nu^i\nu^j}{g(\widetilde{x})}$$

where

$$\widehat{g}^{ij}(\widetilde{x}, \widetilde{\nu}) = \frac{1}{g(\widetilde{x})} \in (\widetilde{\nu}^2)\eta^{ij} \tag{97}$$

It is easy to show the relations

$$g_{ij}g^{ih} = \delta^h_j$$

and

$$\widehat{g}_{ij}\widehat{g}^{ih} = \delta^h_j \tag{98}$$

In deducing the first of the above relations (98) we have used the following formula

$$z^m\delta(z) = 0 \tag{99}$$

for $m > 0$.
The covariant and contravariant tensors are related as follows:

$$\phi_i = g_{ij}\phi^j$$

and

$$\phi^i = g^{ij}\phi_j \tag{100}$$

But for ν^i and ν_i, we can make with $\widehat{g}^{ij}$ and $\widehat{g}_{ij}$ instead of Finsler metric tensors g_{ij} and g^{ij}. That is,

$$\nu_i = \widehat{g}_{ij}\nu^j$$

and

$$\nu^i = \widehat{g}^{ij}\nu_j \tag{101}$$

This is evident from their expressions given in (82) and (97).
Now, let us define, coefficients $\widehat{\gamma}_{\alpha\beta\lambda}$ and $\widehat{\gamma}^\delta_{\alpha\beta}$ as

$$\widehat{\gamma}_{\alpha\beta\lambda} = \frac{1}{2}\left[\frac{\partial \widehat{g}_{\alpha\beta}}{\partial x^\lambda} + \frac{\partial \widehat{g}_{\beta\lambda}}{\partial x^\alpha} - \frac{\partial \widehat{g}_{\alpha\lambda}}{\partial x^\beta}\right] \tag{102}$$

and

$$\widehat{\gamma}^\delta_{\alpha\beta} = \widehat{g}^{\delta\lambda}\widehat{\gamma}_{\alpha\lambda\beta} \tag{103}$$

Then the relation between $\gamma_{\alpha\beta\lambda}$ and $\widehat{\gamma}_{\alpha\beta\lambda}$ can be found to be

$$\gamma_{\alpha\beta\lambda} = \widehat{\gamma}_{\alpha\beta\lambda} + 2\delta(\widetilde{\nu}^2)\left[\frac{\partial \widehat{g}(\widetilde{x})}{\partial x^\lambda}\eta_{\alpha'\alpha}\eta_{\beta'\beta}\nu^{\alpha'}\nu^{\beta'} + \frac{\partial \widehat{g}(\widetilde{x})}{\partial x^\alpha}\eta_{\alpha'\lambda}\eta_{\beta'\beta}\nu^{\alpha'}\nu^{\beta'} - \frac{\partial \widehat{g}(\widetilde{x})}{\partial x^\beta}\eta_{\alpha'\alpha}\eta_{\beta'\lambda}\nu^{\alpha'}\nu^{\beta'}\right] \tag{104}$$

Consequently, we find

$$\gamma_{\alpha\beta\lambda}\nu^\alpha\nu^\lambda = \widehat{\gamma}_{\alpha\beta\lambda}\nu^\alpha\nu^\lambda \tag{105}$$

Again,

$$\gamma^{\delta}_{\alpha\lambda}\nu^{\alpha}\nu^{\lambda} = g^{\delta\beta}\gamma_{\alpha\beta\lambda}\nu^{\alpha}\nu^{\lambda} = \left[\widehat{g}^{\delta\beta} - \frac{4\delta(\widetilde{\nu}^2)\nu^{\delta}\nu^{\beta}}{g(\widetilde{x})}\right]\widehat{\gamma}_{\alpha\beta\lambda}\nu^{\alpha}\nu^{\lambda}$$

or,

$$\gamma^{\delta}_{\alpha\lambda}\nu^{\alpha}\nu^{\lambda} = \widehat{\gamma}^{\delta}_{\alpha\lambda}\nu^{\alpha}\nu^{\lambda} - \frac{4\delta(\widetilde{\nu}^2)\nu^{\delta}}{g(\widetilde{x})}\widehat{\gamma}_{\alpha\beta\lambda}\nu^{\alpha}\nu^{\beta}\nu^{\lambda} \tag{106}$$

Now,

$$\delta(\widetilde{\nu}^2)\widehat{\gamma}_{\alpha\beta\lambda}\nu^{\alpha}\nu^{\beta}\nu^{\lambda} = \frac{1}{2}\delta(\widetilde{\nu}^2)\left[\frac{\partial(\widehat{g}_{\alpha\beta}\nu^{\alpha}\nu^{\beta})}{\partial x^{\lambda}}\nu^{\lambda} + \frac{\partial(\widehat{g}_{\beta\lambda}\nu^{\beta}\nu^{\lambda})}{\partial x^{\alpha}}\nu^{\alpha} - \frac{\partial(\widehat{g}_{\alpha\lambda}\nu^{\alpha}\nu^{\lambda})}{\partial x^{\beta}}\nu^{\beta}\right]$$

Since, $\widehat{g}_{\alpha\beta}\nu^{\alpha}\nu^{\beta} = F^2(\widetilde{x},\widetilde{\nu}) = g(\widetilde{x}) \in (\widetilde{\nu}^2)\widetilde{\nu}^2$, we have

Now,

$$\delta(\widetilde{\nu}^2)\widehat{\gamma}_{\alpha\beta\lambda}\nu^{\alpha}\nu^{\beta}\nu^{\lambda} = \frac{1}{2} \in (\widetilde{\nu}^2)\widetilde{\nu}^2\delta(\widetilde{\nu}^2)\left[\frac{\partial g(\widetilde{x})}{\partial x^{\lambda}}\nu^{\lambda} + \frac{\partial g(\widetilde{x})}{\partial x^{\alpha}}\nu^{\alpha} - \frac{\partial g(\widetilde{x})}{\partial x^{\beta}}\nu^{\beta}\right] = 0$$

[since, $\widetilde{\nu}^2\delta(\widetilde{\nu}^2) = 0$.]
Thus, from (106) and (95) we arrive at the following relation

$$\Gamma^{i}_{jk}\nu^{j}\nu^{k} = \gamma^{i}_{jk}\nu^{j}\nu^{k} = \widehat{\gamma}^{i}_{jk}\nu^{j}\nu^{k} \tag{107}$$

Consequently, the geodesic of this Finslerian microspace can be expressed by the following form:

$$\frac{d\nu^i}{ds} + \widehat{\gamma}^{i}_{jk}\nu^{j}\nu^{k} = 0 \tag{108}$$

or, equivalently,

$$\frac{d^2x^i}{ds^2} + \widehat{\gamma}^{i}_{jk}\frac{dx^j}{ds}\frac{dx^k}{ds} = 0 \tag{109}$$

It is to be noted that $\widehat{\gamma}^{i}_{jk}\nu^{j}$ are Christoffel symbols of second kind as calculated from $\widehat{g}_{ij}$ and $\widehat{g}^{ij}$ which are not actually Finsler metric tensors defining the fundamental function of the Finsler space. Of course, they are related to the Finsler metric tensor. It is evident from above that the autoparallel curve or geodesic in this Finsler space will be the same if one deals with $\widehat{\gamma}^{i}_{jk}\nu^{j}$ computed from $\widehat{g}_{ij}$ and $\widehat{g}^{ij}$. This is of course, not true in general for other types of Finsler space.
A non linear connection can be defined (Miron and Anastasiei (1987)) in order to obtain a covariant basis for the module of the vector fields. In local coordinates the transformation matrix is given by $X^{*\mu}_{\lambda} = \frac{\partial x'^{\mu}}{\partial x^{\lambda}}$ for the transformation $x'^{\mu} = x'^{\mu}(x^{\lambda})$ of the base space (the underlying manifold). The transformation for the directional variables $\widetilde{\nu}$ is given by

$$\nu'^{\mu} = X^{*\mu}_{\lambda}\nu^{\lambda} \tag{110}$$

[cf. equations (18) and (19)].

Now, the covariant basis is $\left(\frac{\delta}{\delta x^{\mu}}, \frac{\partial}{\partial \nu^{\mu}}\right)$ where

$$\frac{\delta}{\delta x^{\mu}} = \frac{\partial}{\partial x^{\mu}} - N^{\lambda}_{\mu}\frac{\partial}{\partial \nu^{\lambda}} \tag{111}$$

Here, the matrix N^{λ}_{μ} which is the local representation of the non linear connection is given by

$$N^{\lambda}_{\mu} = \frac{1}{2}\frac{\partial}{\partial \nu^{\lambda}}\left(\gamma^{\mu}_{\alpha\beta}\nu^{\alpha}\nu^{\beta}\right) = \frac{1}{2}\frac{\partial}{\partial \nu^{\lambda}}\left(\widehat{\gamma}^{\mu}_{\alpha\beta}\nu^{\alpha}\nu^{\beta}\right) \tag{112}$$

[by using (107)]

and N^{λ}_{μ} satisfy the following transformation law

$$N'^{\alpha}_{\beta} = X^{*\alpha}_{\lambda}X^{\mu}_{\beta}N^{\lambda}_{\mu} + X^{*\alpha}_{\lambda}\frac{\partial X^{\lambda}_{\beta}}{\partial x'^{\mu}}\nu'^{\mu} \tag{113}$$

where $X^{\mu}_{\lambda} X^{*\lambda}_{\alpha} = \delta^{\mu}_{\alpha}$.
With these, the basis $\left(\frac{\delta}{\delta x^{\mu}}, \frac{\partial}{\partial \nu^{\mu}}\right)$ satisfies the required transformation law

$$\left(\frac{\delta}{\delta x'^{\mu}}, \frac{\partial}{\partial \nu'^{\mu}}\right) = \left(X^{\lambda}_{\mu}\frac{\delta}{\delta x^{\lambda}}, X^{\lambda}_{\mu}\frac{\partial}{\partial \nu^{\lambda}}\right) \tag{114}$$

Also, the connection is defined as

$$F^{\lambda}_{\alpha\beta} = \frac{1}{2} g^{\mu\lambda}\left(\frac{\delta g_{\mu\alpha}}{\delta x^{\beta}} + \frac{\delta g_{\beta\mu}}{\delta x^{\alpha}} - \frac{\delta g_{\alpha\beta}}{\delta x^{\mu}}\right) \tag{115}$$

so as to have a covariant derivative of a general vector $V^{\mu}(\widetilde{x}, \widetilde{\nu})$

$$V^{\lambda}_{;\beta} = \frac{\delta V^{\lambda}}{\delta x^{\beta}} + F^{\lambda}_{\alpha\beta} V^{\alpha}$$

$$V'^{\mu}_{;\alpha} = X^{*\mu}_{\lambda} X^{\beta}_{\alpha} V^{\lambda}_{;\beta} \tag{116}$$

and to make the space to be metric, i.e., $g_{\alpha\beta_{;\mu}} = 0$ where

$$g_{\alpha\beta_{;\mu}} = \frac{\delta g_{\alpha\beta}}{\delta x^{\mu}} - g_{\lambda\beta} F^{\lambda}_{\alpha\mu} - g_{\alpha\lambda} F^{\lambda}_{\beta\mu} \tag{117}$$

With the C-tensor of Cartan defined in (38), the triad $\left(N^{\mu}_{\lambda}, F^{\mu}_{\alpha\beta}, C^{\mu}_{\alpha\beta}\right)$ is known as the Cartan connection of the Finsler space.
Now,

$$\frac{\delta g_{\alpha\beta}}{\delta x^{\mu}} = \frac{\partial g_{\alpha\beta}}{\partial x^{\mu}} - N^{\delta}_{\mu}\frac{\partial g_{\alpha\beta}}{\partial \nu^{\delta}} = \frac{\partial g_{\alpha\beta}}{\partial x^{\mu}} - 2N^{\delta}_{\mu} C_{\alpha\beta\delta}$$

[by (38)]

Then,

$$F^{\lambda}_{\alpha\beta} = \frac{1}{2} g^{\mu\lambda}\left(\frac{\partial g_{\mu\alpha}}{\partial x^{\beta}} + \frac{\partial g_{\beta\mu}}{\partial x^{\alpha}} - \frac{\partial g_{\alpha\beta}}{\partial x^{\mu}}\right) - g^{\mu\lambda}\left(N^{\delta}_{\beta} C_{\mu\alpha\delta} + N^{\delta}_{\alpha} C_{\beta\mu\delta} - N^{\delta}_{\mu} C_{\alpha\beta\delta}\right) = \gamma^{\lambda}_{\alpha\beta} - N^{\delta}_{\beta} C^{\lambda}_{\alpha\delta} - N^{\delta}_{\alpha} C^{\lambda}_{\beta\delta} + g^{\mu\lambda} N^{\delta}_{\mu} C_{\alpha\beta\delta} \tag{118}$$

Therefore, we have (by using (39)) the following relation

$$F^{\lambda}_{\alpha\beta}\nu^{\beta} = \gamma^{\lambda}_{\alpha\beta}\nu^{\beta} - N^{\delta}_{\beta} C^{\lambda}_{\alpha\delta}\nu^{\beta} = \gamma^{\lambda}_{\alpha\beta}\nu^{\beta} - \frac{1}{2}\frac{\partial}{\partial \nu^{\beta}}\left(\widehat{\gamma}^{\delta}_{ij}\nu^{i}\nu^{j}\right) C^{\lambda}_{\alpha\delta}\nu^{\beta} = \gamma^{\lambda}_{\alpha\beta}\nu^{\beta} - \widehat{\gamma}^{\delta}_{ij}\nu^{i}\nu^{j} C^{\lambda}_{\alpha\delta} \tag{119}$$

[using (112)]
Consequently, we have

$$F^{\lambda}_{\alpha\beta}\nu^{\alpha}\nu^{\beta} = \gamma^{\lambda}_{\alpha\beta}\nu^{\alpha}\nu^{\beta} = \widehat{\gamma}^{\lambda}_{\alpha\beta}\nu^{\alpha}\nu^{\beta} \tag{120}$$

From (116) it follows that

$$\frac{\delta \nu^{\lambda}}{\delta x^{\beta}} + F^{\lambda}_{\alpha\beta}\nu^{\alpha} = 0 \tag{121}$$

[since $\widetilde{\nu}$ is independent of $\widetilde{x}$] which gives

$$N^{\lambda}_{\beta} = F^{\lambda}_{\alpha\beta}\nu^{\alpha} \tag{122}$$

The dual basis is given by $(dx^\mu, \delta\nu^\mu)$ with

$$\delta\nu^\mu = d\nu^\mu + N^\mu_\lambda dx^\lambda$$

and the geodesic equation is expressed as

$$\frac{\delta\nu^\mu}{\delta s} = \frac{d\nu^\mu}{ds} + N^\mu_\alpha \frac{dx^\alpha}{ds} = \frac{d\nu^\mu}{ds} + N^\mu_\alpha \nu^\alpha = 0 \tag{123}$$

That is,

$$\frac{\delta\nu^\mu}{\delta s} = 0 = \frac{d\nu^\mu}{ds} + F^\mu_{\alpha\beta}\nu^\alpha\nu^\beta = \frac{d\nu^\mu}{ds} + \widehat{\gamma}^\mu_{\alpha\beta}\nu^\alpha\nu^\beta \tag{124}$$

REFERENCES

1. Antonelli , P.L. , Ingarden, R.S. and Matsumoto, M. (1993).
The Theory of Sprays and Finsler Spaces with Applications in Physics and Biology,
Kluwer Academic Publishers, Dordrecht, Netherland .

2. Asanov, G.S. (1985). Finsler Geometry, Relativity and Gauge theories, D. Reidel, Dordrecht, Netherland .

3. Berwald, L. (1941). Mathematica (Timisoara) **17**, 34.

4. Berwald, L. (1947). Ann. Math. (2) **48**, 755.

5. Blokhintsev, D.I. (1978). Preprint of the Joint Institute for Nuclear Research, Dubna; No. E2 - 11297.

6. Carathéodory, C. (1935). Variationsrechnung und partielle Differetialgleichungen erster Ordnung, Teubner Press, Leipzig.

7. Cartan, E. (1934). Les Espaces de Finsler, Actualités 79, Hermann, Paris.

8. De, S.S. (1997). Int.J.Theor.Phys., **36**, 89

9. De, S.S. (1989). In Hadronic Mechanics and Nonpotential Interaction, M. Mijatovic, ed., Nova Science Publishers, Inc., New York, p-37.

10. Finsler, P. (1918). Über Kurven und Flächen in Allgemeinen Räumen, Dissertation, University of Göttingen

11. Kropina, V.K. (1959). On Projective Finsler spaces with a metric of some special form, (in Russian), Naučn. Dokl. Vysš. Školy. Fis. Mat. **2**,38-42

12. Matsumoto, M. (1986). Foundation of Finsler Geometry and Special Finler Spaces, Kaiseisha Press, Shigaken, Japan.

13. Miron, R. and Anastasiei, M. (1987). Fibrate Vectoriale, Spatii Lagrange, Aplicatii in Theoria Relativitatii, Academiei Republicii Socialiste Romania, Bucharest, Romania.

14. Riemann, G.F.B. (1854). Uber die Hypothesen welche der Geometrie zu Grunde liegen, Habilitation thesisd, Yniversity of Göttingen

15. Rund, H. (1959). The Differential Geometry of Finsler Spaces, Springer - Verlag, Berlin.

CHAPTER TWO:

WAVE EQUATIONS IN FINSLER SPACE

2.1: CLASSICAL WAVE FUNCTION

In the previous chapter we have discussed the special Finsler space of hadronic matter extension, the microlocal space time which is manifested below the length scale of the order of a fundamental length. The macrospace of common experience and also the large scale spacetime of the Universe appear as the 'average' spacetimes of the microspace. Thus, the 'reality' of macrospaces on the larger length scales than the fundamental length scale is only the averaged entity over the hindmost Finslerian spacetime which is anisotropic in character.

Now, we should find the wave or field equation of fundamental particles in this Finslerian microspace. For the free particles supposed to be leptons the classical field $\psi(\widetilde{x},\widetilde{\nu})$ should depend on the directional variables $\widetilde{\nu}$ apart from its dependence on the position coordinates $\widetilde{x}$ as it is on 'entity' of the Finsler space. In fact, all the geometric objects of a Finsler space, such as metric tensor, the connection coefficients, depend on the position coordinates as well as on the directional arguments. The classical wave (field) equation for $\psi(\widetilde{x},\widetilde{\nu})$ can be obtained if we admit the following conjecture (De, 1997):

The equivalence property is to be satisfied by $\psi(\widetilde{x},\widetilde{\nu})$along the neighbouring points in the microdomain (Finsler space) on the autoparallel curve (the geodesic) which is the curve whose tangent vectors result from each other by successive infinitesimal parallel displacements of the type

$$d\nu^i = -\gamma^i_{hj}(\widetilde{x},\widetilde{\nu})\nu^h dx^j \tag{125}$$

(cf. equations (44) and (107))

This property can be stated as " the infinitesimal change of ψ along the autoparallel curve is proportional to the wave (or field) function itself ", that is,

$$\delta\psi = [\psi(\widetilde{x}+d\widetilde{x},\widetilde{\nu}+d\widetilde{\nu}) - \psi(\widetilde{x},\widetilde{\nu})] \propto \psi(\widetilde{x},\widetilde{\nu})$$

or

$$\psi(\widetilde{x}+d\widetilde{x},\widetilde{\nu}+d\widetilde{\nu}) - \psi(\widetilde{x},\widetilde{\nu}) = \epsilon\frac{mc}{\hbar}\psi(\widetilde{x},\widetilde{\nu}) \tag{126}$$

where $(\widetilde{x}+d\widetilde{x},\widetilde{\nu}+d\widetilde{\nu})$ and $(\widetilde{x},\widetilde{\nu})$ lie on the autoparallel curve.

Here, the mass term m appears as the constant of proportionality and may be regarded as the 'inherent' mass of the particle. Also, c is the velocity of light, $\hbar = \frac{h}{2\pi}$, h is the Planck constant and ϵ is a real parameter such that $0 < \epsilon \leq l$. From (126) we have (to the first orders in $d\widetilde{x}$ and $d\widetilde{\nu}$)

$$(dx^\mu\partial_\mu + d\nu^l\partial_l{}')\psi(\widetilde{x},\widetilde{\nu}) = \epsilon\frac{mc}{\hbar}\psi(\widetilde{x},\widetilde{\nu}) \tag{127}$$

where $\partial_\mu \equiv \frac{\partial}{\partial x^\mu}$ and $\partial_l{}' \equiv \frac{\partial}{\partial \nu^l}$.

Using (125), we have from above equation

$$dx^\mu(\partial_\mu - \gamma^l{}_{h\mu}(\widetilde{x},\widetilde{\nu})\nu^h\partial_l{}')\psi(\widetilde{x},\widetilde{\nu}) = \epsilon\frac{mc}{\hbar}\psi(\widetilde{x},\widetilde{\nu}) \tag{128}$$

Now, we have from the equation(112) for the nonlinear connection $(N^{\nu}{}_{\mu})$ (in local representation)

$$N^{\mu}{}_{\nu} = \frac{1}{2}\frac{\partial}{\partial\nu^{\nu}}(\gamma_{\alpha\beta}{}^{\mu}\nu^{\alpha}\nu^{\beta})$$

From this it follows (because of homogeneity property)

$$N^{\mu}{}_{\nu}\nu^{\nu} = \gamma_{\alpha\beta}{}^{\mu}\nu^{\alpha}\nu^{\beta} \tag{129}$$

where $\nu^{\alpha} = \frac{dx^{\alpha}}{ds}$.

Consequently, we have

$$N^{\mu}{}_{\nu}dx^{\nu} = \gamma_{\alpha\beta}{}^{\mu}\nu^{\alpha}dx^{\beta} \tag{130}$$

Then, the equation (128) for ψ becomes

$$dx^{\mu}(\partial_{\mu} - N^{l}{}_{\mu}\partial_{l}{}')\psi(\widetilde{x},\widetilde{\nu}) = \epsilon\frac{mc}{\hbar}\psi(\widetilde{x},\widetilde{\nu}) \tag{131}$$

or, in terms of "covariant" $\frac{\delta}{\delta x^{\mu}}$ given in (111),

$$dx^{\mu}\frac{\delta\psi(\widetilde{x},\widetilde{\nu})}{\delta x^{\mu}} = \epsilon\frac{mc}{\hbar}\psi(\widetilde{x},\widetilde{\nu}) \tag{132}$$

Thus, (as dx^{μ} transforms covariantly like a vector) $\psi(\widetilde{x},\widetilde{\nu})$ behaves like a scalar under the general coordinate transformations, that is,

$$\psi'(\widetilde{x'},\widetilde{\nu'}) - \psi(\widetilde{x},\widetilde{\nu}) \tag{133}$$

where $x'^{\mu} = x'^{\mu}(x^{\nu})$ and $\nu'^{\mu} = X^{*\mu}_{\nu}\nu^{\nu}$
with $X^{*\mu}_{\nu} = \frac{\partial x'^{\mu}}{\partial x^{\nu}}$.

It is to be noted that the equation (132) for the classical wave (or field) function is form invariant under the general coordinate transformation. Also, the parameter ϵ depends on the differential dx^{μ} (and consequently on $d\nu^{\mu}$); that is, ϵ becomes smaller or larger according as the two neighbouring points on the autoparallel curve are closer or farther to each other respectfully. Thus, one can identify the real parameterϵ with the Finslerian arc distance element ds between these points on the autoparallel curve i.e. the geodesic distance element. Then, we get the classical wave (field) equation in the following form:

$$\nu^{\mu}\frac{\delta\psi(\widetilde{x},\widetilde{\nu})}{\delta x^{\mu}} = \frac{mc}{\hbar}\psi(\widetilde{x},\widetilde{\nu}) \tag{134}$$

The equations (132) and (134) for the field $\psi(\widetilde{x},\widetilde{\nu})$ are derived from the property of equivalence of it on the autoparallel curve at the neighbouring points (x^{μ},ν^{μ}) and $(x^{\mu}+dx^{\mu},\nu^{\mu}+d\nu^{\mu})$ for which $\eta_{\mu\nu}dx^{\mu}dx^{\nu} < 0$. On the contrary, for $\eta_{\mu\nu}dx^{\mu}dx^{\nu} \geq 0$, the equivalence property should be taken as

$$\delta\psi = \psi(\widetilde{x}+d\widetilde{x},\widetilde{\nu}+d\widetilde{\nu}) - \psi(\widetilde{x},\widetilde{\nu}) = -i\epsilon\frac{mc}{\hbar}\psi(\widetilde{x},\widetilde{\nu}) \tag{135}$$

which leads to the following equation for $\psi(\widetilde{x},\widetilde{\nu})$:

$$dx^{\mu}\frac{\delta\psi(\widetilde{x},\widetilde{\nu})}{\delta x^{\mu}} = -i\epsilon\frac{mc}{\hbar}\psi(\widetilde{x},\widetilde{\nu}) \tag{136}$$

for $\eta_{\mu\nu}dx^{\mu}dx^{\nu} \geq 0$.
Thus, the wave (or field) equation takes the following form :

$$idx^{\mu}\frac{\delta\psi(\widetilde{x},\widetilde{\nu})}{\delta x^{\mu}} = \epsilon\frac{mc}{\hbar}\sqrt{\theta(d\widetilde{x}^{2})}\psi(\widetilde{x},\widetilde{\nu}) \tag{137}$$

where

$$d\widetilde{x}^2 = \eta_{\mu\nu} dx^\mu dx^\nu$$

and

$$\theta(z) = 1 \quad for \quad z \geq 0$$
$$= -1 \quad for \quad z < 0$$

(137*a*)

As before, if we identify ϵ with the arc distance element ds then the above equation for $\psi(\widetilde{x}, \widetilde{\nu})$ becomes

$$i\nu^\mu \frac{\delta\psi(\widetilde{x}, \widetilde{\nu})}{\delta x^\mu} = \frac{mc}{\hbar} \sqrt{\theta(\widetilde{\nu^2})}\psi(\widetilde{x}, \widetilde{\nu}) \tag{138}$$

where $\nu^\mu = \frac{dx^\mu}{ds}$ and $\widetilde{\nu}^2 = \eta_{\mu\nu}\nu^\mu\nu^\nu$.

2.2: QUANTIZED SPACE-TIME

From the classical field or wave functions we now transit to the quantum field or wave functions and to their equation on the basis of quantized space-time. Synder(1947), and Yang(1947) discussed long ago such theories of quantized space-time. Later, Kadyshevsky(1959,1962)and Tamm(1965) developed the theory. Other contributors include Leznov(1967);Kirzhnits and Cheehin(1967), Blokhintsev(1973), Prugovečki(1984) and Namsrai(1985).

In the theory of quantized space-time there is no usual conceptual meaning of definite space-time point. In fact, the components of the operators of coordinates are not commuted. Blokhintsev(1973) made a general statement of this problem as follows:

The usual (c-number) coordinates of points, (x^0, x^1, x^2, x^3) of space-time, which form a differential manifold $M_4(\widetilde{x})$ (with a certain metric) are changed by linear operators $(\hat{x^0}, \hat{x^1}, \hat{x^2}, \hat{x^3})$ noncommuting with one another. This leads to the question concerning the "measurable" numerical coordinates of a point event and the ordering of events in the "operational space" $M_4(\widetilde{\hat{x}})$. The reasonable answer to this question can be provided by admitting a mapping of the operational space on a space of eigenvalues of $\hat{x}$ or of functions $f(\hat{x})$ which form a complete set of variables. This set should be sufficient for ordering points in the four-dimensional Pseudo-Euclidean space. Toward such consideration Blokhintsev postulated the space $H(\phi)$ of eigenfunctions ϕ of the complete set at each point of space$M_4(\widetilde{x})$.He also considered several examples of which the following will be relevant for our present purpose. There he made the quantum generalization of Finsler space.In fact, the Finsler space which he considered is the usual Minkowski four-dimensional space as pointed out in section 1.6 of chapter 1. The length element ds for this case is expressed as

$$ds = N_\mu dx^\mu \tag{139}$$

where the vector N_μ is a zero-order form in $d\widetilde{x}$. This form is different for time-like, space-like directions and light cone, having three possible values,$\widetilde{N^2} = \pm 1, 0$.

Now, the quantum generalization of this Finsler space is achieved through the change of coordinate differentials dx^μ in (139) by the finite operators

$$\Delta\hat{x}^\mu = a\gamma^\mu \tag{140}$$

where γ^μ are the Dirac matrices and a is a certain length. The operator of interval is taken as follows:

$$\Delta\hat{s} = N_\mu \Delta\hat{x}^\mu \tag{141}$$

for $\widetilde{N}^2 = 1$ and $\widetilde{N}^2 = 0$.
and

$$\Delta\hat{s} = \frac{1}{i} N_\mu \Delta\hat{x}^\mu \tag{142}$$

for $\widetilde{N}^2 = -1$.
Evidently, it follows from (140) that

$$[\Delta\hat{x}^\mu, \Delta\hat{x}^\nu] = 2ia^2 \hat{I}^{\mu\nu} \tag{143}$$

where $\hat{I}^{\mu\nu}$ is the four-dimensional spin operator. The space determined by the relations (140), (141) and (142) will be called $\Gamma_4(\hat{x})$ - space.

It follows from (143) that the eigenvalues of operators $\Delta\hat{x}^0, \Delta\hat{x}^1, \Delta\hat{x}^2, \Delta\hat{x}^3$ do not form the complete set. This set can be built out of the eigenvalues of the interval operator $\Delta\hat{s}$ and unit vector $\widetilde{N}$. By solving the equation for the eigen functions ϕ_λ and eigenvalues of the operator $\hat{\sigma}(\widetilde{N}) = \frac{1}{a}\Delta\hat{s}(\widetilde{N})$; that is

$$\hat{\sigma}(\widetilde{N})\phi_\lambda = \lambda\phi_\lambda \tag{144}$$

it is possible to find out the eigenvalues λ. They are given by

$$\lambda = \pm\sqrt{\widetilde{N}^2} \tag{145}$$

for $\widetilde{N}^2 > 0$ and

$$\lambda = \pm\sqrt{-\widetilde{N}^2} \tag{146}$$

for $\widetilde{N}^2 < 0$.
Thus, the eigenvalues of $\Delta\hat{s}$ are given by $\Delta s = \pm a$ or 0.
Regarding the geometrical field ϕ it should be noted that it cannot be interpreted as a probability since for the tachyon states (146) the invariant $\widetilde{\phi}_\lambda \phi_\lambda = 0$ holds.
It follows from (141), (146) and (143) that the interval operators $\Delta s(\widetilde{N}')$ and $\Delta s(\widetilde{N}'')$ for the two nonparallel directions $\widetilde{N}'$ and $\widetilde{N}''$ do not commute :

$$[\Delta\hat{s}(\widetilde{N}'), \Delta\hat{s}(\widetilde{N}'')] = a^2\gamma^\mu\gamma^\nu(\widetilde{N}' \times \widetilde{N}'')_{\mu\nu} \tag{147}$$

(the symbol $\times$ represents the vector product).
Hence, each point of the quantized space $\Gamma_4(\hat{x})$ can be crossed only by one (though arbitrary) straight line.
Regarding the ordering of events, Blokhintsev has chosen the sign for the interval in accordance with the concept of time τ and distance l. For the time-like interval $\hat{s} = \hat{\tau}$, $\widetilde{N}^2 = 1$, at each point, the rule

$$\lambda = \pm 1, \phi_\lambda \equiv \phi_\pm(\pm\widetilde{N}) \tag{148}$$

gives two values of τ, that is, $\tau = \pm a$, whereas for the space-like interval $\hat{s} = \hat{l}, \widetilde{N}^2 = -1$,

$$\lambda = +1, \phi_\lambda \equiv \phi_+(\pm\widetilde{N}) \tag{149}$$

only one sign is admitted, that is, $l = a$.
Thus, in accordance with this choice, at each point in the space-like direction there can be only one ray $(\widetilde{N})$, while in time-like direction there can be two rays $(\pm\widetilde{N})$. Thereby the ordering of events is determined in the space $\Gamma_4(\hat{x})$. It is realized in the same way as in the Minkowski space with the help of interval s and

unit vector $\widetilde{N}$. The important difference is that only one line (for $\widetilde{N}^2$ =-1) are admitted at each point. The eigenvalue of interval for $\widetilde{N}^2$ =1 coincides with time τ in the reference frame, where $\widetilde{N} = (1,0,0,0)$, and for $\widetilde{N}^2$ =-1 with length l in the frame where $\widetilde{N} = (0, \vec{N})$. As to the interval $\Delta s = 0, \widetilde{N}^2 = 0$, it defines neither length nor time because at $\Delta s = 0$ operators $\hat{x}^0$ and $\hat{x}^h$ (h = 1,2,3) do not commute with $\Delta\hat{s}$ in any reference frame.Therefore the seat of points separated by the light cone $\widetilde{N}^2 = 0$ is undermined. Similar quantized space-time has been considered by Namsrai(1985) for the internal space-time I_4 of the space-time $\hat{R}_4 = E_4 + I_4$ where E_4 is the external space-time.
The coordinates $\hat{x}^\mu \epsilon \hat{R}_4 (\mu = 0,1,2,3)$ are given as
$\hat{x}^\mu = x^\mu + r^\mu, x^\mu \epsilon E_4$ and $r^\mu \epsilon I_4$.
The quantization of space-time is realized in two possible ways :

$$\hat{x}^\mu = x^\mu + l\gamma^\mu \tag{150}$$

or

$$\hat{x}^\mu = x^\mu + il\gamma^\mu \tag{151}$$

where l is a fundamental length. Thus, $\hat{R}_4$ is quantized at small distances. For the first and second cases we have, respectively,

$$[\hat{x}^\mu, \hat{x}^\nu] = 2il^2\sigma^{\mu\nu} \qquad and \qquad [\hat{x}^\mu, \hat{x}^\nu] = \frac{2}{i}l^2\sigma^{\mu\nu} \tag{152}$$

where

$$\sigma^{\mu\nu} = \frac{1}{2i}(\gamma^\mu\gamma^\nu - \gamma^\nu\gamma^\mu) \tag{153}$$

A mathematical procedure has been prescribed there to provide a passage to the large scale from small one. This procedure, the averaging of coordinates $r^\mu = l\gamma^\mu$ or $il\gamma^\mu$ of the internal space I_4, is "to trace the γ matrices". For example, for the first realization,

$$< \hat{x}^\mu >_{\hat{R}_4} = x^\mu, < \hat{x}^\mu \hat{x}^\nu > = x^\mu x^\nu + 4l^2 g^{\mu\nu} \quad , \quad < \hat{s}^2 > = < \hat{x}_0^2 > - < \hat{\vec{x}}^2 > = s_0^2 + 16l^2$$

where $s_0^2 = x_0^2 - \vec{x}^2$.
For the second case,

$$< \hat{x}^\mu \hat{x}^\nu > = x^\mu x^\nu - 4l^2 g^{\mu\nu} \quad , \quad < \hat{s}^2 > = s_0^2 - 16l^2$$

It is to be noted that out of the two realization we have to choose only one realization for quantization of space-time. In Blokhintsev's approach the quantization corresponds only to the first case of Namsrai's approach although there is an ambiguity in the operator of interval $\Delta\hat{s}$ in the former one. In fact as it is evident from (141) and (142) that $\Delta\hat{s}$ differs for $\widetilde{N}^2 = 1$ and $\widetilde{N}^2 = 0$ from that for $\widetilde{N}^2 = -1$.

2.3: MICRODOMAIN AS QUANTIZED FINSLER SPACE

Now the quantum generalization of microlocal space-time of hadronic matter extension which is a Finsler space will be made. The fundamental function of this space is given by (73). For quantization we first write the interval of this Finsler space in the following form :

$$ds_F = F(\widetilde{x}, \widetilde{dx}) = L_\mu dx^\mu \tag{154}$$

where L_μ is a zero-order form in $\nu^\mu \equiv \frac{dx^\mu}{ds_F}$.
In fact, L_μ and L^μ are given by (for nonnull ds_F)

$$L^\mu = g(\widetilde{x})\theta(\widetilde{\nu}^2)\nu^\mu \;\; and \;\; L_\mu = g(\widetilde{x})\theta(\widetilde{\nu}^2)\eta_{\mu\nu}\nu^\nu \tag{155}$$

where $\widetilde{\nu}^2 = \eta_{ij}\nu^i\nu^j$ and $\theta(\widetilde{\nu}^2)$ is given as in (137a).
For null ds_F we can take

$$L^\mu = kg(\widetilde{x})\theta(\widetilde{\nu}^2)\frac{dx^\mu}{dt} \;\; and \;\; L_\mu = kg(\widetilde{x})\theta(\widetilde{\nu}^2)\eta_{\mu\nu}\frac{dx^\nu}{dt} \tag{156}$$

where t is a parameter and k is (-1)p-homogeneous form in the differential dx^μ. Also, k is supposed to be positive definite for all directions (c.f. section 1.6 of chapter 1).
Then, for nonnull ds_F we have

$$\widetilde{L}^2 = \eta_{\mu\nu} L^\mu L^\nu = g^2(\widetilde{x})\theta^2(\widetilde{\nu}^2)\eta_{\mu\nu}\nu^\nu\nu^\mu = g(\widetilde{x})\theta(\widetilde{\nu}^2)F(\widetilde{x},\widetilde{\nu}) = g(\widetilde{x})\theta(\widetilde{\nu}^2) \tag{157}$$

since $F(\widetilde{x},\widetilde{\nu}) = F(x^\mu, \frac{dx^\mu}{ds_F}) = 1$.
For null ds_F we have

$$\widetilde{L}^2 = \eta_{\mu\nu} L^\mu L^\nu = k^2 g^2(\widetilde{x})\theta^2(\widetilde{\nu}^2)\eta_{\mu\nu}\frac{dx^\mu}{dt}\frac{dx^\nu}{dt} = 0 \tag{158}$$

This property of L^μ expresses the Finslerian character of the space-time given by the interval ds_F in (154).
Now, the quantum generalization of this Finsler space is made through change of coordinate differentials dx^μ in (154) by the finite operators

$$\Delta\hat{x}^\mu = \epsilon\sqrt{\theta(\widetilde{\nu}^2)}\gamma^\mu(\widetilde{x}) \tag{159}$$

where $\gamma^\mu(\widetilde{x})(\mu = 0,1,2,3)$ are Dirac matrices for the associated curved space (Riemannian) of the Finsler space (c.f.section 1.8). These matrices are related to the flat space (Dirac) matrices through the vierbein $V^\mu_\alpha(\widetilde{x})$ by

$$\gamma^\mu(\widetilde{x}) = V^\mu_\alpha(\widetilde{x})\gamma^\alpha \tag{160}$$

For the present case the vierbein fields $V^\mu_\alpha(\widetilde{x})$ and $V^\alpha_\mu(\widetilde{x})$ are given as

$$V^\mu_\alpha(\widetilde{x}) = \frac{1}{\sqrt{g(\widetilde{x})}}\delta^\mu_\alpha and V^\alpha_\mu(\widetilde{x}) = \sqrt{g(\widetilde{x})}\delta^\alpha_\mu \tag{161}$$

From (154), the operator of interval is obtained as

$$\Delta\hat{s_F} = L_\mu \Delta\hat{x}^\mu \tag{162}$$

Also, from (159),(160) and (161) it follows that

$$[\Delta\hat{x}^\mu, \Delta\hat{x}^\nu] = 2i\frac{\epsilon^2\theta(\widetilde{\nu}^2)}{g(\widetilde{x})}\sigma^{\mu\nu} \tag{163}$$

where $\sigma^{\mu\nu}$ are given by (153).
Now, defining the operator $\hat{\sigma}(\widetilde{L})$ as before we have

$$\hat{\sigma}(\widetilde{L}) = \frac{\Delta\hat{s_F}}{\epsilon} = \sqrt{\theta(\widetilde{\nu}^2)}L_\mu\gamma^\mu(\widetilde{x}) = \sqrt{\frac{\theta(\widetilde{\nu}^2)}{g(\widetilde{x})}}L_a\gamma^a \tag{164}$$

where γ^a are flat space Dirac matrices. Also, it follows that (using (157) and (158))

$$\hat{\sigma}^2 = \frac{\theta(\widetilde{\nu}^2)}{g(\widetilde{x})}\widetilde{L}^2 = 1 for \widetilde{L}^2 \neq 0 \tag{165}$$

$$\hat{\sigma}^2 = 0 for \widetilde{L}^2 = 0 \tag{166}$$

Thus, the eigenvalues of $\hat{\sigma}$ are ± 1 and 0. Consequently, those of the interval operator $\Delta\hat{s_F}$ are $\pm\epsilon$ and 0. The ordering of events in this space can be determined as in the case of Minkowski(Finsler) space discussed above.

2.4: FIELD EQUATION IN FINSLER SPACE

In section 2.1, we have derived the equation for the classical field (or wave) function $\psi(\widetilde{x},\widetilde{\nu})$ from the property of equivalence of it along the neighbouring points in the microdomain on the autoparallel curve.

Now, if the microdomain which is a Finsler space is quantized then we might derive the quantum field equation in this space. The quantization is admitted in two steps . We start from equivalence property (c.f.section 2.1) which can be written as

$$\psi = dx^\mu \partial_\mu \psi(\widetilde{x}, \widetilde{\nu}) + d\nu^l \partial'_l \psi(\widetilde{x}, \widetilde{\nu}) = -i\epsilon \frac{mc}{\hbar} \sqrt{\theta(\widetilde{\nu}^2)}\psi(\widetilde{x}, \widetilde{\nu}) \tag{167}$$

where $\partial_\mu \equiv \frac{\partial}{\partial x^\mu}$ and $\partial'_l \equiv \frac{\partial}{\partial \nu^l}$.
As the first step, the differentials dx^μ are quantized to $\Delta\hat{x}^\mu$ as given in (159). Consequently, the classical field (or wave) function $\psi(\widetilde{x}, \widetilde{\nu})$ transforms into a spinor, i.e.,

$$\psi(\widetilde{x}, \widetilde{\nu}) \longrightarrow \begin{pmatrix} \psi_1(\widetilde{x}, \widetilde{\nu}) \\ \psi_2(\widetilde{x}, \widetilde{\nu}) \\ \psi_3(\widetilde{x}, \widetilde{\nu}) \\ \psi_4(\widetilde{x}, \widetilde{\nu}) \end{pmatrix} \tag{168}$$

Then, the equation for the spinor $\psi_i(\widetilde{x}, \widetilde{\nu})$ becomes

$$\epsilon\sqrt{\theta(\widetilde{\nu}^2)}\gamma^\mu_{\alpha\beta}(\widetilde{x})\partial_\mu\psi_\beta(\widetilde{x}, \widetilde{\nu}) + d\nu^l \partial'_l \psi_\alpha(\widetilde{x}, \widetilde{\nu}) = -i\epsilon \frac{mc}{\hbar} \sqrt{\theta(\widetilde{\nu}^2)}\psi_\alpha(\widetilde{x}, \widetilde{\nu}) \tag{169}$$

In the second step he differentials $d\nu^l$ are quantized by noting first the fact that since th neighbouring points $(\widetilde{x}, \widetilde{\nu})$ and $(\widetilde{x} + d\widetilde{x}, \widetilde{\nu} + d\widetilde{\nu})$ lie on the autoparallel curve of the Finsler space the relation (125) (c.f. relations (44) and (107) also) between the differentials $d\nu^l$ and dx^μ holds. Therefore, the quantized differentials $\Delta\hat{\nu}^l$ are given by

$$\Delta\hat{\nu}^l = -\epsilon\sqrt{\theta(\widetilde{\nu}^2)}\gamma^l_{h\mu}(\widetilde{x}, \widetilde{\nu})\nu^h \gamma^\mu(\widetilde{x}) \tag{170}$$

and consequently the field (wave) function becomes a 'bispinor' (this nomenclature is only formal and for convenience) $\psi_{\alpha\beta}(\widetilde{x}, \widetilde{\nu})$. Then the resulting equation for the field (or wave) function $\psi_{\alpha\beta}(\widetilde{x}, \widetilde{\nu})$ in the Finsler space is given by

$$\epsilon\sqrt{\theta(\widetilde{\nu}^2)}\gamma^\mu_{\alpha\beta'}(\widetilde{x})\partial_\mu\psi_{\beta'\beta}(\widetilde{x}, \widetilde{\nu}) - \epsilon\sqrt{\theta(\widetilde{\nu}^2)}\gamma^\mu_{\beta\beta'}(\widetilde{x})\gamma^l_{h\mu}(\widetilde{x}, \widetilde{\nu})\nu^h \partial'_l \psi_{\alpha\beta'}(\widetilde{x}, \widetilde{\nu}) = -i\epsilon \frac{mc}{\hbar} \sqrt{\theta(\widetilde{\nu}^2)}\psi_{\alpha\beta}(\widetilde{x}, \widetilde{\nu}) \tag{171}$$

or,

$$i\hbar\gamma^\mu_{\alpha\beta'}(\widetilde{x})\partial_\mu\psi_{\beta'\beta}(\widetilde{x}, \widetilde{\nu}) - i\hbar\gamma^\mu_{\beta\beta'}(\widetilde{x})\gamma^l_{h\mu}(\widetilde{x}, \widetilde{\nu})\nu^h \partial'_l \psi_{\alpha\beta'}(\widetilde{x}, \widetilde{\nu}) = mc\psi_{\alpha\beta}(\widetilde{x}, \widetilde{\nu}) \tag{172}$$

This equation can also be written in the following compact form :

$$i\hbar\gamma^\mu(\widetilde{x})(\partial_\mu - \gamma^l_{h\mu}(\widetilde{x}, \widetilde{\nu})\nu^h \partial'_l)\psi(\widetilde{x}, \widetilde{\nu}) = mc\psi(\widetilde{x}, \widetilde{\nu}) \tag{173}$$

where it is to be remembered that the first and the second operators on the L.H.S. operate on the first and the second indices of the bispinor $\psi(\widetilde{x}, \widetilde{\nu})$ respectively.

2.5: EXTERNAL ELECTROMAGNETIC FIELD

The wave equation for a particle in an external electromagnetic field can be deducted from a similar property of the fields on the autoparallel curve of the Finsler space as for the free field case. This property is expressed by the following relation

$$\delta(\psi\chi) + \frac{ie}{\hbar c}[A_\mu(\widetilde{x}, \widetilde{\nu})dx^\mu]\psi\chi = -i\epsilon \frac{mc}{\hbar} \sqrt{\theta(\widetilde{\nu}^2)}\psi\chi \tag{174}$$

where $\chi(\tilde{x},\tilde{\nu})$ is a scalar function which may be regarded as a phase factor for the field $\psi(\tilde{x},\tilde{\nu})$. The vector field $A_\mu(\tilde{x},\tilde{\nu})$ represents the external electromagnetic field. The relation (174) may be compared with the corresponding relation (167) for the free field, which describes the equivalence property of the free field on the autoparallel curve. Now, we have from (174),

$$dx^\mu \left[\delta_\mu + \frac{ie}{\hbar c}A_\mu(\tilde{x},\tilde{\nu}) + \frac{1}{\chi}\partial_\mu\chi\right]\psi + d\nu^l\left[\partial_l' + \frac{1}{\chi}\partial_l'\chi\right]\psi = -i\epsilon\frac{mc}{\hbar}\sqrt{\theta(\tilde{\nu}^2)}\psi \tag{175}$$

As before, we can use the relation (125) between $d\nu^l$ and dx^μ to find

$$dx^\mu\left[\delta_\mu + \frac{ie}{\hbar c}A_\mu(\tilde{x},\tilde{\nu}) + \frac{1}{\chi}\partial_\mu\chi - \gamma^l_{h\mu}(\tilde{x},\tilde{\nu})\nu^h(\partial_l' + \frac{1}{\chi}\partial_l'\chi)\right]\psi = -i\epsilon\frac{mc}{\hbar}\sqrt{\theta(\tilde{\nu}^2)}\psi \tag{176}$$

or, in terms of nonlinear connections (N^ν_μ) (c.f. (130))

$$dx^\mu\left[\delta_\mu + \frac{ie}{\hbar c}A_\mu(\tilde{x},\tilde{\nu}) + \frac{1}{\chi}\partial_\mu\chi - N^l_\mu(\partial_l' + \frac{1}{\chi}\partial_l'\chi)\right]\psi = -i\epsilon\frac{mc}{\hbar}\sqrt{\theta(\tilde{\nu}^2)}\psi \tag{177}$$

setting $\chi = exp[\frac{i}{\hbar c}\phi(\tilde{x},\tilde{\nu})]$ we have

$$dx^\mu\left[\delta_\mu + \frac{ie}{\hbar c}\left(A_\mu(\tilde{x},\tilde{\nu}) + \frac{1}{e}\partial_\mu\phi - \frac{1}{e}N^l_\mu\partial_l'\phi\right) - N^l_\mu\partial_l'\right]\psi = -i\epsilon\frac{mc}{\hbar}\sqrt{\theta(\tilde{\nu}^2)}\psi \tag{178}$$

or, in terms of "covariant" $\frac{\delta}{\delta x^\mu} = \partial_\mu - N^l_\mu\partial_l'$ [as givn in (111)] we have

$$dx^\mu\left[\frac{\delta}{\delta x^\mu} + \frac{ie}{\hbar c}\left(A_\mu(\tilde{x},\tilde{\nu}) + \frac{1}{e}\frac{\delta\phi}{\delta x^\mu}\right)\right]\psi = -i\epsilon\frac{mc}{\hbar}\sqrt{\theta(\tilde{\nu}^2)}\psi \tag{179}$$

This equation should be regarded as the "classical" field (or wave) equation for $\psi(\tilde{x},\tilde{\nu})$. Here $A_\mu(\tilde{x},\tilde{\nu}) + \frac{1}{e}\frac{\delta\phi}{\delta x^\mu}$ represents the covariant four-vector, the external electromagnetic field, in the Finsler space. Obviously, $A_\mu(\tilde{x},\tilde{\nu}) + \frac{1}{e}\frac{\delta\phi}{\delta x^\mu}$ is the gauge transformation of the electromagnetic field and manifests the gauge invariance of it in the Finsler space.

When the field $A_\mu(\tilde{x},\tilde{\nu})$ and the scalar function $\chi(\tilde{x},\tilde{\nu})$ are separable as follows

$$A_\mu(\tilde{x},\tilde{\nu}) = A_\mu(\tilde{x}) - \gamma^l_{h\mu}(\tilde{x},\tilde{\nu})\nu^h\tilde{A}_l(\tilde{\nu}) \tag{180}$$

(we shall see later that $\gamma^l_{h\mu}(\tilde{x},\tilde{\nu})$ can be made separable for the Finsler space of hadronic matter extension) and

$$\chi(\tilde{x},\tilde{\nu}) = \chi(\tilde{x})\tilde{\chi}(\tilde{\nu}) \tag{181}$$

We have

$$dx^\mu\left[\frac{\delta}{\delta x^\mu} + \frac{ie}{\hbar c}\left(A_\mu(\tilde{x}) - N^l_\mu\tilde{A}_l(\tilde{\nu}) + \frac{1}{e}\partial_\mu\phi(\tilde{x}) - \frac{1}{e}N^l_\mu\partial_l'\tilde{\phi}(\tilde{\nu})\right)\right]\psi = -i\epsilon\frac{mc}{\hbar}\sqrt{\theta(\tilde{\nu}^2)}\psi \tag{182}$$

where $\phi(\tilde{x}) = \frac{\hbar c}{i}\ln\chi(\tilde{x})$ and $\tilde{\phi}(\tilde{\nu}) = \frac{\hbar c}{i}\ln\tilde{\chi}(\tilde{x})$ i.e.

$$\phi(\tilde{x},\tilde{\nu}) = \phi(\tilde{x}) + \tilde{\phi}(\tilde{\nu}) \tag{183}$$

It is to be noted that along the autoparallel curve of the Finsler space

$$A_\mu(\tilde{x},\tilde{\nu})dx^\mu = A_\mu(\tilde{x})dx^\mu - N^l_\mu\tilde{A}_l(\tilde{\nu})dx^\mu = A_\mu(\tilde{x})dx^\mu + \tilde{A}_l(\tilde{\nu})d\nu^l \tag{184}$$

where we have used the relation (125).
Now, if we assume

$$\tilde{A}_l(\widetilde{\nu}) + \frac{1}{e}\partial_l'\tilde{\phi}(\widetilde{\nu}) = 0 \tag{185}$$

the above equation reduces to the following equation for the classical fields :

$$dx^\mu \left[\frac{\delta}{\delta x^\mu} + \frac{ie}{\hbar c}\bar{A}_\mu(\widetilde{x})\right]\psi = -i\epsilon\frac{mc}{\hbar}\sqrt{\theta(\widetilde{\nu}^2)}\psi \tag{186}$$

where $\bar{A}_\mu(\widetilde{x}) = A_\mu(\widetilde{x}) + \frac{1}{e}\partial_\mu\phi(\widetilde{x})$ is the usual gauge transformation in the associated curved space (Riemannian) of the Finsler space.
In order to find out the quantum field (or wave) equation for this case,we begin with the equation (183), (184) and (185)

$$dx^\mu \left[\partial_\mu + \frac{ie}{\hbar c}\left(A_\mu(\widetilde{x}) + \frac{1}{e}\partial_\mu\phi(\widetilde{x})\right)\right]\psi + d\nu^l\partial_l'\psi = -i\epsilon\frac{mc}{\hbar}\sqrt{\theta(\widetilde{\nu}^2)}\psi \tag{187}$$

As before, the Finslerian microdomain may be quantized and in fact, this quantization is accomplished in two steps in which the differentials dx^μ and $d\nu^l$ are transformed into $\Delta\hat{x}^\mu$ and $\Delta\hat{\nu}^l$ respectively. Consequently, the wave function $\psi(\widetilde{x},\widetilde{\nu})$ transforms into a bispinor $\psi_{\alpha\beta}$. From (187), we obtain the resulting equation for the field (or wave) function $\psi_{\alpha\beta}$ for a particle in an external electromagnetic field $\bar{A}_\mu(\widetilde{x})$ as

$$\gamma^\mu_{\alpha\alpha'}(\widetilde{x})(i\hbar\partial_\mu - \frac{e}{c}\bar{A}_\mu(\widetilde{x}))\psi_{\alpha'\beta}(\widetilde{x},\widetilde{\nu}) - i\hbar\gamma^\mu_{\beta\beta'}(\widetilde{x})\gamma^l_{h\mu}(\widetilde{x},\widetilde{\nu})\nu^h\partial_l'\psi_{\alpha\beta'}(\widetilde{x},\widetilde{\nu}) = mc\psi_{\alpha\beta}(\widetilde{x},\widetilde{\nu}) \tag{188}$$

or, as before, in compact form

$$\gamma^\mu(\widetilde{x})(i\hbar\partial_\mu - \frac{e}{c}\bar{A}_\mu(\widetilde{x}))\psi(\widetilde{x},\widetilde{\nu}) - i\hbar\gamma^\mu(\widetilde{x})\gamma^l_{h\mu}(\widetilde{x},\widetilde{\nu})\nu^h\partial_l'\psi(\widetilde{x},\widetilde{\nu}) = mc\psi(\widetilde{x},\widetilde{\nu}) \tag{189}$$

where the action of the oprators on the bispinor $\psi(\widetilde{x},\widetilde{\nu})$ is to be understood as in the free particle case.

2.6: ELECTROMAGNETIC FIELD IN THE FINSLER SPACE

In the above section, the external electromagnetic field has been introduced.In the following, the electromagnetic field(the photon field) in the Finsler space will be presented and its equation will be derived from a similar consideration as in the case of a lepton (spinor) field. From the property of the classical field in the Finsler space it is possible, on quantization of microdomain, to obtain photon field equations similar to Bargmann-Wigner equations for photon field in the Minkowski space-time(Bargmann and Wigner,1948;Dirac,1936). From these Bargmann-Wignertype equations of this field in the Finsler space it will be shown that in the case of "separable" field function the "$\widetilde{x}$-part" of the field represents the electromagnetic field for the associated curved space (the pseudo Riemannian space or the Minkowski space as the case may be) of the Finsler space. In fact, this part of field function satisfies the usual field equation for the free field case(as well as for the case of interacting field as we shall see later).
Let us begin with the "classical" field $\psi_{(i',j')}(\widetilde{x},\widetilde{\nu})$ which for each pair of indices (i',j') satisfies the "equvalence property" encountered earlier in the previous sections (c.f. section 2.1). This property of the field is to be satisfied by it on the autoparallel curve of the Finsler space (the microdomain). For a free field this is given as (on the autoparallel curve) [see also, De (2000b)].

$$\delta\psi_{(i',j')}(\widetilde{x},\widetilde{\nu}) = dx^\mu\partial_\mu\psi_{(i',j')}(\widetilde{x},\widetilde{\nu}) + d\nu^l\partial_l'\psi_{(i',j')}(\widetilde{x},\widetilde{\nu}) = 0 \tag{190}$$

for each pair (i',j').
From this equation, by proceeding as before,that is, through the quantization of the microdomain in two steps, namely, $(i)dx^\mu \to \Delta\hat{x}^\mu$ and $(ii)d\nu^l \to \Delta\hat{\nu}^l$.

We arrive at the following equation for the quantum field $\psi_{(i,j),(i',j')}(\widetilde{x},\widetilde{\nu})$, which is a bispinor for each (i',j') :

$$\gamma^{\mu}(\widetilde{x})(\partial_{\mu}-\gamma^{l}_{h\mu}(\widetilde{x},\widetilde{\nu})\nu^{h}\partial'_{l})\psi_{(i',j')}(\widetilde{x},\widetilde{\nu})=0 \tag{191}$$

Here, $\psi_{(i',j')}(\widetilde{x},\widetilde{\nu})$ is now a bispinor for each (i',j') and the previous convention regarding the action of the operators on the bispinor has been adopted here also . That is, in "full form" the above equation is

$$\gamma^{\mu}_{ik}(x)\partial_{\mu}\psi_{(k,j),(i',j')}(\widetilde{x},\widetilde{\nu})-\gamma^{\mu}_{jk}(\widetilde{x})\gamma^{l}_{h\mu}(\widetilde{x},\widetilde{\nu})\nu^{h}\partial'_{l}\psi_{(i,k),(i',j')}(\widetilde{x},\widetilde{\nu})=0 \tag{192}$$

for each (i',j').
Now, one could have considered the classical field $\psi_{(i,j)}(\widetilde{x},\widetilde{\nu})$ and followed the same procedure as above (that is, its property on the autoparallel curve and the space-time quantization of the Finslerian microdomain) to arrive at the following equation for the other pair of indices (i',j'):

$$\gamma^{\mu}_{i'k}(x)\partial_{\mu}\psi_{(i,j),(k,j')}(\widetilde{x},\widetilde{\nu})-\gamma^{\mu}_{j'k}(\widetilde{x})\gamma^{l}_{h\mu}(\widetilde{x},\widetilde{\nu})\nu^{h}\partial'_{l}\psi_{(i,j),(i',k)}(\widetilde{x},\widetilde{\nu})=0 \tag{193}$$

for each (i,j)
Thus, $\psi_{(i,j),(i',j')}(\widetilde{x},\widetilde{\nu})$ is completely symmetric with respect to the pairs (i,j) and (i',j'). The equation satisfied by the field for either of these pairs is nothing but the equation satisfied by the lepton (spinor) field (for massless particle) considered in section 2.4. This consideration is analogous to the case of Bargmann-Wigner equations for spin $s\geq\frac{1}{2}$ particles, where, of course, the equations(Dirac equations for each index) have to be postulated.
Now, when the field function $\psi_{(i,j),(i',j')}(\widetilde{x},\widetilde{\nu})$ is separable as the direct product of two bispinors $\hat{\psi}_{ii'}(\widetilde{x})$ and $\hat{\phi}_{j,j'}(\widetilde{\nu})$ (represented as 4×4 matrices), one depending on the co ordinates x^{μ} and the other on the directional variables ν^{μ} of the Finsler space, that is,

$$[\psi_{(i,j),(i',j')}(\widetilde{x},\widetilde{\nu})]=[\hat{\psi}_{ii'}(\widetilde{x})]\times[\hat{\phi}_{jj'}(\widetilde{\nu})] \tag{194}$$

where $\hat{\psi}(\widetilde{x})=[\hat{\psi}_{ii'}(\widetilde{x})]$ and $\hat{\phi}(\widetilde{\nu})=[\hat{\phi}_{jj'}(\widetilde{\nu})]$ are completely symmetric with respect to their respective indices , we have from (192)

$$\gamma^{\mu}_{ik}(\widetilde{x})\partial_{\mu}\hat{\psi}_{ki'}(\widetilde{x})\times\hat{\phi}_{jj'}(\widetilde{\nu})-\hat{\psi}_{ii'}(\widetilde{x})\times\gamma^{\mu}_{jk}(\widetilde{x})\gamma^{l}_{h\mu}(\widetilde{x},\widetilde{\nu})\nu^{h}\partial'_{l}\hat{\phi}_{kj'}(\widetilde{\nu})=0 \tag{195}$$

for each (i',j') .

It should be noted that the Dirac matrices $\gamma^{\mu}(\widetilde{x})$ for the associated curved space of the Finsler space are connected with the flat space Dirac matrices γ^{α} through the vierbeins $V^{\mu}_{\alpha}(\widetilde{x})$ which are given by (161). Also, the coefficients $\gamma^{l}_{h\mu}(\widetilde{x},\widetilde{\nu})$ are separable as functions of the coordinates and the directional arguments of the Finsler space which we are interested with. Now, if the bispinor $\hat{\phi}(\widetilde{\nu})$ satisfies the following equation

$$\gamma^{\mu}_{jk}(\widetilde{x})\gamma^{l}_{h\mu}(\widetilde{x},\widetilde{\nu})\nu^{h}\partial'_{l}\hat{\phi}_{kj'}(\widetilde{\nu})=0 \tag{196}$$

Then we have from (195),

$$\gamma^{\mu}_{ik}(\widetilde{x})\partial_{\mu}\hat{\psi}_{ki'}(\widetilde{x})\times\hat{\phi}_{jj'}(\widetilde{\nu})-\hat{\psi}_{ii'}(\widetilde{x})\times 0=0 \tag{197}$$

and consequently,

$$\gamma^{\mu}_{ik}(\widetilde{x})\partial_{\mu}\hat{\psi}_{ki'}(\widetilde{x})=0, \tag{198}$$

for each i'.
For the flat space(i.e., for Minkowski space-time) this equation becomes

$$\gamma^{\alpha}_{ik}\partial_{\alpha}\hat{\psi}_{ki'}(\widetilde{x})=0, \tag{199}$$

for each i'.
Similarly, from (193)we can arrive at the equation

$$\gamma^{\alpha}_{i'k}\partial_{\alpha}\hat{\psi}_{ik}(\widetilde{x})=0, \tag{200}$$

for each i.
These equations,(199) and (200) are the usual Bargmann -Wigner equations for massless spin 1 particle, that is, for the photon field. These equations can be written as (in matrix form)

$$\gamma \cdot \partial \hat{\psi}(\widetilde{x}) = 0 \tag{201}$$

$$\hat{\psi}(\widetilde{x})\gamma^{\top} \cdot \overleftarrow{\partial} = 0 \tag{202}$$

Since$\hat{\psi}(\widetilde{x})$ is symmetric (i.e., $\hat{\psi}^{\top}(\widetilde{x}) = \hat{\psi}(\widetilde{x})$), it can be expressible as (Bargmann and Wigner,1948;see also Lurie,1968)

$$\hat{\psi}(\widetilde{x}) = \frac{1}{2}\sigma^{ab} C F_{ab}(\widetilde{x}) \tag{203}$$

where $\sigma^{ab} = \frac{i}{2}(\gamma^a\gamma^b - \gamma^b\gamma^a)$ and C is the charge conjugation matrix; $C = i\gamma^2\gamma^0$. It should be noted that $\sigma^{ab}C$ is symmetric. Then by addition of the two equations of (2.6.11) and using the expression of $\hat{\psi}(\widetilde{x})$ from (201) and (202), we have

$$[\gamma^\alpha, \sigma^{ab}] C \partial_\alpha F_{ab}(\widetilde{x}) = 0$$

where we have used the relation $C^{-1}\gamma^\mu C = -\gamma^{\mu^{\top}}$. Now, since $[\gamma^\alpha, \sigma^{ab}] = 2i(g^{\alpha a}\gamma^b - g^{\alpha b}\gamma^a)$ where $g^{ab} = diag.(1, -1, -1, -1)$. We obtain ultimately

$$\gamma^a C \partial^b F_{ab}(\widetilde{x}) = 0$$

This equation leads to the Maxwell's equations

$$\partial^b F_{ab}(\widetilde{x}) = 0 \tag{204}$$

In terms of the electromagnetic potential field $A_\mu(\widetilde{x})$ given as

$$F_{ab}(\widetilde{x}) = \partial_a A_b(\widetilde{x}) - \partial_b A_a(\widetilde{x}) \tag{205}$$

and with Lorentz condition

$$\partial^a A_a(\widetilde{x}) = 0 \tag{206}$$

We have from (204),

$$\Box \, A_a(\widetilde{x}) = 0 \tag{207}$$

where $\Box \ = \partial^\alpha\partial_\alpha$. Thus, the "$(\widetilde{x})$ -part " of the field of the Finsler space gives rise to the usual electromagnetic field which satisfies the Maxwell equations in the Minkowski space-time. Of course, the Finslerian electromagnetic fields satisfy Bargmann-Wigner type equations given in (192) and (193).

2.7: ELECTROMAGNETIC INTERACTION

We now introduce electromagnetic interaction with the spinor (lepton) fields [see alsao De (2000 a, b)]. It will be done from the properties of the classical fields on the autoparallel curve of the Finsler space and then by proceeding to the quantum transitions of these fields in the process of quantization of the space-time of the microdomain. These properties of the classical fields on the autoparallel curve are expressed by the following equations :

$$\delta(\psi\chi) + \frac{ie}{\hbar c}[A_\mu dx^\mu]\psi\chi = -i\epsilon\frac{mc}{\hbar}\sqrt{\theta(\widetilde{\nu}^2)}\psi\chi \tag{208}$$

and

$$\delta\psi_{(i',j')}(\widetilde{x}, \widetilde{\nu}) = -i\epsilon e\sqrt{\theta(\widetilde{\nu}^2)}\widetilde{C}_{(i',j')}(\widetilde{x}, \widetilde{\nu}) \tag{209}$$

for each (i', j')
where $\psi(\widetilde{x}, \widetilde{\nu})$ and $\psi_{(i',j')}(\widetilde{x}, \widetilde{\nu})$ are "classical" fields. The field $\psi_{(i',j')}(\widetilde{x}, \widetilde{\nu})$ gives rise, on quantization of space-time of microdomain, to the electromagnetic field. $\widetilde{C}_{(i',j')}(\widetilde{x}, \widetilde{\nu})$, for each (i', j'), is proportional

to the change of the field $\psi_{(i',j')}(\widetilde{x},\widetilde{\nu})$ on the autoparallel curve (as evident from (209)) and becomes a bispinor for each (i',j') on quantization of microdomain. This field is, in fact, connected with the interacting spinor field and characterization of it will be made later. The equation (208) is the same as that in (174) for the external electromagnetic field. Thus, by space-time quantization one can arrive from (208) at the same equations in (188) and (189). Therefore, we shall consider the equation (209) presently. As before, on the autoparallel curve we have

$$dx^{\mu}\partial_{\mu}\psi_{(i',j')}(\widetilde{x},\widetilde{\nu}) + d\nu^{l}\partial'_{l}\psi_{(i',j')}(\widetilde{x},\widetilde{\nu}) = -i\epsilon e\sqrt{\theta(\widetilde{\nu}^2)}\widetilde{C}_{(i',j')}(\widetilde{x},\widetilde{\nu}) \tag{210}$$

for each (i',j')

Now, we make the quantum transition, that is, quantize the Finslerian space-time of microdomain in two steps, as before, namely

$$(i)\ dx^{\mu} \rightarrow \Delta\hat{x}^{\mu} = \epsilon\sqrt{\theta(\widetilde{\nu}^2)}\gamma^{\mu}(\widetilde{x}) \tag{211}$$

$$(ii)\ d\nu^{l} \rightarrow \Delta\hat{\nu}^{l} = -\epsilon\sqrt{\theta(\widetilde{\nu}^2)}\gamma^{l}_{h\mu}(\widetilde{x},\widetilde{\nu})\nu^{h}\gamma^{\mu}(\widetilde{x}) \tag{212}$$

In this process $\psi_{(i',j')}(\widetilde{x},\widetilde{\nu})$ for each (i',j') becomes a bispinor $\psi_{(i,j),(i',j')}(\widetilde{x},\widetilde{\nu})$.

Also, $\widetilde{C}_{(i',j')}(\widetilde{x},\widetilde{\nu}) \rightarrow \widetilde{C}_{(i',j')}(\widetilde{x},\widetilde{\nu})$ (a bispinor or a 4×4 matrix for each (i',j')) $= [\widetilde{C}_{(i,j),(i',j')}(\widetilde{x},\widetilde{\nu})]$

Consequently, the equation (210) changes to

$$(\gamma^{\mu}(\widetilde{x}))_{ik}\partial_{\mu}\psi_{(k,j),(i',j')}(\widetilde{x},\widetilde{\nu}) - (\gamma^{\mu}(\widetilde{x}))_{jk}\gamma^{l}_{h\mu}(\widetilde{x},\widetilde{\nu})\nu^{h}\partial'_{l}\psi_{(i,k),(i',j')}(\widetilde{x},\widetilde{\nu}) = -ie\widetilde{C}_{(i,j),(i',j')}(\widetilde{x},\widetilde{\nu}) \tag{213}$$

for each (i',j')

Similarly, starting from the classical field $\psi_{(i,j)}(\widetilde{x},\widetilde{\nu})$ one can arrive at the following equation for each (i,j) :

$$(\gamma^{\mu}(\widetilde{x}))_{i'k}\partial_{\mu}\psi_{(i,j),(k,j')}(\widetilde{x},\widetilde{\nu}) - (\gamma^{\mu}(\widetilde{x}))_{j'k}\gamma^{l}_{h\mu}(\widetilde{x},\widetilde{\nu})\nu^{h}\partial'_{l}\psi_{(i,j),(i',k)}(\widetilde{x},\widetilde{\nu}) = -ie\widetilde{C}_{(i,j),(i',j')}(\widetilde{x},\widetilde{\nu}) \tag{214}$$

Both $\psi_{(i,j),(i',j')}(\widetilde{x},\widetilde{\nu})$ and $\widetilde{C}_{(i,j),(i',j')}(\widetilde{x},\widetilde{\nu})$ are completely symmetric with respect to the pairs of indices (i,j) and (i',j') as in the free-field case considered in the previous section. The set of equations (188), (213) and (214) are the equations that are satisfied by the interacting fields in the Finsler space.

Now, the specification of $\widetilde{C}(\widetilde{x},\widetilde{\nu}) = [\widetilde{C}_{(i,j),(i',j')}(\widetilde{x},\widetilde{\nu})]$ will be made through the direct products of the fields represented as metrices. In case $[\psi_{(i,j),(i',j')}(\widetilde{x},\widetilde{\nu})]$ is separable as

$$[\psi_{(i,j),(i',j')}(\widetilde{x},\widetilde{\nu})] = [\hat{\psi}_{ii'}(\widetilde{x})] \times [\hat{\phi}_{jj'}(\widetilde{\nu})] \tag{215}$$

$\widetilde{C}(\widetilde{x},\widetilde{\nu})$ can be specified as

$$[\widetilde{C}(\widetilde{x},\widetilde{\nu})] = [\xi_{ii'}(\widetilde{x})] \times [\hat{\phi}_{jj'}(\widetilde{\nu})] \tag{216}$$

where $\hat{\psi}(\widetilde{x}) = [\hat{\psi}_{ii'}(\widetilde{x})]$, $\hat{\phi}(\widetilde{\nu}) = [\hat{\phi}_{jj'}(\widetilde{\nu})]$ and $\xi(\widetilde{x}) = [\xi_{ii'}(\widetilde{x})]$ are completely symmetric with respect to their respective indices. $\xi(\widetilde{x})$ is yet to be specified. With these separations, we have from (213)

$$[(\gamma^{\mu}(\widetilde{x}))_{ik}\partial_{\mu}\hat{\psi}_{ki'}(\widetilde{x})]\times[\hat{\phi}_{jj'}(\widetilde{\nu})] - [\hat{\psi}_{ii'}(\widetilde{x})]\times[(\gamma^{\mu}(\widetilde{x}))_{jk}\gamma^{l}_{h\mu}(\widetilde{x},\widetilde{\nu})\nu^{h}\partial'_{l}\hat{\phi}_{kj'}(\widetilde{\nu})] = [-ie\xi_{ii'}(\widetilde{x})]\times[\hat{\phi}_{jj'}(\widetilde{\nu})] \tag{217}$$

for each (i',j')

Similarly, from (214) we have

$$[(\gamma^{\mu}(\widetilde{x}))_{i'k}\partial_{\mu}\hat{\psi}_{ik}(\widetilde{x})]\times[\hat{\phi}_{jj'}(\widetilde{\nu})] - [\hat{\psi}_{ii'}(\widetilde{x})]\times[(\gamma^{\mu}(\widetilde{x}))_{j'k}\gamma^{l}_{h\mu}(\widetilde{x},\widetilde{\nu})\nu^{h}\partial'_{l}\hat{\phi}_{jk}(\widetilde{\nu})] = [-ie\xi_{ii'}(\widetilde{x})]\times[\hat{\phi}_{jj'}(\widetilde{\nu})] \tag{218}$$

for each (i,j)

Now, as before, if the bispinor $\hat{\phi}(\widetilde{\nu})$ satisfies the following equations

$$(\gamma^{\mu}(\widetilde{x}))_{jk}\gamma^{l}_{h\mu}(\widetilde{x},\widetilde{\nu})\nu^{h}\partial'_{l}\hat{\phi}_{kj'}(\widetilde{\nu}) = 0 \tag{219}$$

and

$$(\gamma^{\mu}(\widetilde{x}))_{j'k}\gamma^{l}_{h\mu}(\widetilde{x},\widetilde{\nu})\nu^{h}\partial'_{l}\hat{\phi}_{jk}(\widetilde{\nu}) = 0 \tag{220}$$

then we have from (217) and (218) the following equations for $\hat{\psi}(\tilde{x})$:

$$(\gamma^{\mu}(\tilde{x}))_{ik}\partial_{\mu}\hat{\phi}_{ki'}(\tilde{x}) + ie\xi_{ii'}(\tilde{x}) = 0 \tag{221}$$

and

$$(\gamma^{\mu}(\tilde{x}))_{i'k}\partial_{\mu}\hat{\phi}_{ik}(\tilde{x}) + ie\xi_{ii'}(\tilde{x}) = 0 \tag{222}$$

Also, we remember that $\gamma^{\mu}(\tilde{x})$ are connected with the flat space Dirac matrices γ^{α} by the vierbeins $V^{\mu}_{\alpha}(\tilde{x})$ and also that $\gamma^{l}_{h\mu}(\tilde{x},\tilde{\nu})$ are separable for the Finsler space of our consideration. Then, for the flat space (i.e., for Minkowski space-time), the above equations become

$$(\gamma^{\alpha})_{ik}\partial_{\alpha}\hat{\phi}_{ki'}(\tilde{x}) + ie\xi_{ii'}(\tilde{x}) = 0 \tag{223}$$

and

$$(\gamma^{\alpha})_{i'k}\partial_{\alpha}\hat{\phi}_{ik}(\tilde{x}) + ie\xi_{ii'}(\tilde{x}) = 0 \tag{224}$$

Now, as before, the symmetric bispinor(represented as a $4X4$ matrix)$\hat{\psi}(\tilde{x})$ can be expressed as

$$\hat{\psi}(\tilde{x}) = \frac{1}{2}\sigma^{ab}CF_{ab}(\tilde{x}) \tag{225}$$

and $\xi(\tilde{x})$ which is also symmetric can be specified as

$$\xi(\tilde{x}) = \gamma^{\alpha}C(\bar{\psi}(\tilde{x})\gamma_{\alpha}\psi(\tilde{x})) \tag{226}$$

where C is the charge conjugation matrix. It should be noted that $\gamma^{\alpha}C$ is symmetric. Here, $\psi(\tilde{x})$ and $\bar{\psi}(\tilde{x})$ are respectively the "$\tilde{x}$ -parts" of the separated (lepton) field $\psi(\tilde{x},\tilde{\nu})$ and its adjoint field $\bar{\psi}(\tilde{x},\tilde{\nu})$ which is defined as

$$\bar{\psi}(\tilde{x},\tilde{\nu}) = \gamma^{0}\psi(\tilde{x},\nu)^{\dagger}\gamma^{0} \tag{227}$$

$$\psi(\tilde{x},\tilde{\nu}) = \gamma^{0}\bar{\psi}(\tilde{x},\tilde{\nu})^{\dagger}\gamma^{0} \tag{228}$$

(This definition of adjoint field is intended for the bispinor field in the Finsler space of our consideration) In fact, the field $\psi(\tilde{x},\tilde{\nu})$ (represented as a $4X4$ matrix) is separated as

$$\psi(\tilde{x},\tilde{\nu}) = \psi(\tilde{x})X\phi(\tilde{\nu})^{T} = \phi(\tilde{\nu})^{T}X\psi(\tilde{x}) = \psi(\tilde{x})\phi(\nu)^{T} \tag{229}$$

where $\psi(\tilde{x})$ and $\phi(\tilde{\nu})$ are spinors and are represented as $4X1$ matrices. Thus, the adjoint bispinor $\bar{\psi}(\tilde{x},\tilde{\nu})$ is given by

$$\begin{aligned}\bar{\psi}(\tilde{x},\tilde{\nu}) &= \gamma^{0}(\psi(\tilde{x})\phi(\tilde{\nu})^{T})^{\dagger}\gamma^{0} = \gamma^{0}\phi^{*}(\tilde{\nu})\psi(\tilde{x})^{\dagger}\gamma^{0} = (\phi(\tilde{\nu})^{\dagger}\gamma^{0})^{T}(\psi(\tilde{x})^{\dagger}\gamma^{0}) \\ &= (\bar{\phi}(\tilde{\nu}))^{T}\bar{\psi}(\tilde{x}) = (\bar{\phi}(\tilde{\nu}))^{T}X\bar{\psi}(\tilde{x}) = \bar{\psi}(\tilde{x})X(\bar{\phi}(\tilde{\nu}))^{T}\end{aligned} \tag{230}$$

(however, $(\gamma^{0})^{T} = \gamma^{0}$) where $\bar{\psi}(\tilde{x})$ and $\bar{\phi}(\tilde{\nu})$ are adjoint spinors of $\psi(x)$ and $\phi(\tilde{\nu})$ respectively. We shall show later that $\psi(\tilde{x})$ satisfies the usual Dirac equation (for a particle in an electromagnetic field) in flat space and also that $\phi(\tilde{\nu})$ which can be solved gives rise to an additional quantum number. In the subsequent chapter, we shall also show that the " $\tilde{\nu}$-parts" of the fields, i.e. $\phi(\tilde{\nu})$ and $\bar{\phi}(\tilde{\nu})$, in fact, contribute to the " $\tilde{\nu}$-part" of the separated $\widetilde{C}(\tilde{x},\tilde{\nu})$.
With the expressions of $\hat{\psi}(\tilde{x})$ and $\xi(\tilde{x})$ as in (225) and (226) we can write the equations (224) as

$$\gamma^{\alpha}\partial_{\alpha}\hat{\psi}(\tilde{x}) + ie\gamma^{\alpha}C(\bar{\psi}(\tilde{x})\gamma_{\alpha}\psi(\tilde{x})) = 0 \tag{231}$$

$$\hat{\psi}(\tilde{x})(\gamma^{\alpha})^{T}\overleftarrow{\partial}_{\alpha} + ie\gamma^{\alpha}C(\bar{\psi}(\tilde{x})\gamma_{\alpha}\psi(\tilde{x})) = 0 \tag{232}$$

Adding these equations we get

$$\gamma^{\alpha}\partial_{\alpha}\hat{\psi}(\tilde{x}) + \hat{\psi}(\tilde{x})(\gamma^{\alpha})^{T}\overleftarrow{\partial}_{\alpha} + 2ie\gamma^{\alpha}C(\bar{\psi}(\tilde{x})\gamma_{\alpha}\psi(\tilde{x})) = 0 \tag{233}$$

and using the relations $C^{-1}\gamma^{\mu}C = -(\gamma^{\mu})^T$ and $[\gamma^{\alpha}, \sigma^{ab}] = 2i(g^{\alpha a}\gamma^b - g^{\alpha b}\gamma^a)$ where $g^{ab} = diag.(1, -1, -1, -1)$, we can arrive at the following equation

$$\partial^b F_{ab}(\widetilde{x}) = e(\bar{\psi}(\widetilde{x})\gamma_a\psi(\widetilde{x})) \tag{234}$$

In terms of the electromagnetic potential field $A_{\mu}(\widetilde{x})$ defined in (205) and with the Lorentz condition (206), the above equation can also be written as

$$\Box\, A_{\mu}(\widetilde{x}) = -e(\bar{\psi}(\widetilde{x})\gamma_{\mu}\psi(\widetilde{x})) \tag{235}$$

The equations (234) and (235) are the equations satisfied by the interacting fields in the Minkowski space-time.

2.8: GENERAL COORDINATE TRANSFORMATION

In section 2.1, it has been shown that the classical field $\psi(\widetilde{x}, \widetilde{\nu})$ behaves like a scalar under general coordinate transformations, that is,

$$\psi(\widetilde{x}, \widetilde{\nu}) = \psi'(\widetilde{x}', \widetilde{\nu}') \tag{236}$$

where $x')^{\mu} \equiv x'^{\mu}x^{\nu}$ and $\nu'^{\mu} = X^{*\mu}_{\nu}\nu^{\nu}$ with $X^{*\mu}_{\nu} = \frac{\partial(x')^{\mu}}{\partial x^{\nu}}$.
The equation for the field $\psi(\widetilde{x}, \widetilde{\nu})$ is given in (2.1.8) or in more appropriate form (137). It is form-invariant under general coordinate transformations. Here, we write this form-invariant equation (137) as follows :

$$idx^{\mu}\frac{\delta\psi(\widetilde{x}, \widetilde{\nu})}{\delta x^{\mu}} = \epsilon\frac{mc}{\hbar}\sqrt{\theta(\widetilde{\nu^2})}\psi(\widetilde{x}, \widetilde{\nu}) \tag{237}$$

where the covariant $\frac{\delta}{\delta x^{\mu}}$ is given in (111), which is

$$\frac{\delta}{\delta x^{\mu}} = \partial_{\mu} - N^l_{\mu}\partial'_l \tag{238}$$

The nonlinear connection (N^l_{μ})(in local representation) is given by

$$N^{\mu}_{\nu} = \frac{1}{2}\frac{\partial}{\partial\nu^{\nu}}(\gamma^{\mu}_{\alpha\beta}\nu^{\alpha}\nu^{\beta}) \tag{239}$$

and also

$$N^{\mu}_{\nu}\nu^{\nu} = \gamma^{\mu}_{\alpha\beta}\nu^{\alpha}\nu^{\beta} \tag{240}$$

In 'quantum transition' via quantization of space-time of Finslerian microdomain one arrives at the quantum field (or wave) equation for the field which becomes a bispinor through this process. The equation for the bispinor field is given in (171) and it is rewritten as

$$i\hbar\left(\gamma^{\mu}(\widetilde{x}\,)\frac{\partial\psi(\widetilde{x}\,,\widetilde{\nu}\,)}{\partial x^{\mu}} - N^l_{\mu}(\widetilde{x}\,,\widetilde{\nu}\,)\psi(\widetilde{x}\,,\widetilde{\nu}\,)\frac{\overleftarrow{\partial}\,\gamma^{\mu T}(\widetilde{x}\,)}{\partial\nu^l}\right) = mc\psi(\widetilde{x}\,,\widetilde{\nu}\,) \tag{241}$$

It should be noted that for the Finsler space which we are regarding as the space of extended hadrons, $\gamma^l_{h\mu}(\widetilde{x}\,,\widetilde{\nu}\,) = \gamma^l_{h\mu}(\widetilde{x}\,)$, independent of $\widetilde{\nu}$. Consequently,

$$N^l_{\mu}(\widetilde{x}\,,\widetilde{\nu}\,) = \gamma^l_{h\mu}(\widetilde{x}\,)\nu^h \tag{242}$$

Under the general coordinate transformation (236), the equation $\psi'(\widetilde{x}\,',\widetilde{\nu}\,')$, which is form invariant is given by

$$idx'^{\mu}\frac{\delta\psi'(\widetilde{x}\,',\widetilde{\nu}\,')}{\delta x'^{\mu}} = \epsilon\frac{mc}{\hbar}\sqrt{\theta(\widetilde{\nu}\,'^{\,2})}\psi'(\widetilde{x}\,',\widetilde{\nu}\,') \tag{243}$$

where $\widetilde{\nu}\,'^{\,2} = \xi_{lm}\nu\,'^{\,l}\nu\,'^{\,m}$ with $\xi_{lm} = \eta_{ij}X^i_l X^j_m$.
Here, (X^i_l) is given by

$$X^i_l = \frac{\partial x^i}{\partial x\,'^{\,l}} \tag{244}$$

Also,

$$X^i_l\; X^{*\,l}_j = X^{*\,i}_l\; X^l_j = \delta_{ij}$$

Clearly,

$$\theta(\widetilde{\nu}^{\,2}) = \theta(\widetilde{\nu}\,'^{\,2}) \tag{245}$$

From this equation (243), as before, on quantization of spacetime in the transformed coordinates, we can arrive at the following equation for the bispinor $\psi'(\widetilde{x}\,',\widetilde{\nu}\,')$:

$$i\hbar\left(\gamma^\mu(\widetilde{x}\,')\frac{\partial\psi'(\widetilde{x}\,',\widetilde{\nu}\,')}{\partial x'^\mu} - N^{\prime l}_\mu(\widetilde{x}\,',\widetilde{\nu}\,')\psi'(\widetilde{x}\,',\widetilde{\nu}\,')\frac{\overleftarrow{\partial}\gamma^{\mu T}(\widetilde{x}\,')}{\partial\nu'^l}\right) = mc\psi'(\widetilde{x}\,',\widetilde{\nu}\,') \tag{246}$$

Here, the bispinor $\psi(\widetilde{x}\,,\widetilde{\nu}\,)$ is not, in general, a scalar under general coordinate transformations. In fact, for the first step of quantization (that is, $dx^\mu \longrightarrow \Delta\hat{x}^\mu$), $\psi(\widetilde{x}\,,\widetilde{\nu}\,)$ changes to a spinor $\psi_\alpha(\widetilde{x}\,,\widetilde{\nu}\,)$. Thus, there is a mapping f_1 of the classical field onto the spinor space as follows:

$$f_1 \;:\; \psi(\widetilde{x}\,,\widetilde{\nu}\,) \longrightarrow \psi_\alpha(\widetilde{x}\,,\widetilde{\nu}\,)$$

or,

$$\psi_\alpha(\widetilde{x}\,,\widetilde{\nu}\,) = f_1\psi(\widetilde{x}\,,\widetilde{\nu}\,)$$

In the second step of quantization (that is, $d\nu^\mu \longrightarrow \Delta\hat{\nu}^\mu$) the spinor $\psi_\alpha(\widetilde{x}\,,\widetilde{\nu}\,)$ is mapped onto the bispinor $\psi_{\alpha\beta}(\widetilde{x}\,,\widetilde{\nu}\,)$, that is,

$$f_2 \;:\; \psi_\alpha(\widetilde{x}\,,\widetilde{\nu}\,) \longrightarrow \psi_{\alpha\beta}(\widetilde{x}\,,\widetilde{\nu}\,)$$

Therefore,

$$f_1 f_2 \;:\; \psi(\widetilde{x}\,,\widetilde{\nu}\,) \longrightarrow \psi_{\alpha\beta}(\widetilde{x}\,,\widetilde{\nu}\,) \tag{247}$$

is the mapping of the classical field onto the bispinor space, induced by the 'quantization'. The inverse mappings are

$$f_2^{-1} \;:\; \psi_{\alpha\beta}(\widetilde{x}\,,\widetilde{\nu}\,) \longrightarrow \psi_\alpha(\widetilde{x}\,,\widetilde{\nu}\,)$$
$$f_1^{-1} \;:\; \psi_\alpha(\widetilde{x}\,,\widetilde{\nu}\,) \longrightarrow \psi(\widetilde{x}\,,\widetilde{\nu}\,)$$

That is,

$$f_2^{-1}f_1^{-1} \;:\; \psi_{\alpha\beta}(\widetilde{x}\,,\widetilde{\nu}\,) \longrightarrow \psi(\widetilde{x}\,,\widetilde{\nu}\,) \tag{248}$$

Similarly, the corresponding mappings in the transformed Finsler space are given by

$$\begin{aligned} g_1 \;&:\; \psi'(\widetilde{x}\,',\widetilde{\nu}\,') \longrightarrow \psi\,'_\alpha(\widetilde{x}\,',\widetilde{\nu}\,') \\ g_2 \;&:\; \psi'_\alpha(\widetilde{x}\,',\widetilde{\nu}\,') \longrightarrow \psi\,'_{\alpha\beta}(\widetilde{x}\,',\widetilde{\nu}\,') \\ g_1 g_2 \;&:\; \psi'(\widetilde{x}\,',\widetilde{\nu}\,') \longrightarrow \psi\,'_{\alpha\beta}(\widetilde{x}\,',\widetilde{\nu}\,') \end{aligned} \tag{249}$$

$$\begin{aligned} g_2^{-1} \;&:\; \psi'_{\alpha\beta}(\widetilde{x}\,',\widetilde{\nu}\,') \longrightarrow \psi\,'_\alpha(\widetilde{x}\,',\widetilde{\nu}\,') \\ g_1^{-1} \;&:\; \psi'_\alpha(\widetilde{x}\,',\widetilde{\nu}\,') \longrightarrow \psi\,'(\widetilde{x}\,',\widetilde{\nu}\,') \\ g_2^{-1}g_1^{-1} \;&:\; \psi\,'_{\alpha\beta}(\widetilde{x}\,',\widetilde{\nu}\,') \longrightarrow \psi\,'(\widetilde{x}\,',\widetilde{\nu}\,') \end{aligned} \tag{250}$$

Since$\psi(\widetilde{x}\,,\widetilde{\nu}\,)$ is a scalar we must have

$$g_2^{-1}g_1^{-1}(\psi\,'_{\alpha\beta}(\widetilde{x}\,',\widetilde{\nu}\,')) = f_2^{-1}f_1^{-1}(\psi_{\alpha\beta}(\widetilde{x}\,,\widetilde{\nu}\,))$$

or,

$$\psi'_{\alpha\beta}(\widetilde{x}',\widetilde{\nu}') = g_1 g_2 f_2^{-1} f_1^{-1}(\psi_{\alpha\beta}(\widetilde{x},\widetilde{\nu})) \tag{251}$$

For the case $f_1 = g_1$, $f_2 = g_2$ one has

$$\psi'_{\alpha\beta}(\widetilde{x}',\widetilde{\nu}') = \psi_{\alpha\beta}(\widetilde{x},\widetilde{\nu}) \tag{252}$$

that is, the bispinor field behaves like a scalar under such coordinate transformation. This can occur if the coordinate transformation is linear. In general case of coordinate transformations the form invariance of the field equation (for bispinor) should be understood in the following manner:
The classical fields are scalars under general coordinate transformations and their equations are form invariant. By ' spacetime quantization' and consequent mappings of the classical fields onto the bispinor fields one arrives at the corresponding form invariant field equations. The bispinors are related by the mapping of the type (251). The inverse mappings in (248) , (250) and (251) on the bispinors can be interpreted as the mappings associated with 'classical-ization' of the spacetime. The relation between the bispinors (the mappings) given by (251) is, in general, of nonlinear type. The present consideration is schematically depicted in the figure 1. However, it is possible that the field (or wave) function in the

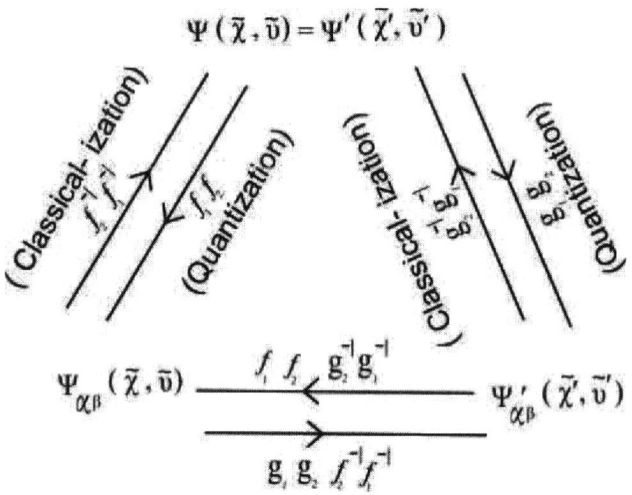

Figure 1: Form invariance of classical and quantum field equations. Relation between bispinors.

coordinate space x^μ (the associated Riemann space) is a scalar under general coordinate transformations. In fact, the field of the coordinate space is the '$\widetilde{x}$'-part of the bispinor $\psi_{\alpha\beta}(\widetilde{x},\widetilde{\nu})$ if it is separable as the direct product of spinors considered as matrices. On the other hand, if the bispinor is not separable then the field in the '$\widetilde{x}$-space' can be defined as follows:

$$\psi_\alpha(\widetilde{x}) = \int \psi_{\alpha\beta}(\widetilde{x},\widetilde{\nu})\chi_\beta(\widetilde{\nu})d^4\nu \tag{253}$$

where $\chi_\beta(\widetilde{\nu})$ is (spinor) probability density function. It is to be noted that the bispinors are not separable in all transformed spaces, that is, the bispinors obtained under general coordinate transformations may not be separable. For the Finsler space of extended hadrons, that we are considering the bispinors in that space are, however, separable. This is due to the fact that we are considering here a special coordinate system for the spacetime of the anisotropic Finslerian microdomain.

REFERENCES

1. Bargmann,V and Wigner, E.P. (1948) Proceedings of the National Academy of Sciences of the USA, **34**, 211
2. Blokhintsev, D.I. (1973).Space and Time in the Microworld. D. Reidel, Dordrecht, Holland
3. De, S.S. (1997). Int.J.Theor.Phys., **36**, 89
4. De, S.S. (2000a). Int.J.Theor.Phys., **41**, 1291
5. De, S.S. (2000b). Int.J.Theor.Phys., **41**, 1307
6. Dirac, P.A.M (1936) Proceedings of the Royal Society of London, Series A, **155**, 447
7. Kadyshevsky,V.G. (1959), Soviet Journal, JETP, **41**, 1885
8. Kadyshevsky,V.G. (1962), Doklad Akademii Nauk, USSR, **147**, 588
9. Kirzhnits, D.A. and Chechin,V.A (1967), In the Proceedings of the International Symposium on Nonlocal Quantum Field Theory, JINR, Preprint P2-3590, Dubna, USSR
10. Leznov, A.N (1967) In the Proceedings of the International Symposium on Nonlocal Quantum Field Theory, JINR, Preprint P2-3590, Dubna, USSR
11. Lurié, D. (1968). Particles and Fields, Interscience Publishers, New York.
12. Namsrai, Kh (1985). Nonlocal Quantum Field Theory and Stochastic Quantum Mechanics, D. Reidel, Dordrecht, Holland
13. Prugovecki,E (1984) Stochastic Quantum Mechanics and Quantum Spacetime, D. Reidel, Dordrecht, Holland
14. Snyder, H.S (1947) Phys. Rev. **71**, 38; ibid, **72**, 68
15. Tamn, I.E. (1965) In the Proceedings of the International Conference on Elementary Particles, Kyoto
16. Yang, C.N. (1947) Phys. Rev. **72**, 874

CHAPTER THREE:

FIELD EQUATIONS IN THE ASSOCIATED RIEMANNIAN SPACES

3.1: ASSOCIATED RIEMANNIAN SPACE

In section two of chapter one, the associated Riemannian space has been defined in relation to the (α, β)-metric Finsler space. The Finsler space which we are considering as the microspace in, in fact, a (α, β)-metric Finsler space. In section eight of the same chapter, we have discussed that the associated Riemannian space is the macrospace which may either be the Minkowski spacetime (the laboratory spacetime) or the background spacetime of the Universe (the Robertson Walker spacetime). The metric tensors of these spaces have, in fact, been shown to be the 'averaged' metric tensors from that of the Finslerian microspace. Thus, the associated Riemannian space is identical with the averaged microspace (Finslerian) and the corresponds to the macrospace.

In chapter two, the the wave(or field)equatoins in the Finsler space for the free field (lepton) as well as in interaction with external electromagnetic field have been deduced. Also, the field equations for free electromagnetic field have been deduced. For interaction with the lepton the field equations have also been deduced. For the later cases, the equations for the associated Riemannian space (and for the Minkowski space-time) have been found by decomposing the Finslerian fields as the direct products of the fields of the "$\widetilde{x}$ - space" (the associated Riemannian space) and the fields dependent on the directional variables of the Finsler space. It is to be noted that with the decomposition of fields we firstly deduced equations for the field in the associated (pseudo) Riemannian space and then by using vierbeins the equations in local inertial frame(that is, the Minkowski space-time) have been obtained. Alternatively, the same equations could have been derived from those of the pseudo Riemannian space by making the parameters b_k in the metric tensors (c.f. equation (86)) tend to zero. Actually, the local inertial frame is equivalent to the curved pseudo-Riemannian space with $b_k \to 0$.Presently, we shall find out the equations for the lepton field in the associated Riemanniam space from the equations of the Finslerian(lepton) field, which were constructed in the previous chapter.

3.2: DIRAC EQUATION IN ROBERTSON - WALKER SPACETIME

For the Finlerian space we are considering the function $g(x^{\sim})$ in the metric tensor can be chosen as (c.f. equation (87)

$$g(\widetilde{x}) \equiv F(t) = exp(\pm b_0 x^0)$$
$$or$$
$$(b_0 x^0)^n$$
$$or$$
$$(1 + b_0 x^0)^n \tag{254}$$

where

$$x^0 = ct$$

In this case, the connection coefficients $\gamma^l_{h\mu}(\widetilde{x},\widetilde{\nu})$ are separated as

$$\gamma^l_{h\mu}(\widetilde{x},\widetilde{\nu}) = \zeta(t)\gamma^l_{h\mu}(\widetilde{\nu}) \tag{255}$$

where

$$2b_0 c\zeta(t) = \frac{F'(t)}{F(t)} \tag{256}$$

In fact, in view of the relation (120) it can easily be seen that $\gamma^l_{h\mu}$ are independent of the directional arguments for this Finsler space. Actually, if we calculate

$$G^l_{hk} = \frac{\partial^2 G^l}{\partial\nu^h \partial\nu^k} \tag{257}$$

where $G^l = \frac{1}{2}\gamma^l_{ij}(\widetilde{x},\widetilde{\nu})\nu^i\nu^j$, it will be found that G^l_{hk} are independent of $\widetilde{\nu}$. Such types of Finsler spaces are called affinely-connected spaces or Berwald spaces (Rund, 1959).

Now, the Finslerian(lepton) field $\psi(\widetilde{x},\widetilde{\nu})$, a bispinor, satisfies the equation(172), which can be written in the following form:

$$i\hbar\left[\gamma^\mu(\widetilde{x})\partial_\mu\psi(\widetilde{x},\widetilde{\nu}) - \zeta(t)\gamma^l_{h\mu}\nu^h\left(\psi(\widetilde{x},\widetilde{\nu})\overleftarrow{\partial}'_l\gamma^{\mu T}(\widetilde{x})\right)\right] = mc\psi(\widetilde{x},\widetilde{\nu}) \tag{258}$$

where $\psi(\widetilde{x},\widetilde{\nu})$ is represented as a 4×4 matrix. Here, the Dirac matrices $\gamma^\mu(\widetilde{x})$ for the curved space-time manifold(the associated Riemannian space or the $\widetilde{x}$-space) are connected with the flat space Dirac matrices γ^a through the vierbein fields V^μ_a by the relations

$$\gamma^\mu(\widetilde{x}) = V^\mu_a(\widetilde{x})\gamma^a \tag{259}$$

and

$$\gamma_\mu(\widetilde{x}) = V^a_\mu(\widetilde{x})\gamma_a \tag{260}$$

where $V^a_\mu(\widetilde{x})$ are the inverse vierbein fields, such that

$$V^\mu_a(\widetilde{x})V^b_\mu(\widetilde{x}) = \delta^b_a \tag{261}$$

In the present case, the vierbein fields are diagonal and they are given by

$$V^\mu_a(\widetilde{x}) = [F(t)]^{\frac{-1}{2}}\delta^\mu_a$$

and

$$V^a_\mu(\widetilde{x}) = [F(t)]^{\frac{1}{2}}\delta^a_\mu$$

Writing $[F(t)]^{\frac{-1}{2}} = e(t)$, we have

$$V^\mu_a(\widetilde{x}) = e(t)\delta^\mu_a$$

and

$$V^a_\mu(\widetilde{x}) = \frac{1}{e(t)}\delta^a_\mu \tag{262}$$

Then, equation (258) becomes

$$i\hbar e(t)[\gamma^\mu\partial_\mu\psi(\widetilde{x},\widetilde{\nu}) - \zeta(t)\psi(\widetilde{x},\widetilde{\nu})\overleftarrow{\partial}'_l\gamma^{\mu^T}\gamma^l_{h\mu}\nu^h] = mc\psi(\widetilde{x},\widetilde{\nu}) \tag{263}$$

Now, let us decompose $\psi(\widetilde{x}, \widetilde{\nu})$ in the following way (De,1997):

$$\psi(\widetilde{x}, \widetilde{\nu}) = \psi_1(\widetilde{x}) \times \phi^T(\widetilde{\nu}) + \psi_2(\widetilde{x}) \times \phi^{c^T}(\widetilde{\nu}) \tag{264}$$

where the spinors $\psi_1(\widetilde{x})$ and $\psi_2(\widetilde{x})$ are eigenstates of γ^0 with eigenstates +1 and -1 respectively and $\phi(\widetilde{\nu})$ and $\phi^c(\widetilde{\nu})$ satisfy, respectively, the following equations

$$i\hbar\gamma^\mu\gamma^l_{h\nu}\nu^h\partial'_l\phi(\widetilde{\nu}) = \left(Mc - \frac{3i\hbar b_0}{2})\phi(\widetilde{\nu}\right) \tag{265}$$

$$i\hbar\gamma^\mu\gamma^l_{h\nu}\nu^h\partial'_l\phi^c(\widetilde{\nu}) = \left(Mc + \frac{3i\hbar b_0}{2})\phi^c(\widetilde{\nu}\right) \tag{266}$$

Then,it is easily seen that the field $\psi(\widetilde{x}, \widetilde{\nu})$ satisfies the Dirac equation in the '"$\widetilde{x}$-space", the associated Riemannian space. This Riemannian space is the space-time conformal to the Minkowski flat space. The equation for $\psi(\widetilde{x}, \widetilde{\nu})$ is given as

$$i\hbar\gamma^\mu\partial_\mu\psi(\widetilde{x}, \widetilde{\nu}) + \frac{3i\hbar b_0}{2}\zeta(t)\gamma^0\psi(\widetilde{x}, \widetilde{\nu}) = \frac{c}{e(t)}(m + M\zeta(t)e(t))\psi(\widetilde{x}, \widetilde{\nu}) \tag{267}$$

Consequently, the Dirac equation for the Robertson-Walker space-time can be obtained by a pure-time transformation(c.f.equations(90) and (92)). Here, the additional mass term M appears as the constant in the process of separation of the equation(263) and this can be considered as a manifestation of the anisotropic Finslerian character of the microdomain. In a curved space time with the metric tensor $g_{\mu\nu} = \eta_{\mu\nu}exp[\lambda(\widetilde{x})]$ that is, a space-time conformal to Minkowski space-time, the Dirac equation for a particle has been considered by Bediaga, et al(1989), Mielke(1981) and also by Yajima and Kimura(1985). Of course, in the present case it is evident that the mass of the particle is time-dependent if M is nonzero. In fact for nonzero M the mass of the particle is $m+M\zeta(t)e(t)$ where m can be regarded as the "inherent" mass of the particle. We shall see later that if the time-dependent part of mass is expressed in terms of cosmological time this term is found to be dominant in the very early universe and had a significant role in that era. Although, the time-dependent part has an extremely small value compared to the inherent mass in the present epoch of the universe and thus the inherent mass becomes the mass of the particle to the extremely high order of accuracy. Another important fact about the field and its equation (that is, the Dirac equation in curved space-time) is that the field in this case is$\widetilde{\nu}$-dependent and, in fact,is a bispinor. The "$\widetilde{\nu}$-part" in the decomposition of $\psi(\widetilde{x}, \widetilde{\nu})$ can give rise to an additional quantum number if it is the field of the constituent-particle in the hadron configuration and this will be considered in the subsequent chapter. On the other hand, for the cases other than this one the spinorial character of the field with the $\widetilde{\nu}$-variable dependence has no other physical significance(than that of the time-dependent mass). The usual field for the curved space-time(the$\widetilde{x}$-space) can be obtained by the following "averaging procedure"

$$\psi(\widetilde{x}) = \int \psi(\widetilde{x}, \widetilde{\nu})\chi(\widetilde{\nu})d^4\nu \tag{268}$$

where $\chi(\widetilde{\nu})$ is a (spinor) probability density or a weight function. Clearly, the field(spinor)$\psi(\widetilde{x})$ satisfies the Dirac equation in curved space-time, that is, the equation(267).

The Dirac equation for the local inertial frame (the Minkowski flat space) can, of course, be recovered from(267) by using the vierbeins $V^\alpha_\mu(X)$ which connect the curved space-time with the flat one in normal coordinates. The vierbeins are given by $V_\mu^\alpha(X) = \left(\frac{\partial y_X^\alpha}{\partial X^\mu}\right)_{x=X}$ $(\alpha = 0,1,2,3)$ where y_X^α(with index α referring to the local inertial frame) are normal coordinates at a point X. In the present case, the index μ is associated with the conformal space-time to the Minkowski space-time, which by pure-time transformation becomes the Robertson-Walker space-time.It is to be noted that if we keep y_X^α fixed then the effect of changing x^μ is given by

$$V_\mu^\alpha \longrightarrow \frac{\partial x^\nu}{\partial x^\mu} V_\nu^\alpha \tag{269}$$

On the other hand, y_X^α may be changed by Lorentz transformation $\Lambda_\beta^\alpha(X)$ and in this case the vierbeins $V_\mu^\alpha(X)$ are changed to $\Lambda_\beta^\alpha(X)V_\alpha^\beta(X)$ which leave the metric of the curved space-time invariant. The vierbeins for the present case are given in (262).
With these vierbeins we obtain from(267) the Dirac equation in local inertial frame in normal coordinate system as

$$i\hbar\gamma^\alpha\partial_\alpha\psi(\widetilde{x},\widetilde{\nu}) = mc\psi(\widetilde{x},\widetilde{\nu}) \tag{270}$$

if one neglects the extremely small second terms both from the left and right hand sides of (267). Of course, the mass term $M\zeta(t)e(t)$ may be retained here and although it has no effect on the mass of the particle in the present epoch of the universe its dominance in the very early epoch is very significant(De,1993 a;b). We shall return to it later.

Actually, in laboratory space-time(Minkowskian) we can decompose the field as in (229), that is,

$$\psi(\widetilde{x},\widetilde{\nu}) = \psi(\widetilde{x}) \times \phi^T(\widetilde{\nu}) = \psi(\widetilde{x})\phi^T(\widetilde{\nu}) \tag{271}$$

where $\phi(\widetilde{\nu})$ satisfies the following equation

$$i\hbar\gamma^\mu P^l_{h\mu}(\widetilde{\nu})\nu^h\partial_l\phi(\widetilde{\nu}) = Mc\phi(\widetilde{\nu}) \tag{272}$$

It is to be noted that this corresponds to the decomposition of $\psi(\widetilde{x},\widetilde{\nu})$ as given in (264) for the case $\phi(\widetilde{\nu}) = \phi^c(\widetilde{\nu})$ and here, of course, the space-time is a flat one ($b_0 \longrightarrow 0$). Now with this decomposition it is easy to see that (from equation (172))$\psi(\widetilde{x}$ satisfies the Dirac equation in flat space-time:

$$i\hbar\gamma^\mu\partial_\mu\psi(\widetilde{x}) = mc\psi(\widetilde{x}) \tag{273}$$

Here, of course, $b_0 \longrightarrow 0$(and consequently $F(t) \longrightarrow 1$) makes $M \longrightarrow 0$, as we shall later see that M is proportional to b_0. The bispinor $\psi(\widetilde{x},\widetilde{\nu})$ which satisfies the Dirac equation(270) in the local inertial frame (the Minkowski space-time) reduces to a spinor through the averaging procedure (268) and that spinor satisfies the Dirac equation in local frame with or without the time-dependent mass term. Thus, the time-dependence of particle masses is the manifestation of the Finslerian character of the space-time whose underlying manifold (the$\widetilde{x}$-space or the associated Riemannian space) is a curved one. In the present case, this space is a Robertson-Walker space-time (obtained through pure-time transformation) which is the background space-time of our universe.

3.3: HOMOGENEOUS SOLUTIONS FOR $\phi(\widetilde{\nu})$ AND $\phi^C(\widetilde{\nu})$

We seek a class of solutions for $\phi(\widetilde{\nu})$ and $\phi^C(\widetilde{\nu})$ which are homogeneous of degree zero. In fact, the metric tensors of the Finsler space and the fundamental function are homogeneous functions of degree zero and one, respectively, in the directional arguments. Therefore, one can argue that only this class of homogeneous solutions is physically relevant. For the Finsler space that we are considering, the equations for $\phi(\widetilde{\nu})$ and $\phi^C(\widetilde{\nu})$ for such type of solutions become (from (266))

$$i\hbar b_0 \sum_{l=1}^{3} \gamma^l \left(\nu^l \frac{\partial}{\partial \nu^0} + \nu^0 \frac{\partial}{\partial \nu^l} \right) \phi(\widetilde{\nu}) = -Mc\phi(\widetilde{\nu}) + \frac{3i\hbar b_0}{2} \phi(\widetilde{\nu}) \tag{274}$$

and

$$i\hbar b_0 \sum_{l=1}^{3} \gamma^l \left(\nu^l \frac{\partial}{\partial \nu^0} + \nu^0 \frac{\partial}{\partial \nu^l} \right) \phi^c(\widetilde{\nu}) = -Mc\phi^c(\widetilde{\nu}) - \frac{3i\hbar b_0}{2} \phi^c(\widetilde{\nu}) \tag{275}$$

The general form of the solutions which are homogeneous of degree zero is given as

$$\phi(\widetilde{\nu}) = \left[f_1 \left(\frac{\overrightarrow{\nu}^2}{\nu^{0 2}} \right) + \frac{i \overrightarrow{\gamma} . \overrightarrow{\nu}}{\nu^0} f_2 \left(\frac{\overrightarrow{\nu}^2}{\nu^{0 2}} \right) \right] \omega^b \tag{276}$$

where ω^b is a four-component(arbitrary) spinor independent of $\widetilde{\nu}$. Here, f_1 and f_2 are functions of the homogeneous variable(of degree zero in the directional variable $\widetilde{\nu}$) $x = \frac{\overrightarrow{\nu}^2}{\nu^{02}}$. These functions satisfy the following coupled equations

$$2f_1'(x)(1-x) = ikf_2(x) \tag{277}$$

$$(x-3)f_2(x) + 2x(x-1)f_2'(x) = -ikf_1(x) \tag{278}$$

where k is a constant which may be real or complex. It is, in fact given by

$$i\hbar b_0 k = -Mc + \frac{3i\hbar b_0}{2} \tag{279}$$

For $\phi^c(\widetilde{\nu})$ whose form of the solutions is also given by(276), the constant k in (277) and (278) is now replaced by k' which is given as

$$i\hbar b_0 k' = -Mc - \frac{3i\hbar b_0}{2} \tag{280}$$

Obviously,

$$Re[k] = \frac{3}{2}$$

and

$$Re[k'] = -\frac{3}{2}$$

$$Im[k] = Im[k'] = \frac{Mc}{\hbar b_0} \tag{281}$$

For the case $M = 0$, $Im[k] = Im[k'] = 0$, that is, k and k' are real. For the case of the equation (272),the solution for $\phi(\widetilde{\nu})$ is also of the form(276), but in this case k is purely imaginary and is given by $k = \frac{iMc}{\hbar b_0}$. Now, it is easy to see that $f_1(x)$ and $f_2(x)$, which satisfy the coupled first order differential equations (277) and (278), individually satisfy the following second order differential equations

$$4(x-1)^2 \left[xf_1''(x) + \frac{3}{2} f_1'(x) \right] + k^2 f_1(x) = 0 \tag{282}$$

$$4(x-1)^2 \left[xf_2''(x) + \frac{5}{2} f_2'(x) \right] + [k^2 - 2(1-x)] f_2(x) = 0 \tag{283}$$

We can find the solution of the equation(282) for $f_1(x)$ and consequently for $f_2(x)$ from (277) and (278) or directly from (283), by transforming the equation(282) by the substitution $y = x^{-\frac{1}{2}}$.

We get the following equation for f_1:

$$\frac{d^2 f_1}{dy^2} + \frac{k^2 f_1}{(1-y^2)^2} = 0 \qquad (y^2 \neq 1) \tag{284}$$

Again, by the substitution

$$f_1 = (1-y^2)^{\frac{1}{2}} v \qquad (y^2 \neq 1) \tag{285}$$

We can find the following equation for v:

$$(1-y^2)\frac{d^2 v}{dy^2} - 2y\frac{dv}{dy} + \frac{k^2-1}{1-y^2} v = 0 \tag{286}$$

The solutions of this equation are associated Legendre functions. In fact, associated Legendre functions of the first kind $P_0^{\pm\mu}(\pm y) = P_{-1}^{\pm\mu}(\pm y)$ and those functions of the second kind $Q_0^{\pm\mu}(\pm y), Q_{-1}^{\pm\mu}(\pm y)$ are the required solutions for v where $\mu = (1-k^2)^{\frac{1}{2}}$.
For the case $M = 0$, we have $k^2 = k'^2 = \frac{9}{4}$ and consequently $\mu = \frac{\sqrt{5}i}{2}$. Now, the solutions for v can be identified for $y^2 > 1$ *and* $y^2 < 1$. In fact, for $y^2 > 1$, one can make the substitution $f_1 = (y^2-1)^{\frac{1}{2}} v$ in the equation(284) to arrive at the same equation(286) for v . For $y^2 > 1$, the solution for v can be taken to be $Q_0^{\mu}(y)$ whose integral representation is given by(Erdelyi, 1953).

$$Q_0^{\mu}(y) = \frac{1}{2} e^{\pi\mu i}\Gamma(\mu+1)(y^2-1)^{-\frac{\mu}{2}} \int_0^{\pi} (y+\cos t)^{\mu-1} \sin t dt \qquad for \quad Re(\mu+1) > 0 \tag{287}$$

By making the substitution $y + cost = \xi$ the above integral can be computed and we find

$$v = Q_0^{\mu}(y) = \frac{1}{2\mu} e^{\pi\mu i}\Gamma(\mu+1)(y^2-1)^{-\frac{\mu}{2}} \left[(y+1)^{\mu} - (y-1)^{\mu}\right] \tag{288}$$

Thus, apart from the inessential constant one can take the solution for f_1 for $y^2 > 1$ $(i.e., \quad y > 1 \quad or \quad y < -1)$ as

$$f_1 = \frac{(y^2-1)^{\frac{1}{2}}}{2\mu} \left[\left(\frac{y+1}{y-1}\right)^{\frac{\mu}{2}} - \left(\frac{y-1}{y+1}\right)^{\frac{\mu}{2}}\right] \tag{289}$$

For $y^2 < 1$, (i.e., for $-1 < y < 1$), it is easy to see that the solution for v is

$$v = \frac{1}{2\mu} \left[\left(\frac{y+1}{y-1}\right)^{\frac{\mu}{2}} - \left(\frac{y-1}{y+1}\right)^{\frac{\mu}{2}}\right] \tag{290}$$

and consquenly

$$f_1 = (1-y^2)^{\frac{1}{2}} v \tag{291}$$

The above solutions for f_1 can be expressed in the following alternative forms:

$$f_1 = \frac{(y^2-1)^{\frac{1}{2}}}{d} \sin\left[\frac{d}{2}\ln\left(\frac{y+1}{y-1}\right)\right] \qquad for y > 1 \quad or \quad y < -1 \tag{292}$$

$$f_1 = \frac{(1-y^2)^{\frac{1}{2}}}{d} \sin\left[\frac{d}{2}\ln\left(\frac{1+y}{1-y}\right)\right] \qquad for \quad -1 < y < 1 \tag{293}$$

where $d = \frac{\sqrt{5}}{2}$
Now, f_2 can be found by using (277) and (278) and from the solutions for f_1 as in (292) and (293). We can find

$$f_2 = \frac{2}{ik}(1-x) f_1'(x) = \frac{(1-y^2)y}{ik}\frac{df_1}{dy} \tag{294}$$

Therefore,

$$f_2 = \frac{y(y^2-1)^{\frac{1}{2}}}{ik}\left[\cos\left\{\frac{d}{2}\ln\left(\frac{y+1}{y-1}\right)\right\} - \frac{y}{d}\sin\left\{\frac{d}{2}\ln\left(\frac{y+1}{y-1}\right)\right\}\right] \qquad for\ y>1 \quad or \quad y<-1 \tag{295}$$

$$f_2 = \frac{y(1-y^2)^{\frac{1}{2}}}{ik}\left[\cos\left\{\frac{d}{2}\ln\left(\frac{1+y}{1-y}\right)\right\} - \frac{y}{d}\sin\left\{\frac{d}{2}\ln\left(\frac{1+y}{1-y}\right)\right\}\right] \qquad for\ -1<y<1 \tag{296}$$

Here, the homogeneous variable y of degree zero in the directional variable $\widetilde{\nu}$ is given as

$$y = \frac{\nu^0}{(\overrightarrow{\nu}^2)^{\frac{1}{2}}} \tag{297}$$

Thus, for this case of $M=0$ the solution for $\phi(\widetilde{\nu})$ from (276) can be written as

$$\phi(\widetilde{\nu}) = F(\nu^0, \overrightarrow{\nu})\omega^b \tag{298}$$

where

$$F(\nu^0, \overrightarrow{\nu}) = f_1 + \frac{i\overrightarrow{\gamma}.\overrightarrow{\nu}}{\nu_0} f_2$$

$$-\sqrt{\left(\frac{\nu_0^2}{\overrightarrow{\nu}^2}-1\right)}\left[\frac{1}{d}\sin\left\{\frac{d}{2}\ln\left(\frac{\nu^0+\sqrt{\overrightarrow{\nu}^2}}{\nu^0-\sqrt{\overrightarrow{\nu}^2}}\right)\right\} + \frac{\overrightarrow{\gamma}.\overrightarrow{\nu}}{k\sqrt{\overrightarrow{\nu}^2}}\cos\left\{\frac{d}{2}\ln\left(\frac{\nu^0+\sqrt{\overrightarrow{\nu}^2}}{\nu^0-\sqrt{\overrightarrow{\nu}^2}}\right)\right\} \frac{\nu_0}{d\sqrt{\overrightarrow{\nu}^2}}\sin\left\{\frac{d}{2}\ln\left(\frac{\nu^0+\sqrt{\overrightarrow{\nu}^2}}{\nu^0-\sqrt{\overrightarrow{\nu}^2}}\right)\right\}\right]$$

$$for \quad \nu^o > \sqrt{\overrightarrow{\nu}^2} \quad or \quad \nu^o < -\sqrt{\overrightarrow{\nu}^2} \tag{299}$$

and

$$F(\nu^0, \overrightarrow{\nu}) =$$

$$\sqrt{\left(1-\frac{\nu_0^2}{\overrightarrow{\nu}^2}\right)}\left[\frac{1}{d}\sin\left\{\frac{d}{2}\ln\left(\frac{\nu^0+\sqrt{\overrightarrow{\nu}^2}}{\sqrt{\overrightarrow{\nu}^2}-\nu^0}\right)\right\} + \frac{\overrightarrow{\gamma}.\overrightarrow{\nu}}{k\sqrt{\overrightarrow{\nu}^2}}\cos\left\{\frac{d}{2}\ln\left(\frac{\nu^0+\sqrt{\overrightarrow{\nu}^2}}{\sqrt{\overrightarrow{\nu}^2}-\nu^0}\right)\right\} - \frac{\nu_0}{d\sqrt{\overrightarrow{\nu}^2}}\sin\left\{\frac{d}{2}\ln\left(\frac{\nu^0+\sqrt{\overrightarrow{\nu}^2}}{\sqrt{\overrightarrow{\nu}^2}-\nu^0}\right)\right\}\right]$$

$$for \quad -\sqrt{\overrightarrow{\nu}^2} < \nu^o < \sqrt{\overrightarrow{\nu}^2} \tag{300}$$

The solution for $\phi^c(\widetilde{\nu})$ can be obtained from(299) and (300) by changing $k=\frac{3}{2}$ to $k'=-\frac{3}{2}$. This change of constants from k to k' can also be achieved by changing $\overrightarrow{\nu}$ to $-\overrightarrow{\nu}$ (keeping the constant k unchanged). That is, the solution for $\phi^c(\widetilde{\nu})$ can be written as

$$\phi^c(\widetilde{\nu}) = F(\nu^0, -\overrightarrow{\nu})\omega_c^b \tag{301}$$

where ω_c^b is an arbitrary four-component spinor independent of $\widetilde{\nu}$ and we can choose it as

$$\omega_c^b = i\gamma^2(\omega^b)^* \tag{302}$$

It is to noted that that the solutions for $\phi(\widetilde{\nu})$ and $\phi^c(\widetilde{\nu})$ at $y^2=1$, i.e., $\nu^0=\pm(\overrightarrow{\nu}^2)^{\frac{1}{2}}$ can be obtained by taking the limits of $F(\nu^0,\overrightarrow{\nu})$ as given in (299) and (300). The left and right limits at $\nu^0=\pm(\overrightarrow{\nu}^2)^{\frac{1}{2}}$ are obviously zero. Therefore, $\phi(\widetilde{\nu})$ and $\phi^C(\widetilde{\nu})$ tend to zero at $\nu^0=\pm(\overrightarrow{\nu}^2)^{\frac{1}{2}}$. this result is also in consistent with the relations (277) and (278). It is easy to see that $\phi(\widetilde{\nu})$ (and also $\phi^C(\widetilde{\nu})$) is finite as $\overrightarrow{\nu}^2 \longrightarrow 0 (i.e., y \longrightarrow \infty)$. In fact, $\phi(\widetilde{\nu}) \longrightarrow \omega^b as\ \overrightarrow{\nu}^2 \longrightarrow 0$, i.e., as $\widetilde{\nu}$ tends to directional variable

$(\nu^0, 0, 0, 0)$. Also, as the direction variable $\widetilde{\nu}$ tends to $(0, \overrightarrow{\nu})$, that is, as $y \longrightarrow 0$,

$$\phi(\widetilde{\nu}) \longrightarrow \frac{\overrightarrow{\gamma}.\overrightarrow{\nu}}{k(\overrightarrow{\nu}^2)^{\frac{1}{2}}}\omega^b \quad and \quad \phi^c(\widetilde{\nu}) \longrightarrow -\frac{\overrightarrow{\gamma}.\overrightarrow{\nu}}{k(\overrightarrow{\nu^2})^{\frac{1}{2}}}\omega^b$$

The following relations are true for $F(\nu^0, \overrightarrow{\nu})$:

$$F(\nu^0, -\overrightarrow{\nu}) = -F(-\nu^0, \overrightarrow{\nu}) \tag{303}$$

$$F(-\nu^0, -\overrightarrow{\nu}) = -F(\nu^0, \overrightarrow{\nu}) \tag{304}$$

$$i\gamma^2 F^*(\nu^0, \overrightarrow{\nu}) = F(\nu^0, -\overrightarrow{\nu})i\gamma^2 \tag{305}$$

$$\gamma^2[F(\nu^0, \overrightarrow{\nu})]^T = F(\nu^0, \overrightarrow{\nu})\gamma^2 \tag{306}$$

where the following representation of γ-matrices is being used:

$$\gamma^0 = \begin{pmatrix} 1 & 0 \\ 0 & -1 \end{pmatrix}, \quad \overrightarrow{\gamma} = \begin{pmatrix} 0 & \overrightarrow{\sigma} \\ -\overrightarrow{\sigma} & 0 \end{pmatrix}, \quad \sigma^1 = \begin{pmatrix} 0 & 1 \\ 1 & 0 \end{pmatrix}, \quad \sigma^2 = \begin{pmatrix} 0 & -i \\ i & 0 \end{pmatrix}, \quad \sigma^3 = \begin{pmatrix} 1 & 0 \\ 0 & -1 \end{pmatrix} \tag{307}$$

We also have the relation

$$\phi^c(\overrightarrow{\nu}) = F(\nu^0, -\overrightarrow{\nu})\omega_c^b = F(\nu^0, -\overrightarrow{\nu})i\gamma^2(\omega^b)^* = i\gamma^2 F^*(\nu^0, \overrightarrow{\nu})(\omega^b)^* = i\gamma^2\phi^*(\overrightarrow{\nu}) \tag{307a}$$

[*using equations* (302) *and* (305)]

For $M \neq 0$, i.e., for $Im[k] = Im[k'] \neq 0$, the solution of(286) is also given by $v = Q_0^\mu(y)$ but in this case μ is a complex number. It is given as

$$\mu = \left[1 - \left(\frac{3}{2} + id\right)^2\right]^{\frac{1}{2}} \qquad where \quad Im[k] = Im[k'] = d \tag{308}$$

The imaginary part of k or k' is arbitrary whereas the real part of it is fixed. One can choose the imaginary part to the same order with the real part (which is equal to $\frac{3}{2}$). Presently, we choose

$$Im[k] = Im[k'] = d = 3/\sqrt{13} \tag{309}$$

This gives

$$\mu = 1 - \frac{9i}{2\sqrt{13}} \tag{310}$$

With this μ, the solution for f_1 is given by (289) for $y > 1$ *or* $y < -1$ and by (290) and(291) for $-1 < y < 1$. As before, f_2 is related to f_1 by the relation (294). It is to be noted that this choice of value for d can give rise to the finite limits for f_1 *and* f_2 *as* $y \longrightarrow \infty$ and also for $y \longrightarrow 0$. The functions f_1 *and* f_2 *at* $y = \pm 1$ are given by their limits as $y \longrightarrow \pm 1$ from left as well as from right. It is easy to see that these limits are also finite(left and right limits being equal at $y = 1$ and also at $y = -1$) for this value of d and consequently for μ as given by (310).
For the value of d as given in (309) we obtain the following relation, from(281), which connects the mass

M with the parameter b_0(of the space-time):

$$\frac{3\hbar b_0}{\sqrt{13}} = Mc \quad or, \quad (0.832)\hbar b_0 = Mc \tag{311}$$

It is remarked here that the terms containing b_0 and M may be neglected in order to obtain the usual field equations in the local inertial frame of the "$\overrightarrow{x}$-space" (the flat space) but it will be seen in the next chapter that the $\overrightarrow{\nu}$-dependent $\phi(\overrightarrow{\nu})$ (for the case$M = 0$) can give rise to an additional quantum number that can generate the internal symmetry of hadrons.

3.4: ELECTROMAGNETIC INTERACTION

We have considered electromagnetic interaction in section 2.5 and 2.7. The first of the field equations in section 2.7 for electromagnetic interaction with the lepton field is of the same form as that(i.e., equation(188)) for the interaction with the external electromagnetic field. The other field equation(213) and(214) for the electromagnetic fields have already been considered in section 2.7 to obtain the field equations in the associated Riemannian space. Now, we shall consider the equation(188) in order to find the field equation in the "$\widetilde{x} - space$" [see also De (2001); De (2002)].

Firstly, we point out that if the bispinor field $\psi(\widetilde{x}, \widetilde{\nu})$ represents a particle field then the antiparticle field function $\psi^c(\widetilde{x}, \widetilde{\nu})$ must satisfy the following equation:

$$\gamma^\mu(\widetilde{x})[i\hbar\partial_\mu + \frac{e}{c}\overline{A}_\mu(\widetilde{x})]\psi^c(\widetilde{x}, \widetilde{\nu}) - \psi^c(\widetilde{x}, \widetilde{\nu})\overleftarrow{\partial'}[i\hbar\gamma^l_{h\mu}(\widetilde{x}, \widetilde{\nu})\nu^h\gamma^{\mu^T}(\widetilde{x})] = mc\psi^c(\widetilde{x}, \widetilde{\nu}) \tag{312}$$

where $\psi^c(\widetilde{x}, \widetilde{\nu})$ is represented as a 4×4 matrix. This equation is obtained from (2.5.14) by the transformation $e \longrightarrow -e$.
The relation between the particle and antiparticle fields can be found to be

$$\psi^c(\widetilde{x}, \widetilde{\nu}) = i\gamma^2\psi^*(\widetilde{x}, \widetilde{\nu})i\gamma^2 \tag{313}$$

In fact, by taking the complex conjugate of the equation(188) for $\psi(\widetilde{x}, \widetilde{\nu})$ and then by left as well as right multiplication by $i\gamma^2$, one can arrive at the equation(312) for $\psi^c(\widetilde{x}, \widetilde{\nu})$ if the relation(313) is being taken into account.
If the bispinor field $\psi(\widetilde{x}, \widetilde{\nu})$ is decomposed, as before, as in(264) and (271), the field $\psi^c(\widetilde{x}, \widetilde{\nu})$ can also be decomposed. For the case of the decomposition (271), that is,

$$\psi(\widetilde{x}, \widetilde{\nu}) = \psi(\widetilde{x})\phi^T(\widetilde{\nu}) = \psi(\widetilde{x}) \times \phi^T(\widetilde{\nu}) \tag{314}$$

we have

$$\begin{aligned}
\psi^c(\widetilde{x}, \widetilde{\nu}) &= i\gamma^2(\psi(\widetilde{x})\phi^T(\widetilde{\nu}))^* i\gamma^2 \\
&= i\gamma^2\psi^*(\widetilde{x})(i\gamma^2\phi^*(\widetilde{\nu}))^T \qquad (since \quad \gamma^{2^T} \; = \; \gamma^2\;) \\
&= (i\gamma^2\psi^*(\widetilde{x})) \times (i\gamma^2\psi^*(\widetilde{x}))^T
\end{aligned} \tag{315}$$

Now, if $\psi(\widetilde{x})$ represents a particle field in the "$\widetilde{x} - space$" then

$$i\gamma^2\psi^*(\widetilde{x}) = \psi^c(\widetilde{x}) \tag{316}$$

is the antiparticle field. This will be evident when we shall find the equation of the fields in this space. Also, $i\gamma^2\phi^*(\widetilde{\nu})$ satisfies the same equation(272) for $\phi(\widetilde{\nu})$ and we can put

$$i\gamma^2\phi^*(\widetilde{\nu}) = \phi^c(\widetilde{\nu}) \tag{317}$$

It is to be noted that $\phi^c(\widetilde{\nu})$ is, in this case, not of the form as given in (301). Then we have from (315) the following decomposition for $\psi^c(\widetilde{x}, \widetilde{\nu})$:

$$\psi^c(\widetilde{x}, \widetilde{\nu}) = \psi^c(\widetilde{x}) \times \phi^{cT}(\widetilde{\nu}) \tag{318}$$

For the decomposition (264) we have

$$\begin{aligned}\psi^c(\widetilde{x}, \widetilde{\nu}) &= i\gamma^2[\psi_1(\widetilde{x})\phi^T(\widetilde{\nu}) + \psi_2(\widetilde{x})\phi^{c^T}(\widetilde{\nu})]^* i\gamma^2 \\ &= i\gamma^2\psi_1^*(\widetilde{x})(i\gamma^2\phi^*(\widetilde{\nu}))^T + i\gamma^2\psi_2^*(\widetilde{x})(i\gamma^2\phi^{c^*})^T \\ &= \psi_1^c(\widetilde{x})(i\gamma^2\phi^*(\widetilde{\nu}))^T + \psi_2^c(\widetilde{x})(i\gamma^2\phi^{c^*})^T\end{aligned} \tag{319}$$

where

$$\psi_1^c(\widetilde{x}) = i\gamma^2\psi_1^*(\widetilde{x}) \qquad and \qquad \psi_2^c(\widetilde{x}) = i\gamma^2\psi_2^*(\widetilde{x}) \tag{320}$$

$\psi_1^c(\widetilde{x})$ and $\psi_2^c(\widetilde{x})$ are, now, the eigenstates of γ^0 with eigenvalues -1 and +1 respectively.Again, $\phi(\widetilde{\nu})$ and $\phi^c(\widetilde{\nu})$ are solutions of the two equations in (274) and they are related by (317). In general case(i.e., for $M \neq 0$) the solution for $\phi^c(\widetilde{\nu})$ is again not of the form (301) when $\phi(\widetilde{\nu})$ is of the form (298). As shown earlier that for the case $M = 0$, the solutions for $\phi(\widetilde{\nu})$ and $\phi^c(\widetilde{\nu})$ which are related by(307a) or (317) are of the form (298) and(301) respectively. It also follows from (317) that

$$i\gamma^2\phi^{c^*}(\widetilde{\nu}) = \phi(\widetilde{\nu}) \tag{321}$$

Consequently, we have the following decomposition for the bispinor $\phi^c(\widetilde{x}, \widetilde{\nu})$:

$$\begin{aligned}\psi^c(\widetilde{x}, \widetilde{\nu}) &= \psi_1^c(\widetilde{x})(\phi^c(\widetilde{\nu}))^T + \psi_2^c(\widetilde{x})\phi^T(\widetilde{\nu}) \\ &= \psi_1^c(\widetilde{x}) \times \phi^{c^T}(\widetilde{\nu}) + \psi_2^c(\widetilde{x}) \times \phi^T(\widetilde{\nu})\end{aligned} \tag{322}$$

Now, from the equation (188) by using the decomposition (264) for $\phi(\widetilde{x}, \widetilde{\nu})$ it is easy to see that the bispinor $\phi(\widetilde{x}, \widetilde{\nu})$ satisfies the following equation:

$$i\hbar\gamma^\mu\partial_\mu\psi(\widetilde{x}, \widetilde{\nu}) - \frac{e}{c}\gamma^\mu\bar{A}_\mu(\widetilde{x})\psi(\widetilde{x}, \widetilde{\nu}) + \frac{3i\hbar b_0}{2}\zeta(t)\gamma^0\psi(\widetilde{x}, \widetilde{\nu}) = \frac{c}{e(t)}(m + M\zeta(t)e(t))\psi(\widetilde{x}, \widetilde{\nu}) \tag{323}$$

where, as before, $\phi(\widetilde{\nu})$ and $\phi^c(\widetilde{\nu})$ satisfy equations(265) or(274). $\zeta(t)$ and $e(t)$ are given by (256) and (262) respectively. In fact, the connection coefficients for this Finsler space are separated as (255). The field equation(323) is for the field in the associated Riemannian space of the Finsler space. By the averaging procedure (267), we, in fact, find the field equation for the field $\psi(\widetilde{x})$ for that space. $\psi(\widetilde{x})$ satisfies the following eqation

$$\gamma^\mu(i\hbar\partial_\mu - \frac{e}{c}\bar{A}_\mu(\widetilde{x}))\psi(\widetilde{x}) + \frac{3i\hbar b_0}{2}\zeta(t)\gamma^0\psi(\widetilde{x}) = \frac{c}{e(t)}(m + M\zeta(t)e(t))\psi(\widetilde{x}) \tag{324}$$

The field equation in local inertial frame (the flat Minkowski space) can be derived from (324) with the use of the vierbeins $V_\mu^\alpha(\widetilde{X})$, given by (262), if one neglects the extremely small terms as before. The equation is

$$\gamma^\mu(i\hbar\partial_\mu - \frac{e}{c}A_\mu(\widetilde{x}))\psi(\widetilde{x}) = c(m + M\zeta(t)e(t))\psi(\widetilde{x}) \tag{325}$$

where $A_\mu(\widetilde{x}) = e(t)\bar{A}_\mu(\widetilde{x})$ (expressed in local coordinates, the normal coordinates). Of course, $A_\mu(\widetilde{x}) \simeq \bar{A}_\mu(\widetilde{x})$ as $F(t) \simeq 1$ in the present epoch of the universe. This is evident if the function $F(t)$ is expressed in terms of the cosmological time T (see section 1.8). The equation (325) is the usual field equation in Minkowski space because of the fact that the additional time-dependent mass term $Mc\zeta(t)e(t)$ is negligible in the present epoch of the universe and also it vanishes for the case $Im[k] = Im[k'] = 0$. For nonzero imaginary parts of k and k', this time-dependent mass term is significant only in the very early era of the universe. It is to be noted that we could have been arrived at the above equation (325) by using the decomposition(314) in which case $\phi(\widetilde{\nu})$ satisfies the equation (272).

Now, in the other field equations(213) and(214) for the electromagnetic interaction from which the field equation in the "$\widetilde{x}-space$" have already been obtained, the field $\widetilde{C}(\widetilde{x},\widetilde{\nu})$ has been specified in (216) and (226).
It is, in fact, given as

$$\widetilde{C}(\widetilde{x},\widetilde{\nu}) = \gamma^{\alpha}C(\bar{\psi}(\widetilde{x})\gamma_{\alpha}\psi(\widetilde{x})) \times \hat{\phi}(\widetilde{\nu}) \tag{326}$$

where $\hat{\phi}(\widetilde{\nu})$ satisfies the equations(222). These equations(222) are of the form (272) for $M=0$. In fact, the bispinor $\hat{\phi}(\widetilde{\nu})$ satisfies (272) (with $M=0$) for each index of it. Thus, the solution for the bispinor $\hat{\phi}(\widetilde{\nu})$ is given by

$$\hat{\phi}(\widetilde{\nu}) = \left(f_1 + \frac{i\overrightarrow{\gamma}\cdot\overrightarrow{\nu}}{\nu^0}f_2\right)\bar{\omega} \tag{327}$$

where $\bar{\omega}$ is an arbitrary 4×4 matrix independent of $\widetilde{\nu}$. Since $\hat{\phi}(\widetilde{\nu})$ is symmetric we can choose $\bar{\omega}$ to be γ^2. Thus,

$$\hat{\phi}(\widetilde{\nu}) = \left(f_1 + \frac{i\overrightarrow{\gamma}.\overrightarrow{\nu}}{\nu^0}f_2\right)\gamma^2 \tag{328}$$

In this case f_1 and f_2 satisfy the following equations

$$(1-x)f_1'(x) = 0 \qquad and \qquad (x-3)f_2(x) + 2x(x-1)f_2'(x) = 0 \tag{329}$$

where $x = \frac{\overrightarrow{\nu}^2}{\nu^{0^2}}$ is a variable (homogeneous function of $\widetilde{\nu}$ with degree zero). The nontrivial and finite (for all values of $\widetilde{\nu}$) solutions for f_1 and f_2 are given by

$$f_1(x) = A \quad and \quad f_2(x) = C\frac{x-1}{x^{\frac{3}{2}}}H(x-1) \tag{330}$$

where

$$\begin{aligned} H(\xi) &= 1 \quad for \quad \xi \geq 0 \\ &= 0 \quad for \quad \xi < 0 \end{aligned} \tag{331}$$

Here, A and C are arbitrary constants and we can choose $A=C=1$. With these solutions for f_1 and f_2, the bispinor $\hat{\phi}(\widetilde{\nu})$ can be found as

$$\hat{\phi}(\widetilde{\nu}) = \hat{F}(\nu^0,\overrightarrow{\nu})\gamma^2 \tag{332}$$

with

$$\hat{F}(\nu^0,\overrightarrow{\nu}) = I + i\frac{\overrightarrow{\gamma}.\overrightarrow{\nu}}{\sqrt{\overrightarrow{\nu}^2}}\left(\frac{\overrightarrow{\nu}^2-\nu^{0^2}}{\overrightarrow{\nu}^2}\right)H\left(\frac{\overrightarrow{\nu}^2-\nu^{0^2}}{\overrightarrow{\nu}^2}\right) \tag{333}$$

Using the representation (307) for $\gamma-matrices$ it is easy to see that the following relations hold

$$\begin{aligned} \gamma^2\hat{F}^T = \hat{F}\gamma^2 \quad or \quad \hat{F}^T\gamma^2 &= \gamma^2\hat{F} \\ \hat{F}^{\dagger} &= \hat{F} \\ q(\widetilde{\nu})\hat{F}^{-1}\gamma^0 &= \gamma^0\hat{F} \\ &where, \end{aligned}$$

$$q(\widetilde{\nu}) = 1 - \left(\frac{\overrightarrow{\nu}^2-\nu^{0^2}}{\overrightarrow{\nu}^2}\right)^2 H\left(\frac{\overrightarrow{\nu}^2-\nu^{0^2}}{\overrightarrow{\nu}^2}\right) \tag{334}$$

Evidently,

$$\hat{\phi}^T(\widetilde{\nu}) = \gamma^{2^T}\hat{F}^T = \gamma^2\hat{F}^T = \hat{F}\gamma^2 = \hat{\phi}(\widetilde{\nu}) \tag{335}$$

Now it is to be noted that the Finslerian(lepton) field $\psi(\widetilde{x},\widetilde{\nu})$ (the bispinor) is separated as in (264) where $\phi(\widetilde{\nu})$ and $\phi^c(\widetilde{\nu})$ satisfy the equations(271) corresponds to the case of flat space that is achieved by tending $b_0 \longrightarrow 0$, i.e., $F(t) \longrightarrow 1$ and in this case $\phi(\widetilde{\nu})$ satisfies the equation(272) for $M = 0$. The equation(272) with nonzero M may also correspond to the flat space. But presently we shall consider the case of $M = 0$ and for this case the solution for $\phi(\widetilde{\nu})$ is given by

$$\phi(\widetilde{\nu}) = \hat{F}(\nu^0, \overrightarrow{\nu})\hat{\omega} \tag{336}$$

where $\hat{\omega}$ represents an arbitrary spinor independent of $\widetilde{\nu}$. This arbitrary spinor $\hat{\omega}$ can always be expressed as the linear combination of four independent spinors $\omega^b (b = 1, 2, 3, 4)$ given by

$$\omega^1 = \begin{pmatrix} 1 \\ 0 \\ 0 \\ 0 \end{pmatrix}, \quad \omega^2 = \begin{pmatrix} 0 \\ 1 \\ 0 \\ 0 \end{pmatrix}, \quad \omega^3 = \begin{pmatrix} 0 \\ 0 \\ 1 \\ 0 \end{pmatrix}, \quad \omega^4 = \begin{pmatrix} 0 \\ 0 \\ 0 \\ 1 \end{pmatrix} \tag{337}$$

Thus, we have four solutions $\phi^b(\widetilde{\nu})$ $(b = 1, 2, 3, 4)$ given as

$$\phi^b(\widetilde{\nu}) = \hat{F}(\nu^0, \overrightarrow{\nu})\omega^b \tag{338}$$

Now, the adjoint bispinor $\bar{\psi}(\widetilde{x},\widetilde{\nu})$ was defined in(227) and corresponding to the separation (271) of the bispinor $\psi(\widetilde{x},\widetilde{\nu})$ the adjoint bispinor is separated as in(229). Consequently, we can calculate the current

$$\sum_{b=1}^{4} \bar{\psi}^b(\widetilde{x},\widetilde{\nu})\gamma_\mu(\widetilde{x})\psi^b(\widetilde{x},\widetilde{\nu}) \equiv \psi^b(\widetilde{x},\widetilde{\nu})\gamma_\mu(\widetilde{x})\psi^b(\widetilde{x},\widetilde{\nu}) \quad (\textit{ using summation convention }) \tag{339}$$

where $\psi^b(\widetilde{x},\widetilde{\nu}) = \psi(\widetilde{x})(\phi^b(\widetilde{\nu}))^T$
and consequently,

$$\bar{\psi}^b(\widetilde{x},\widetilde{\nu}) = (\bar{\phi}^b(\widetilde{\nu}))^T \bar{\psi}(\widetilde{x}) \tag{340}$$

$\bar{\phi}^b(\widetilde{\nu})$ being adjoint spinor, that is, $\bar{\phi}^b(\widetilde{\nu}) = (\phi^b(\widetilde{\nu}))^\dagger \gamma^0$
Thus,

$$\begin{aligned} \bar{\psi}^b(\widetilde{x},\widetilde{\nu})\gamma_\mu(\widetilde{x})\psi^b(\widetilde{x},\widetilde{\nu}) &= (\bar{\phi}^b(\widetilde{\nu}))^T(\phi^b(\widetilde{\nu}))^T(\bar{\psi}(\widetilde{x})\gamma_\mu(\widetilde{x})\psi(\widetilde{x})) \\ &= (\phi^b(\widetilde{\nu})\bar{\phi}^b(\widetilde{\nu}))^T(\bar{\psi}(\widetilde{x})\gamma_\mu(\widetilde{x})\psi(\widetilde{x})) \\ &= (\hat{F}(\nu^0, \overrightarrow{\nu})\omega^b\omega^{b^T}\hat{F}(\nu^0, \overrightarrow{\nu})\gamma^0)^T(\bar{\psi}(\widetilde{x})\gamma_\mu(\widetilde{x})\psi(\widetilde{x})) \end{aligned} \tag{341}$$

(using(3.4.22) and noting that ω^b's are real)

Now, it follows that

$$\omega^b\omega^{b^T} = I \tag{342}$$

and hence

$$\bar{\psi}^b(\widetilde{x},\widetilde{\nu})\gamma_\mu(\widetilde{x})\psi^b(\widetilde{x},\widetilde{\nu}) = \gamma^0(\hat{F}^T(\nu^0, \overrightarrow{\nu}))^2(\bar{\psi}(\widetilde{x})\gamma_\mu(\widetilde{x})\psi(\widetilde{x})) \tag{343}$$

or,

$$\gamma^2\gamma^0\bar{\psi}^b(\widetilde{x},\widetilde{\nu})\gamma_\mu(\widetilde{x})\psi^b(\widetilde{x},\widetilde{\nu}) = \hat{F}(\nu^0, \overrightarrow{\nu})\hat{F}(\nu^0, \overrightarrow{\nu})\gamma^2(\bar{\psi}(\widetilde{x})\gamma_\mu(\widetilde{x})\psi(\widetilde{x})) \tag{344}$$

(by using(3.4.22))

Again, using the third relation in (334) we have from above the following relation:

$$\hat{\phi}(\widetilde{\nu})(\bar{\psi}(\widetilde{x})\gamma_{\mu}(\widetilde{x})\psi(\widetilde{x})) = -\frac{1}{q(\widetilde{\nu})}\gamma^0\hat{F}(\nu^0,\overrightarrow{\nu})\bar{\psi}^b(\widetilde{x},\widetilde{\nu})\gamma_{\mu}(\widetilde{x})\psi^b(\widetilde{x},\widetilde{\nu})$$
$$= -\frac{1}{q(\widetilde{\nu})}\gamma^0\hat{\phi}(\widetilde{\nu})\bar{\psi}^b(\widetilde{x},\widetilde{\nu})\gamma_{\mu}(\widetilde{x})\psi^b(\widetilde{x},\widetilde{\nu}) \qquad (345)$$

By using the vierbein field $V_{\mu}^{\alpha}(\widetilde{x})$ *and* $V_{\alpha}^{\mu}(\widetilde{x})$ it follows from (334) that

$$\widetilde{C}(\widetilde{x},\widetilde{\nu}) = \gamma^{\mu}(\widetilde{x})C(\bar{\psi}(\widetilde{x})\gamma_{\mu}(\widetilde{x})\psi(\widetilde{x})) \times \hat{\phi}(\widetilde{\nu})$$
$$= \gamma^{\mu}(\widetilde{x})C \times \hat{\phi}(\widetilde{\nu})(\bar{\psi}(\widetilde{x})\gamma_{\mu}(\widetilde{x})\psi(\widetilde{x}))$$
$$= \gamma^{\mu}(\widetilde{x})C \times [-\frac{1}{q(\widetilde{\nu})}\gamma^0\hat{\phi}(\widetilde{\nu})\bar{\psi}^b(\widetilde{x},\widetilde{\nu})\gamma_{\mu}(\widetilde{x})\psi^b(\widetilde{x},\widetilde{\nu})] \qquad (346)$$

Similar relation also follows for the bispinor(lepton) field $\psi^b(\widetilde{x},\widetilde{\nu})$ with the separation of it as in(334) in the case of $\phi^b(\widetilde{\nu})$ satisfying equation (272) with nonzero M. Thus, $\tilde{C}(\widetilde{x},\widetilde{\nu})$ is connected with the bispinor(lepton) current (339), with which the electromagnetic field interacts.
It is to be noted that the solution for $\phi^b(\widetilde{\nu})$ that satisfies the equation (272) for nonzero M is given by (276) where f_1 and f_2 are given in (289), (290), (291) and (294). In this case μ is real and is given by

$$\mu - \sqrt{1+\frac{M^2c^2}{\hbar^2 b_0^2}} \qquad (347)$$

It is an interesting fact that this solution is singular at ${\nu^0}^2 - \overrightarrow{\nu}^2 = 0$, although it is finite for all other values of the directional variable $\widetilde{\nu}$. On the other hand, the solution for $\phi^b(\widetilde{\nu})$ for$M = 0$ is finite for all $\widetilde{\nu}$, even at ${\nu^0}^2 - \overrightarrow{\nu}^2 = 0$, as we have seen earlier. Thus, there is an intrinsic connection between the mass M and the directional variables for which ${\nu^0}^2 = \overrightarrow{\nu}^2$. This situation is comparable to the fact that for a massive particle its mass becomes infinite if its velocity tends to that of light. Although the analogy is not exact, one may guess that for the present case $\widetilde{\nu}$ may be the four-velocity of the "internal fluctuations".
We have defined a bispinor(lepton) current in (339). This current is a 4×4matrix. Instead of it one can define a current $j_{\mu}(\widetilde{x},\widetilde{\nu})$ given by

$$j_{\mu}(\widetilde{x},\widetilde{\nu}) = Tr.(\gamma^0\bar{\psi}^b(\widetilde{x},\widetilde{\nu})\gamma_{\mu}(\widetilde{x})\psi^b(\widetilde{x},\widetilde{\nu})) \qquad (348)$$

By using the relation (343), we can relate $j_{\mu}(\widetilde{x},\widetilde{\nu})$ with the "$\widetilde{x}-space$" lepton current as follows:

$$j_{\mu}(\widetilde{x},\widetilde{\nu}) = cTr[(\hat{F}^T(\widetilde{x},\widetilde{\nu}))^2](\bar{\psi}(\widetilde{x})\gamma_{\mu}(\widetilde{x})\psi(\widetilde{x})) = P(\widetilde{\nu})(\bar{\psi}(\widetilde{x})\gamma_{\mu}(\widetilde{x})\psi(\widetilde{x})) \qquad (349)$$

where

$$P(\widetilde{\nu}) = 4c\left[1+\left(\frac{\overrightarrow{\nu}^2-{\nu^0}^2}{\overrightarrow{\nu}^2}\right)^2 H\left(\frac{\overrightarrow{\nu}^2-{\nu^0}^2}{\overrightarrow{\nu}^2}\right)\right] \qquad (350)$$

In deducing (350), the expression for $\hat{F}(\nu^0,\overrightarrow{\nu})$ as in (333) has been used. With this current $j_{\mu}(\widetilde{x},\widetilde{\nu})$ we have the following expression for $\widetilde{C}(\widetilde{x},\widetilde{\nu})$:

$$\widetilde{C}(\widetilde{x},\widetilde{\nu}) = \frac{1}{P(\widetilde{\nu})}\gamma^{\mu}(\widetilde{x})Cj_{\mu}(\widetilde{x},\widetilde{\nu}) \times \hat{\psi}(\widetilde{\nu}) = \gamma^{\mu}(\widetilde{x})C \times \frac{1}{P(\widetilde{\nu})}j_{\mu}(\widetilde{x},\widetilde{\nu})\hat{\phi}(\widetilde{\nu}) \qquad (351)$$

An alternative definition for the current can also be made if we keep in mind the averaging procedure of the bispinor $\psi(\widetilde{x},\widetilde{\nu})$ as in (267). This, in fact, is as follows:

$$J_{\mu}(\widetilde{x},\widetilde{\nu}) = c\chi^{\dagger}(\widetilde{\nu}){\psi^b}^{\dagger}(\widetilde{x},\widetilde{\nu})\gamma^0(\widetilde{x})\gamma_{\mu}(\widetilde{x})\psi^b(\widetilde{x},\widetilde{\nu})\chi(\widetilde{\nu}) \qquad (352)$$

where $\chi(\widetilde{\nu})$ is, as before, a(spinor) probability density or a weight function. It is to be noted that for all these definition of the current the usual continuity equation is satisfied. Now, from(343), it follows that

$$J_\mu(\widetilde{x},\widetilde{\nu}) = [c\chi^\dagger(\widetilde{\nu})\hat{F}^T(\nu^0,\overrightarrow{\nu})\hat{F}^T(\nu^0,\overrightarrow{\nu})\chi(\widetilde{\nu})](\bar{\psi}(\widetilde{x})\gamma_\mu(\widetilde{x})\psi(\widetilde{x})) \tag{353}$$

and consequently,

$$\tilde{C}(\widetilde{x},\widetilde{\nu}) = \gamma^\mu(\widetilde{x})C \times \frac{1}{\zeta(\widetilde{\nu})} J_\mu(\widetilde{x},\widetilde{\nu})\hat{\phi}(\widetilde{\nu}) \tag{354}$$

where

$$\zeta(\widetilde{\nu}) = c\chi^\dagger(\widetilde{\nu})\hat{F}^T(\nu^0,\overrightarrow{\nu})\hat{F}^T(\nu^0,\overrightarrow{\nu})\chi(\widetilde{\nu}) = c\eta^\dagger(\widetilde{\nu})\eta(\widetilde{\nu}) \tag{355}$$

with

$$\eta(\widetilde{\nu}) = \hat{F}^\dagger(\nu^0,\overrightarrow{\nu})\chi(\widetilde{\nu}) \tag{356}$$

Since $\eta^*(\widetilde{\nu}) = \hat{F}^\dagger(\nu^0,\overrightarrow{\nu})\chi^*(\widetilde{\nu}) = \hat{F}(\nu^0,\overrightarrow{\nu})\chi^*(\widetilde{\nu})$ (by using (334)), $\eta^*(\widetilde{\nu})$ is a solution of the equation(272) for $M=0$ if $\chi(\widetilde{\nu})$ is independent of the directional argument $\widetilde{\nu}$.

We have seen in section 3.2 that the bispinor field $\psi^b(\widetilde{x},\widetilde{\nu})$ whose decomposition is given by (264) satisfies the Dirac equation in the "$\widetilde{x}-space$", the associated Riemannian space of the Finsler space. This curved space-time is, in fact, the space-time conformal to the Minkowski flat space. The Dirac equation for the Robertson-Walker space-time can be found through pure-time transformation. The continuity equation for the field $\psi^b(\widetilde{x},\widetilde{\nu})$ that satisfies the equation(266) can be found to be given by

$$\frac{\partial\rho}{\partial t} + div\,\overrightarrow{j} + 3cb_0\zeta(t)\rho = 0 \tag{357}$$

where $(j^\mu) = (c\rho,\overrightarrow{j}) \equiv (j^0,\overrightarrow{j})$ can be defined as above. Particularly, the definition(352) is useful as it is made through the spinor weight function $\chi(\widetilde{\nu})$. This spinor $\chi(\widetilde{\nu})$ makes the bispinor field $\psi^b(\widetilde{x},\widetilde{\nu})$ a spinor field $\phi^b(\widetilde{x},\widetilde{\nu})$ given by

$$\phi^b(\widetilde{x},\widetilde{\nu}) = \psi^b(\widetilde{x},\widetilde{\nu})\chi(\widetilde{\nu}) \tag{358}$$

In fact, the usual field for the "$\widetilde{x}-space$" (curved space-time) is obtained through the averaging procedure

$$\psi^b(\widetilde{x}) = \int \phi^b(\widetilde{x},\widetilde{\nu})d^4\nu \tag{359}$$

(c.f. equation(267))

Now, from (264) it follows that

$$\phi^b(\widetilde{x},\widetilde{\nu}) = \psi_1(\widetilde{x})(\phi^{b^T}(\widetilde{\nu})\chi(\widetilde{\nu})) + \psi_2(\widetilde{x})(\phi^{c^{b^T}}(\widetilde{\nu})\chi(\widetilde{\nu})) = \psi_1(\widetilde{x})\phi_1^b(\widetilde{\nu}) + \psi_2(\widetilde{x})\phi_2^b(\widetilde{\nu}) \tag{360}$$

where

$$\phi_1^b(\widetilde{\nu}) = \phi^{b^T}(\widetilde{\nu})\chi(\widetilde{\nu})\,, \quad \phi_2^b(\widetilde{\nu}) = \phi^{c^{b^T}}(\widetilde{\nu})\chi(\widetilde{\nu}) \tag{361}$$

Here, $\phi_1^b(\widetilde{\nu})$ and $\phi_2^b(\widetilde{\nu})$ are, now, functions (not matrices) of the directional variable $\widetilde{\nu}$. The spinor $\phi^b(\widetilde{x},\widetilde{\nu})$ satisfies the Dirac equation (266) for the curved "$\widetilde{x}-space$". The variable $\widetilde{\nu}$ appears in the field as the parameter only. Also, since $\psi_1(\widetilde{x})$ and $\psi_2(\widetilde{x})$ are eigenstates of γ^0 with eigenvalues +1 and -1 respectively, we can write

$$\psi_1(\widetilde{x}) = \frac{1}{2\phi_1^b(\widetilde{\nu})}(1+\gamma^0)\psi^b(\widetilde{x},\widetilde{\nu}) \tag{362}$$

$$\psi_2(\widetilde{x}) = \frac{1}{2\phi_2^b(\widetilde{\nu})}(1-\gamma^0)\psi^b(\widetilde{x},\widetilde{\nu}) \tag{363}$$

Now, since the current $J_\mu(\widetilde{x},\widetilde{\nu})$ is defined as in (352), we have

$$j^\mu(\widetilde{x},\widetilde{\nu}) = c\phi^{b^\dagger}(\widetilde{x},\widetilde{\nu})\gamma^0\gamma^\mu\phi^b(\widetilde{x},\widetilde{\nu}) \tag{364}$$

Clearly, ρ and j^k $(k = 1, 2, 3)$ are given by

$$\rho = \phi_1^{b^*}(\widetilde{\nu})\phi_1^b(\widetilde{\nu})\psi_1^\dagger(\widetilde{x})\psi_1(\widetilde{x}) + \phi_2^{b^*}(\widetilde{\nu})\phi_2^b(\widetilde{\nu})\psi_2^\dagger(\widetilde{x})\psi_2(\widetilde{x}) = l(\widetilde{\nu})\psi_1^\dagger(\widetilde{x})\psi_1(\widetilde{x}) + m(\widetilde{\nu})\psi_2^\dagger(\widetilde{x})\psi_2(\widetilde{x}) \quad (365)$$

$$j^k = cn(\widetilde{\nu})\psi_1^\dagger(\widetilde{x})\gamma^0\gamma^k\psi_2(\widetilde{x}) + cn^*(\widetilde{\nu})\psi_2^\dagger(\widetilde{x})\gamma^0\gamma^k\psi_1(\widetilde{x}) \quad (366)$$

where

$$l(\widetilde{\nu}) = \phi_1^{b^*}(\widetilde{\nu})\phi_1^b(\widetilde{\nu}) \equiv \sum_b \phi_1^{b^*}(\widetilde{\nu})\phi_1^b(\widetilde{\nu})$$

$$m(\widetilde{\nu}) = \phi_2^{b^*}(\widetilde{\nu})\phi_2^b(\widetilde{\nu}) \equiv \sum_b \phi_2^{b^*}(\widetilde{\nu})\phi_2^b(\widetilde{\nu})$$

and

$$n(\widetilde{\nu}) = \phi_1^{b^*}(\widetilde{\nu})\phi_2^b(\widetilde{\nu}) \equiv \sum_b \phi_1^{b^*}(\widetilde{\nu})\phi_2^b(\widetilde{\nu})$$

(367)

It is interesting to note that the continuity equation(357) for the curved "$\widetilde{x}$ − *space*" can be written in the usual flat space form of the continuity equation, that is,

$$\frac{\partial\hat{\rho}}{\partial t} + div\,\overrightarrow{\hat{j}} = 0 \quad (368)$$

with

$$\hat{j}^\mu = j^\mu exp[\int 3cb_0\zeta(t)dt] = (F(t))^{\frac{3}{2}} j^\mu \quad (369)$$

where $F(t) \equiv g(\widetilde{x})$ is the conformal factor given in (254). In deducing(369) the expression of$\zeta(t)$ from (256) has been used. It follows that

$$\hat{\rho} = (F(t))^{\frac{3}{2}}\rho \quad (370)$$

and $\int \hat{\rho} dV = constant$ in respect of the conformal time t. But $\int \rho dV$ is not a constant and, in fact, is proportional to $(F(t))^{-\frac{3}{2}}$.

REFERENCES

1. Bediaga, I., et al. (1989). Modern Physics Letters A, **4**, 169.
2. De, S.S. (1993a).International Journal of Theoretical Physics, **32**, 1603.
3. De, S.S. (1993b). Communications in Theoretical Physics, **2**, 249.
4. De, S.S.(1997). International Journal of Theoretical Physics, **36**, 89.
5. De, S.S.(2001). International Journal of Theoretical Physics, **40**, 2067.
6. De, S.S.(2002). International Journal of Theoretical Physics, **41**, 1291.
7. De, S.S. : ArXiv: hep-th/0308180.
8. Erdelyi, A., ed.(1953). Higher Transcendental Functions, Mc Graw- Hill, New York.
9. Mielke, E. W.(1981). Journal of Mathematical Physics, **22**, 2034.
10. Rund, H.(1959). The Differential Geometry of Finsler Spaces, Springer, Berlin.
11. Yajima, S., and Kimura, T.(1985). Progress in Theoretical Physics, **74**, 866.

CHAPTER FOUR:

INTERNAL SYMMETRY OF HADRONS, HADRON FIELDS AND INTERACTION

4.1: PARTICLE MASS: A RELATION

In section two of chapter three, we have seen that the mass of the particle (lepton) is epoch dependent and the epoch dependent part of it is a manifestation of the anisotropic Finslerian character of the microdomain. Of course, for some kind of leptons their masses may not be epoch dependent (i.e., the cases $M = 0$). It is seen from (267) that the mass $\overline{m}$ of the lepton is given as

$$\overline{m} = m + M\zeta(t)e(t) \tag{371}$$

where $\zeta(t)$ and $e(t)$ are respectively given by (256) and (262). Here, m is regarded as the inherent mass of the particle and we shall see later that this inherent mass is just the present mass of the particle at a very significant level of accuracy. The time t represents the conformal time and by pure time transformation (cf. equations (90) and (92)) the Roberson-Walker spacetime can be obtained. That is, the following pure time transformation

$$\Omega(t)dt = dT \tag{372}$$

where $\Omega^2(t) = F(t)$
gives the cosmological time. When expressed in terms of T, $\Omega(t)$ is identical with the scale factor R(T) of FRW Universe. Thus,

$$\frac{dT}{dt} = \Omega(t) \equiv R(T) \tag{373}$$

again, we have from (256) and (262)

$$\begin{aligned}\zeta(t)e(t) &= \frac{1}{2b_0c}\frac{F'(t)}{F(t)}\frac{1}{\sqrt{F(t)}} = \frac{1}{b_0c}\frac{1}{\Omega^2}\frac{d\Omega}{dt} = \frac{1}{b_0c}\frac{1}{R^2(T)}\frac{d\Omega}{dT}\frac{dT}{dt} \\ &= \frac{1}{b_0c}\frac{1}{R^2(T)}\frac{dR}{dT}R(T) = \frac{1}{b_0c}\frac{1}{R(T)}\frac{dR}{dT} = \frac{H(T)}{b_0c}\end{aligned} \tag{374}$$

[we have used (273)]

where $H(T) = \frac{1}{R(T)}\frac{dR}{dT}$ is Hubble function. Thus we get,

$$\overline{m} = m + \frac{M}{b_0c}H(T) \tag{375}$$

We have seen in section 3.2 that M is connected with the parameter b_0 by the relation (281) or more specifically by the relation (311). We have discussed there that this specific relation can account for the homogeneous solution of degree zero for $\phi(\widetilde{v})$ for all directional arguments $\widetilde{v}$. Using this relation we find from (375) the following relation for the mass:

$$\overline{m} = m + \frac{0.832\hbar}{c^2}H(T) \tag{376}$$

Again, if we write $M = \lambda m$ and $b_0 c = \tau^{-1}$ then we have from (311),

$$\frac{0.832\hbar}{mc^2} = \lambda\tau = 2\alpha \quad (say) \tag{377}$$

Then we also have the following relation

$$\overline{m} = m(1 + 2\alpha H(T)) \tag{378}$$

where

$$\alpha = \frac{0.416\hbar}{mc^2} \tag{379}$$

The value of α can be computed for the leptons. As for an example, for μ^- whose mass in the natural unit ($\hbar = c = 1$) is 5.37×10^{12} cm^{-1} the computed value of α is 0.26×10^{-23} sec. Again, the age of the Universe, T_0 is connected with τ and the present Hubble constant H_0 by the relation

$$T_0 = \frac{\tau}{3} = \frac{2}{3H_0} \tag{380}$$

With the age of the present Universe to be 15 billion years, i.e., 4.5×10^{17} sec., we can find the value of λ from (377). It is given as

$$\lambda = 3.852 \times 10^{-42} \tag{381}$$

From the relation (378) we find the present mass of μ^- to be given by

$$\overline{m}_{\mu^-} = m_{\mu^-}(1 + 2\alpha H(T_0)) = m_{\mu^-}(1 + 2\alpha H_0) = m_{\mu^-}\left(1 + \frac{4\alpha}{3T_0}\right) = m_{\mu^-}\left(1 + 8 \times 10^{-42}\right) \tag{382}$$

Thus, in the present epoch of the Universe the inherent mass of the particle is the particle mass to an extremely high order of accuracy, as is evident from (382). The epoch dependent part of the particle mass had predominant role in the early epoch of the Universe only, particularly around or before the epoch time α. The cosmological implications of this aspect will be considered in later chapters. It must also be noticed that for not all species of leptons these epoch dependent parts of masses are non zero. Particularly, for the constituent particles (leptons) of the hadrons, it is assumed that $M = 0$, that is, the masses of these constituents are epoch independent. The anistropy of the microdomain for this case of constituent particles is manifested in a different manner that we should see in the following sections. If $M = 0$ for the constituents of the hadrons, we shall, in fact, see that these particles can have additional quantum numbers in the form of internal helicities for them. Thus, we get two fold manifestation of the anisotropy. It might be the epoch dependence of the particle masses or it is in the form of the internal helicity of a particle (constituent) in a hadron configuration. We shall presently consider the later aspect (De, 1997).

4.2: THE OPERATORS D^l AND THEIR COMMUTATION RELATIONS

The equations (274) for the '$\widetilde{\nu}$ - part' wave functions $\phi(\widetilde{v})$ and $\phi(\widetilde{v}^c)$ of the particles in hadron configuration (i.e., the case of $M = 0$) become

$$i\overrightarrow{\gamma}.\overrightarrow{D}(\nu^0, \overrightarrow{\nu})\phi(\widetilde{v}) = \hbar\phi(\widetilde{v})$$

and

$$i\overrightarrow{\gamma}.\overrightarrow{D}(\nu^0,\overrightarrow{\nu})\phi^c(\widetilde{\nu}) = -\hbar\phi^c(\widetilde{v\nu})$$

(383)

where $\overrightarrow{\gamma}.\overrightarrow{D} = \sum_{l=1}^{3}\gamma^l D^l$
with

$$D^l = \frac{2}{3i}\hbar\left(\nu^0\frac{\partial}{\partial\nu^l} + \nu^l\frac{\partial}{\partial\nu^0}\right) \tag{384}$$

The components of the three vector $\overrightarrow{D} = (D^1, D^2, D^3)$ satisfy the following commutation relations:

$$[\nu^0, D^m] = \frac{2}{3}i\hbar\nu^m$$

$$[\nu^l, D^m] = \frac{2}{3}i\hbar\nu^0\delta_{lm} \quad (l \neq 0)$$

$$[q_0, D^m] = \frac{2}{3}i\hbar q_m$$

$$[q_l, D^m] = \frac{2}{3}i\hbar q_0\delta_{lm} \quad (l \neq 0)$$

$$[D^0, D^m] = -\frac{8}{9}i\hbar(\nu^0 q_m + \nu^m q_0) \quad (m \neq 0)$$

$$[D^l, D^m] = -\frac{4}{9}i\hbar(\nu^l q_m - \nu^m q_l) \quad (l, m = 1, 2, 3)$$

(385)

where

$$q_0 = i\hbar\frac{\partial}{\partial\nu^0} = q^0$$

$$q_l = -i\hbar\frac{\partial}{\partial\nu^l} = -q^l \quad (l = 1, 2, 3)$$

(386)

Also, ν^ρ, q^ρ satisfy the following commutation relations:

$$[\nu^\rho, \nu^\sigma] = 0$$

$$[q_\rho, q_\sigma] = [q^\rho, q^\sigma] = 0$$

$$[\nu^\rho, q_\sigma] = i\hbar g_{\rho\sigma}$$

(387)

The last of the commutation relations (385) can also be written as

$$\overrightarrow{D} \times \overrightarrow{D} = -\frac{4}{9}i\hbar(\overrightarrow{\nu} \times \overrightarrow{q}) \tag{388}$$

or,

$$[D^l, D^m] = -\frac{4}{9}i\hbar \in_{lmn} \widehat{J_n} \tag{389}$$

with $\overrightarrow{\widehat{J_n}} = \overrightarrow{\nu} \times \overrightarrow{q}$
Also, we can have the commutation relations

$$[\widehat{J_l}, D^j] = i \in_{ljk} D^k \tag{390}$$

These commutation relations may be compared with those for the 'boost' generators K_j (Perl, 1974) which in the usual coordinate representation are given by

$$K_j = -i\left(t\frac{\partial}{\partial x_j} + x_j\frac{\partial}{\partial t}\right) \tag{391}$$

[in the natural unit, $\hbar = c = 1$]

The commutation relations satisfied by K_j are

$$\begin{aligned} [K_j, P^0] &= iP_j \\ [K_j, P_k] &= -i\delta_{kj}P^0 \\ [J_k, K_j] &= i \in_{kjl} K_l \\ [K_k, K_j] &= -i \in_{kjl} J_l \end{aligned} \tag{392}$$

Thus, in a sense the vector $\overrightarrow{D}$ can be thought as the boost generator in the tangent space of the Finsler space.

4.3: ADDITIONAL QUANTUM NUMBERS OF HADRON-CONSTITUENTS

The wave (or field) function $\psi(\widetilde{x},\widetilde{\nu})$ as separated in (264) has two parts, $\psi_1(\widetilde{x})\times\phi^T(\widetilde{\nu})$ and $\psi_2(\widetilde{x})\times\phi^{cT}(\widetilde{\nu})$. In the 'rest' frame of the particles, these two parts are related as the particle and antiparticle to each other. In fact, for this case of the constituents particles in the hadron configuration, for which $M = 0$, the antiparticle state $\psi^c(\widetilde{x},\widetilde{\nu})$ corresponding to a particle state with $\psi(\widetilde{x},\widetilde{\nu}) = \psi(\widetilde{x}) \times \phi^T(\widetilde{\nu})$ is given by

$$\psi^c(\widetilde{x},\widetilde{\nu}) = i\gamma^2\psi^*(\widetilde{x}) \times (i\gamma^2\phi^*(\widetilde{\nu}))^T = \psi^c(\widetilde{x}) \times \phi^{cT}(\widetilde{\nu}) \tag{393}$$

[cf. discussions of the section 3.4]
Here, $\phi(\widetilde{\nu})$ and $\phi^c(\widetilde{\nu})$ satisfy the equation (383). Also the equation satisfied by $\phi^c(\widetilde{\nu}) \equiv \phi^c(\nu^0,\overrightarrow{\nu})$ is also been satisfied by $\phi(\nu^0,-\overrightarrow{\nu})$ since $D(\nu^0,-\overrightarrow{\nu}) = -D(\nu^0,\overrightarrow{\nu})$.
Thus,

$$\phi^c(\nu^0,\overrightarrow{\nu}) \propto \phi(\nu^0,-\overrightarrow{\nu}) \tag{394}$$

It must be noted that the equations (383) for $\phi(\widetilde{\nu})$ and $\phi^c(\widetilde{\nu})$ do not contain the parameter b_0 (as also of M, since $M = 0$) and thus, when one makes $b_0 \to 0$ in the transition from the comoving coordinates of FRW spacetime to the local inertial frame (the Minkowski flat spacetime) these equations are unaffected and continue to hold. Now, there are two independent solutions for the '$\widetilde{\nu}$ - part' wave function $\phi(\overrightarrow{\nu})$ (suppressing the dependence on ν^0, which will be irrelevant in the subsequent discussions) that satisfy the first equation of (383) which is, in the unit $c = \hbar = 1$,

$$i\overrightarrow{\gamma}.\overrightarrow{D}\phi(\overrightarrow{\nu}) = \phi(\overrightarrow{\nu}) \tag{395}$$

Let us denote them by $\phi(\overrightarrow{\nu})$ and $\overline{\phi}(\overrightarrow{\nu})$. Again,

$$i\overrightarrow{\gamma}.\overrightarrow{D} = -(\gamma^1\gamma^2\gamma^3)(\overrightarrow{\Sigma}.\overrightarrow{D}) = -(\overrightarrow{\Sigma}.\overrightarrow{D})(\gamma^1\gamma^2\gamma^3) \tag{396}$$

where the representation (307) of γ-matrices has been used and

$$\vec{\Sigma} = \begin{pmatrix} \vec{\sigma} & 0 \\ 0 & \vec{\sigma} \end{pmatrix} \tag{397}$$

As the three operators $i\vec{\gamma}.\vec{D}$, $(\gamma^1\gamma^2\gamma^3)$, $(\vec{\Sigma}.\vec{D})$ are mutually commuting, one can construct simultaneous eigenstates of them. Thus, we can take ϕ and $\overline{\phi}$ to be the eigenstates of $(\gamma^1\gamma^2\gamma^3)$ *and* $(\vec{\Sigma}.\vec{D})$. Let ϕ and $\overline{\phi}$ be the eigenstates of $(\gamma^1\gamma^2\gamma^3)$ with eigenvalues +1 and -1. respectively. (It can be easily seen that the eigenvalues of $(\gamma^1\gamma^2\gamma^3)$ *and* $(\vec{\Sigma}.\vec{D})$ are only ± 1). As ϕ and $\overline{\phi}$ are linearly independent eigenstates one can always make them to have eigenvalues +1 and -1 respectively.

Thus, we have

$$\vec{\Sigma}.\vec{D}\phi(\vec{\nu}) = -\phi(\vec{\nu}); \vec{\Sigma}.\vec{D}\phi(\vec{\nu}) = \phi(\vec{\nu}) \tag{398}$$

For the antiparticles '$\widetilde{\nu}$ - part' wave function that satisfies the second of the equations (13) one can similarly have two linearly independent solutions $\phi^c(\vec{\nu})$ and $\overline{\phi}^c(\vec{\nu})$ where $\phi^c = i\gamma^2\phi^*$ and $\overline{\phi}^c = i\gamma^2\overline{\phi}^*$. Obviously,

$$\begin{aligned} \gamma^1\gamma^2\gamma^3\phi^c(\vec{\nu}) &= -\phi^c(\vec{\nu}) \\ \gamma^1\gamma^2\gamma^3\overline{\phi}^c(\vec{\nu}) &= \overline{\phi}^c(\vec{\nu}) \\ \vec{\Sigma}.\vec{D}\phi^c(\vec{\nu}) &= -\phi^c(\vec{\nu}) \\ \vec{\Sigma}.\vec{D}\overline{\phi}^c(\vec{\nu}) &= \overline{\phi}^c(\vec{\nu}) \end{aligned} \tag{399}$$

As we have seen from above discussion leading to the equation (384) that $\phi(\vec{\nu})$ and $\phi(-\vec{\nu})$ satisfy two equations in (383) for $\phi(\vec{\nu})$ and $\phi^c(\vec{\nu})$ respectively and the transformation $\vec{\nu} \rightarrow -\vec{\nu}$ does not change the eigenvalues for the eigenstates of $\gamma^1\gamma^2\gamma^3$, we can always take $\phi(-\vec{\nu}) = \overline{\phi}^c(\vec{\nu})$ and $\overline{\phi}(-\vec{\nu}) = \phi^c(\vec{\nu})$ (with the constant of proportionality to be unity without any loss of generality).
From the solutions of $\phi(\vec{\nu})$ and $\phi^c(\vec{\nu})$ as given explicitly in (298), (299) and (301), these specifications can also be verified. If fact, if we take the solutions ϕ and $\overline{\phi}$ as follows:

$$\begin{aligned} \phi(\vec{\nu}) &= F(\nu^0, \vec{\nu})\omega^b \\ \overline{\phi}(\vec{\nu}) &= F(\nu^0, \vec{\nu})\omega^b_c \end{aligned} \tag{400}$$

where $F(\nu^0, \vec{\nu})$ is given in (299) and ω^b_c is related to ω^b as in (302), then we have

$$\phi^c(\widetilde{\nu}) = i\gamma^2\phi^*(\vec{\nu}) = i\gamma^2F^*(\nu^0, \vec{\nu})\omega^{b^*} = F(\nu^0, -\vec{\nu})i\gamma^2\omega^{b^*} = F(\nu^0, -\vec{\nu})\omega^b_c = \overline{\phi}(-\vec{\nu}) \tag{401}$$

[using (304)]

and

$$\overline{\phi}^c(\widetilde{\nu}) = i\gamma^2\overline{\phi}^*(\vec{\nu}) = i\gamma^2F^*(\nu^0, \vec{\nu})\omega^{b^*}_c = F(\nu^0, -\vec{\nu})i\gamma^2\omega^{b^*}_c = F(\nu^0, -\vec{\nu})\omega^b = \phi(-\vec{\nu}) \tag{402}$$

[using (304)]

In tabular form these can be summarized in the following table:

Table 1:

$\widetilde{\nu}$-part of the wave function	Eigenvalues of $\overrightarrow{\Sigma}.\overrightarrow{D}$	Eigenvalues of $\gamma^1\gamma^2\gamma^3$
Particle $\phi(\overrightarrow{\nu})$	-1	$+1$
Particle $\overline{\phi}(\overrightarrow{\nu})$	$+1$	-1
Antiparticle, $\phi^c(\overrightarrow{\nu}) = \overline{\phi}(-\overrightarrow{\nu})$	-1	-1
Antiparticle, $\overline{\phi}^c(\overrightarrow{\nu}) = \phi(-\overrightarrow{\nu})$	$+1$	$+1$

Here it is emphasized that $\phi(\overrightarrow{\nu})$ and $\phi(-\overrightarrow{\nu})$ represent particle and corresponding antiparticle states (as the $\widetilde{\nu}$-part of the fields) and also that these are both eigenstates of $\gamma^1\gamma^2\gamma^3$ with the same eigenvalue. Likewise, $\overline{\phi}(\overrightarrow{\nu})$ and $\overline{\phi}(-\overrightarrow{\nu})$ are related as the particle and corresponding antiparticle states having the same eigenvalue of $\gamma^1\gamma^2\gamma^3$ since these are also the eigenstates of it. Such choice is necessary because the transformation $\overrightarrow{\nu} \to -\overrightarrow{\nu}$ makes a particle state into an antiparticle state and vice versa but leaves either $\gamma^1\gamma^2\gamma^3\phi = \phi$ or $\gamma^1\gamma^2\gamma^3\phi = -\phi$ unchanged. Thus, either $\gamma^1\gamma^2\gamma^3\phi = \phi$ or $\gamma^1\gamma^2\gamma^3\phi = -\phi$ can be regarded as the 'constraint' on the field (wave) function. A similar situation arises for the Weyl two component theory of neutrinos. The space reflection $\overrightarrow{x} \to -\overrightarrow{x}$ makes a particle or antiparticle state to a 'no physical state'. This is because of the fact that the parity is violated in this theory based on the Minkowski spacetime which is isotropic in character. On the contrary, the spacetime of the present consideration is anisotropic in nature, which is manifested in the transition from particle to antiparticle states or vice versa by the 'reflection' $\overrightarrow{\nu} \to -\overrightarrow{\nu}$ in the tangent space of the anisotropic Finsler space. In fact, by the imposition of any of the above two constraints this physically manifested anisotropic character of the Finslerian spacetime is restored for this case of $M = 0$.
Now, it evident that from the above table that particle and antiparticle states with the $\widetilde{\nu}$-part wave functions $\phi(\overrightarrow{\nu})$ and $\phi(-\overrightarrow{\nu})$ (and also $\overline{\phi}(\overrightarrow{\nu})$ and $\overline{\phi}(-\overrightarrow{\nu})$) are eigenstates of an 'internal helicity' $\overrightarrow{S}.\overrightarrow{D}$ where $S = \frac{1}{2}\overrightarrow{\Sigma}$, the 'internal' spin angular momentum, with opposite eigenvalues $\pm\frac{1}{2}$. Thus, in the hadron configuration the particle and antiparticle are in the internal (spin) angular momentum states having opposite helicities. This relation between the fermion number and internal helicity gives rise to a preferred direction (as one can not make simultaneous eigenstates for the three operators, the three mutually non commuting components of $S = \frac{1}{2}\overrightarrow{\Sigma}$, one direction, say the 3-axis, must be chosen as the preferred direction). Because of this preferred direction in the tangent space of the Finsler space of microdomain, this spacetime is manifestly anisotropic, that is, anisotropic from the physical point of view. This preferred direction ushers in a relation between fermion number and the helicity (internal) states and consequently an additional conserved quantum number arises. This internal quantum number can generate the internal symmetry of hadrons. In the following section this will be considered.

4.4: INTERNAL SYMMETRY OF HADRONS

Let us use the following representation of the γ-matrices

$$\gamma_\mu = \begin{pmatrix} 0 & \sigma_\mu \\ \overline{\sigma}_\mu & 0 \end{pmatrix}$$

$$\gamma_5 = i\gamma_0\gamma_1\gamma_2\gamma_3 = \begin{pmatrix} 1 & 0 \\ 0 & -1 \end{pmatrix}$$

$$\overline{\sigma}^\mu = \sigma_\mu = (1, -\overrightarrow{\sigma}), \sigma^\mu = \overline{\sigma}_\mu = (1, +\overrightarrow{\sigma})$$

$$[\ \mu = 0,1,2,3\ ] \tag{403}$$

where $\overrightarrow{\sigma} = (\sigma^1, \sigma^2, \sigma^3)$ are Pauli's matrices. Now, the wave function in the Finsler space, $\psi(\widetilde{x}, \widetilde{\nu})$ when separated as in (264) satisfies the Dirac equation in Minkowski flat spacetime (also in FRW spacetime). There, the functions $\psi_1(\widetilde{x})$ and $\psi_2(\widetilde{x})$ are eigenstates of γ^0 with eigenvalues +1 and -1 respectively.

Also, one can have two linearly independent eigenstates of γ^0 having the same eigenvalue and therefore there are the following two sets of eigenstates of γ^0 in the above representation:

$$\psi_1 = \begin{pmatrix} \chi_1 \\ \chi_1 \end{pmatrix}, \qquad \psi_2 = \begin{pmatrix} \chi_2 \\ -\chi_2 \end{pmatrix}$$
$$\overline{\psi}_1 = \begin{pmatrix} \overline{\chi}_1 \\ \overline{\chi}_1 \end{pmatrix}, \qquad \overline{\psi}_2 = \begin{pmatrix} \overline{\chi}_2 \\ -\overline{\chi}_2 \end{pmatrix} \tag{404}$$

where, $\chi_1, \chi_2, \overline{\chi}_1, \overline{\chi}_2$ are two component semispinors. From these, one can form the bispinor fields $\psi_1(\widetilde{x}, \widetilde{\nu})$ and $\psi_2(\widetilde{x}, \widetilde{\nu})$ which may be related as particle and antiparticle to each other as follows:

$$\begin{aligned} \psi_1(\widetilde{x}, \widetilde{\nu}) &= \begin{pmatrix} \chi_1 \\ \chi_1 \end{pmatrix} \times \phi^T(\vec{\nu}) + \begin{pmatrix} \chi_2 \\ -\chi_2 \end{pmatrix} \times \overline{\phi}^T(-\vec{\nu}) \\ &= \begin{pmatrix} \chi_1 \times \phi^T(\vec{\nu}) + \chi_2 \times \overline{\phi}^T(-\vec{\nu}) \\ \chi_1 \times \phi^T(\vec{\nu}) - \chi_2 \times \overline{\phi}^T(-\vec{\nu}) \end{pmatrix} \end{aligned} \tag{405}$$

$$\begin{aligned} \psi_2(\widetilde{x}, \widetilde{\nu}) &= \begin{pmatrix} \overline{\chi}_1 \\ \overline{\chi}_1 \end{pmatrix} \times \overline{\phi}^T(\vec{\nu}) + \begin{pmatrix} \overline{\chi}_2 \\ -\overline{\chi}_2 \end{pmatrix} \times \phi^T(-\vec{\nu}) \\ &= \begin{pmatrix} \overline{\chi}_1 \times \overline{\phi}^T(\vec{\nu}) + \overline{\chi}_2 \times \phi^T(-\vec{\nu}) \\ \overline{\chi}_1 \times \overline{\phi}^T(\vec{\nu}) - \overline{\chi}_2 \times \phi^T(-\vec{\nu}) \end{pmatrix} \end{aligned} \tag{406}$$

It is easily seen that $\psi_1(\widetilde{x}, \widetilde{\nu})$ and $\psi_2(\widetilde{x}, \widetilde{\nu})$ are eigenstates of the internal helicity operator with eigenvalues of $-\frac{1}{2}$ and $+\frac{1}{2}$ respectively (cf. Table 1). Also, as pointed out in section 3.2, $\psi_1(\widetilde{x}, \widetilde{\nu})$ and $\psi_2(\widetilde{x}, \widetilde{\nu})$ which satisfy equation (266) are , in fact, the Dirac spinors (in the first indicates of them) as they satisfy the usual Dirac equation in local inertial frame, that is, in Minkowski flat spacetime. Again, as these two fields have opposite internal helicities and are related as particle (fermion) and antiparticle, one can identify the internal helicity with the fermion number.
Budini (1979) has made an important suggestion that one can generate isospin algebra from the conformal reflection group. The simplest conformally covariant spinor field equation postulated as an 0(4,2) covariant equation in a pseudo Euclidean manifold $M^{4,2}$ is of the form

$$\left(\Gamma_a \frac{\partial}{\partial \eta_a} + m\right) \xi(\eta) = 0 \tag{407}$$

where m is a constant matrix and $\xi(\eta)$ is an eight component spinor field. Here, the element of Clifford algebra Γ_a are the basis unit vectors of $M^{4,2}$. In the fundamental representation where Γ_a's are represented by the 8×8 matrices of the form

$$\Gamma_a = \begin{pmatrix} 0 & \Xi \\ H & 0 \end{pmatrix} \tag{408}$$

the conformal spinors $\xi(\eta)$ can be expressed as

$$\xi^s = \begin{pmatrix} \phi_1 \\ \phi_2 \end{pmatrix} \tag{409}$$

in which ϕ_1 and ϕ_2 are Cartan semispinors (Cartan, 1966). In this basis (also called semispinor basis or Cartan basis) the equation (406) becomes equivalent in the Minkowski space $M^{3,1}$ to the following coupled equations:

$$i\gamma^\mu \partial_\mu \phi_1 = -m\phi_2$$
$$i\gamma^\mu \partial_\mu \phi_2 = -m\phi_1 \tag{410}$$

Also, it is possible to obtain from equation (406) a pair of standard Dirac equations in the Minkowski space by a unitary transformation given by

$$C_1\psi = \xi^D = \begin{pmatrix} \psi_1 \\ \psi_2 \end{pmatrix}, \qquad C_1^{-1}\Gamma_\mu C_1 = \Gamma_\mu^D = \begin{pmatrix} \gamma_\mu & 0 \\ 0 & \gamma_\mu \end{pmatrix}$$

where

$$C_1 = \begin{pmatrix} L & R \\ R & L \end{pmatrix}$$

with

$$L = \frac{1}{2}(1+\gamma_5), \qquad R = \frac{1}{2}(1-\gamma_5) \tag{411}$$

in the representation (403) of γ_5. The superscript D stands for the 'Dirac basis'. In this Dirac basis, ψ_1, ψ_2 each satisfies the usual Dirac equation. Here, one should note an important point that reflection group L_4 is of four elements:

$$L_4 = E, \quad S, \quad T, \quad ST = J \tag{412}$$

since 0(3,1) is a subgroup of 0(4,2). Here, E = identity, S= space reflection, T = time reversal and ST = J = strong reflection. In $M^{4,2}$ space, coordinates are taken as η_1, η_2, η_3, η_5, η_0, η_6 with the metric (+, +, +, +, -, -). Here, the reflections

$$S_5 : \eta_5 \longrightarrow \eta_5^{'} = -\eta_5$$
$$T_6 : \eta_6 \longrightarrow \eta_6^{'} = -\eta_6 \tag{413}$$

correspond in the Minkowski space, the inverse radius transformation and the same $\otimes$J. The Abelian group

$$Cp_6 = E, S_5, T_6, S_5T_6 \tag{414}$$

is called the partial conformal reflection group and the total conformal reflection group is given by the direct product

$$C_6 = Cp_6 \otimes L_4 \tag{415}$$

The conformal reflection group is represented in the conformal spinor space by the algebra $U_{4,C}$ which may be called the conformal reflection algebra.

Now, for conformal spinor in the Dirac basis $\xi^D = \begin{pmatrix} \psi_1 \\ \psi_2 \end{pmatrix}$ the Lorentz reflection group L_4 when acting on the Dirac spinor ψ_i is isomorphic to a U_2 algebra whose hermitian elements are given by the matrices 1, γ_0, i $\gamma_0\gamma_5$, γ_5. The transformation S_5, T_6, S_5T_6 that act on the space or time reflection interchanges ϕ_1 and ϕ_2 but transforms ψ_1 and ψ_2 in to themselves. Moreover, conformal reflection (inverse radius transformation) interchanges both $\phi_1 \longleftrightarrow \phi_2$ and $\psi_1 \longleftrightarrow \psi_2$. Also, ψ_1 and ψ_2 may represent physical free massive fermions but ϕ_1 and ϕ_2 do not unless they are massless because they satisfy the coupled equations.

It was Budini's suggestion that one can call a reflection algebra corresponding to a reflection group an internal symmetry algebra for a given field theory.

(a) if the corresponding reflection group when accompanied by the corresponding coordinate reflections is a covariance group for the equation of motion in the Minkowski space.
(b) if it commutes with the Poincare Lie algebra and with the spacetime reflection algebra.
(c) if the transformation induced by the reflection algebra on the fields leave the action of the theory invariant.
If the reflection algebra commutes only with the Poincare algebra but does not commute with the spacetime reflection algebra L_4, the algebra may be termed 'restricted' internal symmetry algebar. Budini (1979) has also shown an important result in the study of the geometry of hadrons that in internal symmetry algebra can be generated from the conformal reflection group which contains as a subgroup of the Lorentz Dirac doublet of the conformal spinor ξ^D correspond.

$$S_5 \to \Gamma_5^D$$
$$T_6 \to \Gamma_6^D$$
$$S_5 T_6 \to \Gamma_5^D \Gamma_6^D \tag{416}$$

Thus, the group Cp_6 can be represented by the Lie algebra $U_{4,C}$ and the corresponding real subalgebra SU_2 may be obtained from the hermitian elements $i\Gamma_6, \Gamma_6, \Gamma_5\Gamma_6$. Then it follows from (44) that the group C_6 is isomorphic to the product

$$U_{2,C} \otimes U_{2,C} = U_{4,C} \tag{417}$$

The following propositions have been proved by Budini:

1. The reflection algebra $U_{2,C}$ corresponding to the partial conformal reflection group Cp_6 in an internal symmetry algebra for the conformal spinor doublets. For massive (but degenerate) components of the doublet, $U_{2,C}$ is maximal.
2. For massless conformal spinors interacting at very short distances, the direct product of the partial conformal reflection group and the strong reflection in the Minkowski space generates a restricted internal symmetry algebra of order eight which can be put in the form $U_{2C,L} \oplus U_{2C,R}$. This $U_{2C,L} \oplus U_{2C,R}$ algebra may be reduced to two independent SU_2 algebras represented by the eight four-dimensional matrices $L \times \sigma_\mu$, $R \times \sigma_\nu$ (where L and R are given by (410)) acting on the two independent doublets of Weyl fields into which the massless conformal spinor or the system of the interacting massive spinors spits at short distances.

Now returning to the bispinor fields as in (405) one can construct spinor fields $\hat{\psi}_1(\tilde{x}, \tilde{\nu})$ and $\hat{\psi}_2(\tilde{x}, \tilde{\nu})$ which also satisfy Dirac equation and thus, are Dirac Spinors in $M^{3,1}$, through the right multiplications by the constant spinors χ^r. In fact, by using the following four linearly independent spinors χ^r $(r = 1, 2, 3, 4)$:

$$\chi^1 = \begin{pmatrix} 1 \\ 0 \\ 0 \\ 0 \end{pmatrix}, \chi^2 = \begin{pmatrix} 0 \\ 1 \\ 0 \\ 0 \end{pmatrix}, \chi^3 = \begin{pmatrix} 0 \\ 0 \\ 1 \\ 0 \end{pmatrix}, \chi^4 = \begin{pmatrix} 0 \\ 0 \\ 0 \\ 1 \end{pmatrix} \tag{418}$$

We can get a set of four spinor doublets in $M^{3,1}$ (each of which represents the $M^{3,1}$ spinor space) as follows:

$$\begin{pmatrix} \widehat{\psi_1^r}(\tilde{x}, \tilde{\nu}) \\ \widehat{\psi_2^r}(\tilde{x}, \tilde{\nu}) \end{pmatrix}^D = \begin{pmatrix} \psi_1(\tilde{x}, \tilde{\nu}) \\ \psi_2(\tilde{x}, \tilde{\nu}) \end{pmatrix} \chi^r \tag{419}$$

$$(r = 1, 2, 3, 4)$$

In order to obtain a spinor field in $M^{4,2}$ from the corresponding one in Minkowski space $M^{3,1}$ we can use the procedure proposed by Dirac (1936). For this, we first note that the coordinates η_a of $M^{4,2}$ and the spacetime coordinates x_μ of $M^{3,1}$ are related as

$$x_\mu = \frac{\eta_\mu}{k} \tag{420}$$

where k is a Lorentz scalar parameter. In Dirac's procedure it is assumed that

$$k = \eta_5 + \eta_6, \eta_a \eta^a = 0 \tag{421}$$

The value of k may be fixed with a conformal covariant procedure (Budini, et al., 1979) and it is fixed as $k = 1$. In this presentation the spinor field $\xi(\eta)$ in $M^{4,2}$ and the spinor field $\xi(\widetilde{x})$ in $M^{3,1}$ are connected as

$$\xi(\widetilde{x}) = k^{-n} e^{-ix^\mu \Pi_\mu} \xi(\eta) \tag{422}$$

with $k = 1$ and

$$\Pi_\mu = S_{\mu 5} + S_{\mu 6} = \frac{i}{2}\Gamma_\mu(\Gamma_5 + \Gamma_6) \tag{423}$$

$$S_{ab} = \frac{i}{4}[\Gamma_a, \Gamma_b] \tag{424}$$

Using this relation (421), we can obtain the conformal spinor in the Dirac basis from the spinor doublet (418) in Minkowski space $M^{3,1}$. These conformal spinors are represented by the doublets $\begin{pmatrix} \psi_1^r \\ \psi_2^r \end{pmatrix}^D$ of Cartan semispinors (in Dirac basis) and the conformal spinors in the fundamental representation can be obtained by the unitary transformation (410). That is

$$\xi^{rS} = \begin{pmatrix} \phi_1^r \\ \phi_2^r \end{pmatrix}^S = C_1 \xi^{rD} = C_1 \begin{pmatrix} \psi_1^r \\ \psi_2^r \end{pmatrix}^D \tag{425}$$

since $C_1^2 = 1$ and the superscript S stands for semispinor basis. These four components Cartan semispinor ϕ_1^r and ϕ_2^r satisfy the coupled equations in the Minkowski space $M^{3,1}$ and consequently the conformal spinor $\xi^r(\eta)$ in a pseudo-Euclidean manifold $M^{4,2}$ satisfies the equation (416). Thus, in the present formalism we can also make use of the excellent argument put forward by Budini to achieve that the direct product of the partial conformal reflection group and strong reflection in the Minkowski space generates a restricted internal symmetry algebra which can be put in the form $U_{2,L} \oplus U_{2,R}$. The elements of the algebra may be represented by the eight four-dimensional matrices $L \times \sigma_\mu$, $R \times \sigma_\nu$ that act on the two independent doublets of the Weyl fields into which the conformal spinor splits and thus, two independent SU_2 algebras are represented by them. Thus, argument of generating internal symmetry algebra by Budini holds good for conformal spinor $\xi^r(\eta)$, for each r, in $M^{4,2}$ or equvalently in $M^{3,1}$ for spinor space $\begin{pmatrix} \widehat{\psi}_1^r(\widetilde{x}, \widetilde{\nu}) \\ \widehat{\psi}_2^r(\widetilde{x}, \widetilde{\nu}) \end{pmatrix}^D$ as given in (418), for each r. Consequently, this SU_2 algebra remains the internal symmetry algebra for the 'bispinor doublet' as constructed below

$$\begin{pmatrix} \psi_1(\widetilde{x}, \widetilde{\nu}) \\ \psi_2(\widetilde{x}, \widetilde{\nu}) \end{pmatrix}^D = \sum_{r=1}^{4} \begin{pmatrix} \widehat{\psi}_1^r(\widetilde{x}, \widetilde{\nu}) \\ \widehat{\psi}_2^r(\widetilde{x}, \widetilde{\nu}) \end{pmatrix}^D \chi^{rT} \tag{426}$$

Here, $\chi_1(\widetilde{x}, \widetilde{\nu})$ and $\chi_2(\widetilde{x}, \widetilde{\nu})$ represent the bispinor fields, as given in (415), which satisfy Dirac equation. In the present formalism this doublet of bispinor fields constructed from Cartan semispinors might be regarded as the constituents of a hadron such that they are in the internal angular momentum state of $S = \frac{1}{2}$. Also, it was pointed out earlier that ψ_1 and ψ_2 are eigenstates of an internal helicity operator with opposite eigenvalues, apart from the fact that they are related as particle and antiparticle. Thus, as in Budini's approach one can treat this doublet $\begin{pmatrix} \psi_1 \\ \psi_2 \end{pmatrix}^D$ of a field with $S_3 = +\frac{1}{2}$ and another with $S_3 = -\frac{1}{2}$, which represent a constituent for the particle and antiparticle configuration in generating the internal symmetry algebra. Moreover, the fixed internal S_3 values for particle and antiparticle states give rise to another quantum number representing the algebra of U_1. These SU_2 and U_1 algebras indicate isospin and hyperchange respectively and consequently one can achieve a Lie group structure $SU_3 \rightarrow SU_2 \times U_1$for the internal symmetry of hadrons. One such consideration has been made by Bandyopadhyay (1989) with the assumption of an internal $l = \frac{1}{2}$ orbital angular momentum by introducing a preferred direction such that l_z values $+\frac{1}{2}$ and $-\frac{1}{2}$ represent particle and antiparticle states. There, the anisotropy is introduced through magnetic monopole and in this space the internal helicity has been connected with the fermion number with only a special choice of the value of a quantum number $\mu = \frac{1}{2}$ where μ denotes the measure of anisotropy. Also, the connection between the internal helicity and the fermion number made by Bandyopadhyay (1989) in a complexified spacetime is valid only for massless particles because the formula for helicity operator used there will not hold good when mass is 'generated' by the imaginary part of the spacetime. On the other hand, in the present formalism the preferred direction arises in an natural way and, in fact, it is reckoned with through the field (or wave) equation in the Finsler space as the physical manifestation of the property of fields in this spacetime below a length scale of the fundamental length l, where hadronic matter is extended . Here, we get an internal helicity operator $\vec{S}.\vec{D} = \frac{1}{2}\vec{\Sigma}.\vec{D}$ for the constituents in the hadronic configuration. The constituents are in the internal (spin) angular momentum ($\vec{S} = \frac{1}{2}\vec{\Sigma}$) state where the two opposite eigenvalues $\pm\frac{1}{2}$ of S_3 (measured along $\vec{D}$) represent respectively the constituents for particle and antiparticle configuration. This angular momentum corresponds to a continuous group (the Lie group structure). In fact, the existence of such continuous (rotational Lie groups) groups which may operate in spaces orthogonal to $M^{3,1}$ was necessitated by Budini et al (1979) and postulated as an ad hoc gauge group. For the present case the existence of a continuous group is not a postulate. Rather it arises naturally and thus, instead of half-orbital angular momentum as postulated by Bandyopadhyay (1989), the spin half angular momentum plays the same role in a natural way in building up the group structure (SU_2)instead of algebra that has been argued excellently by Budini. Finally with the one parameter group U_1 from the additional conserved quantum number that arises in the present consideration we get SU_3 Lie group formalism for the internal symmetry of hadrons.

4.5: MESONS AND BARYONS

With the above mentioned postulate of half orbital angular momentum for the constituents of hadron, Bandyopadhyay et al (1989) have taken a baryonic multiplet corresponding to the internal symmetry group SU_3 representing baryons with spin $\frac{1}{2}$ to arise from the mesonic SU_3 multiplet (with spin zero) by a spinorial constituent having the symmetry group U_1. Here, in the present consideration too, as discussed above , the internal symmetry algebra generated from the partial reflection group can be made to be the internal symmetry group SU_2because the constituents are in the internal spin half angular momentum state. Also, the reflection symmetry represented by these fixed S_3 values for the particle and antiparticle configuration fixes a preferred direction (in the tangent space). Moreover, the internal helicy corresponding to these fixed S_3 values of the constituents can give rise to a conserved quantum number (a conserved 'charge') and is related to the one parameter group U_1. Consequently, one can arrive at the internal SU_3 symmetry group which decomposes into $SU_2 \otimes U_1$. In this scheme we acn get a baryonic state by the introduction of a constituent spinor with a specific internal S_3 value (or a specific internal helicity) in the configuration of a meson. In fact, the two opposite S_3 values for this spinorial constituent correspond to baryons and antibaryons. In meson state, the two constituents of it are in the opposite internal helicity states and consequently bears no signature of internal spin-half angular momentum (or anisotropy of the internal spacetime) outside the configuration. To be more specific, the baryon number is supposed to be twice the sum of internal S_3 values of the spinorial constituents which are not Majorana spinors. On the other hand, the internal quantum number 'strangeness' is the sum of internal S_3 values of the Majorana spinorial constituents in a hadron. We can also get the 'hypercharge' of a hadron, which is the sum of baryon number and the starngeness of it. Thus, the additional non Majorana spinorial constituents appearing in the configuration of a baryon with fixed internal S_3 values (or internal helicities) are responsible for the baryon number of it. Introduction of a spinor (non-Majorana) with specific internal helicity in the configuration of a meson gives rise to a baryon and in fact, we can have spin $\frac{1}{2}$ baryons from spin 0 mesons, spin $\frac{3}{2}$ baryons from spin 1 mesons and so on. In general, spin

J + $\frac{1}{2}$ baryons are constructed from spin J mesons. As the spinor constituents which are responsible for generating baryonic multiplet (of baryons of spin J + $\frac{1}{2}$) having the internal symmetry group SU_3 from the mesonic SU_3 multiplet (of meson spin J) have the symmetry group U_1, we can regard this U_1 group to be the baryon number generating group. Thus

$$SU(3)_{baryons} \subset U(3) = SU(3)_{mesons} \otimes U_1 \tag{427}$$

The meson-baryon mass difference can be derived. This derivation is, in fact, similar to that of Bandyopadhyay et al. (1989). There, the supermultiplets which contain the SU_3 multiplets of mesons and baryons corresponding to spin 0 and $\frac{1}{2}$ respectively as well as for those of spin 1 (vector meson) and spin $\frac{3}{2}$ baryons in the massless case are being constructed. In breaking this supersymmetry the meson-baryon mass difference has been obtained. The symmetry breaking has the group structure

$$U_3 \longrightarrow SU_2 \otimes U_1 \otimes U_1 \tag{428}$$

This generates the mass difference of the numbers of different isomultiplets of the SU_3 multiplet as well as a mass difference between the highest massive meson with hypercharge and the lowest massive baryon.

The supermultiplet for pseudoscalar mesons and spin $\frac{1}{2}$ baryons is taken as

$$\begin{array}{cccccccc} \pi^+ & & K^+ & \overline{K}_0 & p & \Sigma^+ & & \Xi^0 \\ \pi^0 & \eta & & & & \Sigma^0 & \Lambda & \\ \pi^- & & K^0 & K^- & n & \Sigma^- & & \Xi^- \end{array} \tag{429}$$

The mass splitting of $K-\pi$ as well as those for $Y-N$ ($Y=\Lambda,\Sigma$) and $\Xi-Y$ are governed by the decomposition of $SU_3 \longrightarrow SU_2 \otimes U_1$. The group structure $U(3) \longrightarrow SU_2 \otimes U_1 \otimes U_1$ relates the mass split between the highest massive baryon with the hypercharge , K, and the lowest massive baryon N. Since U_1 group appears twice in the mass splitting, once in the Gell-Mann Okubo type split $SU(3)_{baryons} \longrightarrow SU_2 \otimes U_1$ and another U_1 for boson fermion distinguishable, one can have mass difference as twice the mass difference generated by $SU(3)_{baryons} \longrightarrow SU_2 \otimes U_1$ split giving rise to $K-\pi$, $Y(\Lambda,\Sigma)-N$ or $\Xi-Y$ mass difference. Now, if we take the mass difference for Y-N as

$$\frac{3m_\Sigma + m_\Lambda}{4} - m_N \simeq 230Mev,$$

it is found that

$$m_N - m_K \simeq 230Mev = 460Mev \tag{430}$$

which is in agreement with the experimental value. Similarly, one can consider the relation of masses among the vector mesons and spin $\frac{3}{2}$ decuplet baryons. We can have the following relation

$$m_\Delta - m_{K^*} \simeq 2(m_{\Sigma^*} - m_\Delta) \tag{431}$$

It is also in good agreement with the experimental results. It is to be noted that a supermultiplet is feasible only for the case of massless hadrons, and the breaking of this supermultiplet into meson and baryon multiplets (caused by any mass term which, in fact, destroys the conformal symmetry) suggests that the whole spectrum of hadron masses are generated dynamically.

4.6: HADRON FIELDS AND STATES

We shall now construct hadron fields from constituent fields in the anisotropic microdomain regarded as Finslerian [see also De (2002a,b)]. The subatomic particle is of composite type and the constituents are situated at neighboring points of the microdomain. In fact, it is considered here that the constituents lie on a geodesic in this anisotropic spacetime and thus 'every unit field orthogonal to this geodesic is parallel to it' (Chowdhuri, 1981). Actually, this curve is the autoparallel curve whose tangent vectors result from each other by successive infinitesimal parallel displacements which have been considered in chapter two. The neighboring points on the autoparallel curve where the constituents lie (or the corresponding fields depend) are (x^μ, ν^μ), $(x^\mu + dx^\mu, \nu^\mu + d\nu^\mu)$, $(x^\mu - dx^\mu, \nu^\mu - d\nu^\mu)$, ... where dx^μ and ν^μ are quantized. For a pseudo scalar meson whose constituents are a lepton and an antilepton, let $\overline{\psi}^{(-)(\alpha,a)}$ and $\psi^{(-)(\beta,b)}$ be the creation parts of the fields for particle (lepton) and its charge conjugate antiparticle respectively. Here α and β denote the spin indices for the spinor and the indices a and b represent the internal helicities (or the S_3 values). Similarly, the destruction operates for the particle and antiparticle are respectively $\psi^{(+)(\alpha,a)}$ and $\overline{\psi}^{(+)(\beta,b)}$. The field operators $\overline{\psi}^{(-)(\alpha,a)}$ and $\psi^{(-)(\beta,b)}$ are functions of neighboring line support elements $(x^\mu + \frac{1}{2}dx^\mu, \nu^\mu + \frac{1}{2}d\nu^\mu)$, $(x^\mu - \frac{1}{2}dx^\mu, \nu^\mu - \frac{1}{2}d\nu^\mu)$ respectively. Specifically, the field of π^0 meson is constructed from the creation parts of the fields of μ^+ and μ^- or from these of ν_μ and $\overline{\nu}_\mu$; that of π^- from the creation parts of the fields μ^- and ν_μ and so on. Now, remembering the procedure of chapter two in deriving the field equation in Finslerian microdomain, we can have

$$\psi^A((x^\mu \pm \frac{1}{2}dx^\mu, \nu^\mu \pm \frac{1}{2}d\nu^\mu)) = \psi^A(x^\mu, \nu^\mu) \pm \frac{1}{2}i \in \hbar(\gamma^\mu\partial_\mu - \gamma^l_{h\mu}(\widetilde{x}, \widetilde{\nu})\nu^h\gamma^\mu\partial'_l)\psi^A(x^\mu, \nu^\mu)$$
$$= (1 \pm \frac{1}{2} \in mc)\psi^A(x^\mu, \nu^\mu) \qquad (432)$$

where $A = (\alpha, a)$ or (β, b) and m is the mass of the constituent spinor. For neutrino constituents $m = 0$. The parameter $\in$ which is real and positive lies in $(0, l]$, l being a fundamental length. Then, the creation part of the pion field for the macrodomain is taken as

$$\phi^{(-)}_M(\widetilde{x}) = C\int d^4\nu d \in T\chi^\dagger(\widetilde{\nu}, \in)\overline{\psi}^{(-)(\alpha,a)}(\widetilde{x}, \widetilde{\nu})\gamma^5\psi^{(-)(\beta,b)}(\widetilde{x}, \widetilde{\nu})\chi(\widetilde{\nu}, \in) \qquad (433)$$

where C is an appropriate normalization factor. Here, we have used the relation (431). The factors $(1 \pm \frac{1}{2} \in mc)$ have been absorbed into $\chi(\widetilde{\nu}, \in)$ and $\chi^\dagger(\widetilde{\nu}, \in)$ which are respectively a probability density (a spinor or a column vector) and its adjoint in forming the pion field of macrodomain by 'averaging'. The probability density itself may be a function of the parameter $\in$. The time ordered product is represented by the symbol T. Other hadron fields of macrodomain can similarly be constructed. For example, if a hadron is composed of 2n number of lepton and antilepton constituents (some of them be Majorana particles) then the creation part of the hadron field is given as

$$\phi^{(-)}_H(\widetilde{x}) = C\int d^4\nu d \in T\chi^\dagger(\widetilde{\nu}, \in)\overline{\psi}^{(-)(\alpha_1,a_1)}(\widetilde{x}, \widetilde{\nu})\gamma^5\psi^{(-)(\beta_1,b_1)}(\widetilde{x}, \widetilde{\nu})\chi(\widetilde{\nu}, \in)\chi^\dagger(\widetilde{\nu}, \in).$$
$$\overline{\psi}^{(-)(\alpha_2,a_2)}(\widetilde{x}, \widetilde{\nu})\gamma^5\psi^{(-)(\beta_2,b_2)}(\widetilde{x}, \widetilde{\nu})\chi(\widetilde{\nu}, \in)\chi^\dagger(\widetilde{\nu}, \in).............\overline{\psi}^{(-)(\alpha_n,a_n)}(\widetilde{x}, \widetilde{\nu})\gamma^5\psi^{(-)(\beta_n,b_n)}(\widetilde{x}, \widetilde{\nu})\chi(\widetilde{\nu}, \in)(434)$$

For the case of odd number, 2n +1, constituents, the creation part of the hadron field is given by either

$$\phi^{(-)}_H(\widetilde{x}) = C\int d^4\nu d \in T\chi^\dagger(\widetilde{\nu}, \in)\overline{\psi}^{(-)(\alpha_1,a_1)}(\widetilde{x}, \widetilde{\nu})\gamma^5\psi^{(-)(\beta_1,b_1)}(\widetilde{x}, \widetilde{\nu})\chi(\widetilde{\nu}, \in)\chi^\dagger(\widetilde{\nu}, \in) \times$$
$$\overline{\psi}^{(-)(\alpha_2,a_2)}(\widetilde{x}, \widetilde{\nu})\gamma^5\psi^{(-)(\beta_2,b_2)}(\widetilde{x}, \widetilde{\nu})\chi(\widetilde{\nu}, \in)\chi^\dagger(\widetilde{\nu}, \in).............\overline{\psi}^{(-)(\alpha_n,a_n)}(\widetilde{x}, \widetilde{\nu})\gamma^5\psi^{(-)(\beta_n,b_n)}(\widetilde{x}, \widetilde{\nu}) \times$$
$$\chi(\widetilde{\nu}, \in)\chi^\dagger(\widetilde{\nu}, \in)\overline{\psi}^{(-)(\alpha_{n+1},a_{n+1})}(\widetilde{x}, \widetilde{\nu}) \qquad (435)$$

or

$$\phi_H^{(-)}(\widetilde{x}) = C\int d^4\nu d\in T\overline{\psi}^{(-)(\alpha_1,a_1)}(\widetilde{x},\widetilde{\nu})\chi(\widetilde{\nu},\in)\chi^{\dagger}(\widetilde{\nu},\in)\overline{\psi}^{(-)(\alpha_2,a_2)}(\widetilde{x},\widetilde{\nu})\times$$

$$\gamma^5\psi^{(-)(\beta_2,b_2)}(\widetilde{x},\widetilde{\nu})\chi(\widetilde{\nu},\in)\chi^{\dagger}(\widetilde{\nu},\in)\overline{\psi}^{(-)(\alpha_3,a_3)}(\widetilde{x},\widetilde{\nu})\gamma^5\psi^{(-)(\beta_3,b_3)}(\widetilde{x},\widetilde{\nu})\chi(\widetilde{\nu},\in)\chi^{\dagger}(\widetilde{\nu},\in)\times$$

$$\ldots\ldots\ldots\ldots\overline{\psi}^{(-)(\alpha_{n+1},a_{n+1})}(\widetilde{x},\widetilde{\nu})\gamma^5\psi^{(-)(\beta_{n+1},b_{n+1})}(\widetilde{x},\widetilde{\nu})\chi(\widetilde{\nu},\in) \tag{436}$$

In fact, two fields in (435) and (436) correspond to the hadrons which one charge conjugate to each other. In the above expression for the hadron fields of macrodomain, the fields of the constituents at the neighboring line support elements of the Finslerian microdomain are replaced by those at the point $(\widetilde{x},\widetilde{\nu})$ with the absorption of the factors $(1 \pm \frac{1}{2} \in mcj)$, j being positive integers, into the probability density function $\chi(\widetilde{\nu},\in)$ and its adjoint $\chi^{\dagger}(\widetilde{\nu},\in)$. Also, for convenience, we have used same notations $\overline{\psi}^{(-)A}(\widetilde{x},\widetilde{\nu})$ or $\psi^{(-)A}(\widetilde{x},\widetilde{\nu})$ for all types of constituents, that is, for charged muons and neutrinos (Dirac or Majorana). Also, we can replace γ^5 in the above expressions of hadron fields by some other matrices according to transformation properties of the resulting hadron fields in the macrodomain.
We have seen in Chapter two that the separation of the bispinor fields $\psi(\widetilde{x},\widetilde{\nu})$ is possible. In fact, in (228) or in (271), a field $\psi(\widetilde{x},\widetilde{\nu})$ is separated as $\psi(\widetilde{x},\widetilde{\nu}) = \psi(\widetilde{x})\phi^T(\widetilde{\nu})$ where $\psi(\widetilde{x})$ is the field for the macrodomain which is the Minkowskian spacetime. For the present case we can write

$$\overline{\psi}^{(-)(\alpha,a)}(\widetilde{x},\widetilde{\nu}) = \left(\overline{\phi}^{(a)}(\widetilde{\nu})\right)^T\left(\overline{\psi}^{(-)(\alpha)}(\widetilde{x})\right)^T$$

or, rather

$$\overline{\psi}^{(-)(\alpha,a)}(\widetilde{x},\widetilde{\nu}) = \sum_{\overrightarrow{p}}\frac{1}{\sqrt{V}}\sqrt{\left(\frac{m}{E_{\overrightarrow{p}}}\right)}c^{\dagger}_{\overrightarrow{p},\alpha,a}\left(\overline{\phi}^{(a)}(\widetilde{\nu})\right)^T\overline{u}^{\alpha}(\overrightarrow{p})e^{i\widetilde{p}.\widetilde{x}} \tag{437}$$

$$\psi^{(-)(\beta,b)}(\widetilde{x},\widetilde{\nu}) = \psi^{(-)(\beta)}(\widetilde{x})\left(\phi^{(b)}(\widetilde{\nu})\right)^T$$

or, rather

$$\psi^{(-)(\beta,b)}(\widetilde{x},\widetilde{\nu}) = \sum_{\overrightarrow{p'}}\frac{1}{\sqrt{V}}\sqrt{\left(\frac{m}{E_{\overrightarrow{p'}}}\right)}d^{\dagger}_{\overrightarrow{p'},\beta,a}\overline{v}^{\beta}(\overrightarrow{p'})\left(\phi^{(b)}(\widetilde{\nu})\right)^T e^{i\widetilde{p'}.\widetilde{x}} \tag{438}$$

where the four momenta $\widetilde{p}'$ are on the mass shell, that is, $p^0 = \omega_{\overrightarrow{p}} = \sqrt{\overrightarrow{p}^2 + m^2}$ and $p'^0 = \omega_{\overrightarrow{p'}} = \sqrt{\overrightarrow{p'}^2 + m^2}$ in the natural unit $c = \hbar = 1$. $c^{\dagger}_{\overrightarrow{p},\alpha,a}$ and $d^{\dagger}_{\overrightarrow{p'},\beta,a}$ are the creation operators for the particle and antiparticle respectively. Here, we have considered the case the anticommutation relations are

$$\left\{c_{\overrightarrow{p},\alpha,a}, c^{\dagger}_{\overrightarrow{p'},\alpha',a'}\right\} = \left\{d_{\overrightarrow{p},\alpha,a}, d^{\dagger}_{\overrightarrow{p'},\alpha',a'}\right\} = \delta_{\overrightarrow{p}\overrightarrow{p'}}\delta_{\alpha\alpha'}\overrightarrow{p'}\delta_{aa'} \tag{439}$$

where $c_{\overrightarrow{p},\alpha,a}$ and $d_{\overrightarrow{p},\alpha,a}$ are the destruction operators for the particle and antiparticle respectively. All other anticommutators are equal to zero. For the continuous momenta the summations in (437) and (438) are to be replaced by integrations. In fact, one has to replace $\frac{1}{V}\sum_{\overrightarrow{p}}$ by $\frac{1}{(2\pi)^3}\int d^3p$. In this case, the anticommutation relations (439) are to be changed into the following:

$$\left\{c_{\overrightarrow{p},\alpha,a}, c^{\dagger}_{\overrightarrow{p'},\alpha',a'}\right\} = \left\{d_{\overrightarrow{p},\alpha,a}, d^{\dagger}_{\overrightarrow{p'},\alpha',a'}\right\} = \delta^3(\overrightarrow{p} - \overrightarrow{p}')\delta_{\alpha\alpha'}\delta_{aa'}$$

(440)

For the following discussion, if we take momentum to be continuous than a problem of normalization regarding the one particle states $c^{\dagger}_{\overrightarrow{p},\alpha,a} \mid 0>$ and $d^{\dagger}_{\overrightarrow{p},\alpha,a} \mid 0>$ is to be faced with because of the anticommutation relations (440). In fact,

$$< 0 \mid c_{\overrightarrow{p},\alpha,a} c^{\dagger}_{\overrightarrow{p},\alpha,a} \mid 0 >=< 0 \mid \{c_{\overrightarrow{p},\alpha,a}, c^{\dagger}_{\overrightarrow{p},\alpha,a}\} \mid 0 >= \infty$$

But the problem can be resolved by 'smearing' in momentum space. With a square integrable function $f(\overrightarrow{p})$ that is 'concentrated' around a peak value one can construct a state as a linear superposition:

$$c^{\dagger}_{f,\alpha,a} \mid 0 > \;= \int f(\overrightarrow{p}) d^3p \; c^{\dagger}_{\overrightarrow{p},\alpha,a} \mid 0 > \tag{441}$$

Note that $< 0 \mid c_{f,\alpha,a} c^{\dagger}_{f,\alpha,a} \mid 0 >$ becomes finite as $f(\overrightarrow{p})$ is square integrable. The state $c^{\dagger}_{f,\alpha,a} \mid 0 >$ is now interpreted as one particle state. Keeping in mind all these prescriptions regarding the transition to the continuous momentum case, we adopt the formulation with discrete momenta as in (437) and (438). For constituents with zero masses (that is, for neutrinos) the normalization in (437) and (438), that is, $\frac{1}{\sqrt{V}}\sqrt{\left(\frac{m}{E_{\overrightarrow{p}}}\right)}$ is to be replaced by $\frac{1}{\sqrt{V}}$.

Now, with the use of (437) and (438), one can find the hadron fields. We shall find here the pseudoscalar meson field by using these expressions. The creation part of the field follows from (434). It is now given by

$$\begin{aligned}
\phi^{(-)}_M(\widetilde{x}) &= C \sum_{\overrightarrow{p},\overrightarrow{p'}} \frac{1}{V} \sqrt{\left(\frac{m^2}{E_{\overrightarrow{p}} E_{\overrightarrow{p'}}}\right)} T\overline{u}^{(\alpha)}(\overrightarrow{p})\gamma^5 v^{(\beta)}(\overrightarrow{p'}) e^{i(\widetilde{p} \mid \widetilde{p\prime}).\widetilde{x}}. \\
&.c^{\dagger}_{\overrightarrow{p},\alpha,a} d^{\dagger}_{\overrightarrow{p'},\beta,b} \int d^4\nu \; d \in \chi^{\dagger}(\widetilde{\nu}, \in) \overline{\phi}^{(a)^T}(\widetilde{\nu}) \phi^{(b)^T}(\widetilde{\nu}) \chi(\widetilde{\nu}, \in) \\
&= CT\overline{\psi}^{(-)(\alpha,a)}(\widetilde{x})\gamma^5 \psi^{(-)(\beta,b)}(\widetilde{x}) G^{(a,b)}(l)
\end{aligned} \tag{442}$$

where

$$\overline{\psi}^{(-)(\alpha,a)}(\widetilde{x}) = \sum_{\overrightarrow{p}} \frac{1}{\sqrt{V}} \sqrt{\left(\frac{m}{E_{\overrightarrow{p}}}\right)} c^{\dagger}_{\overrightarrow{p},\alpha,a} \overline{u}^{(\alpha)}(\overrightarrow{p}) e^{i\widetilde{p}.\,\widetilde{x}} \tag{443}$$

$$\psi^{(-)(\beta,b)}(\widetilde{x}) = \sum_{\overrightarrow{p'}} \frac{1}{\sqrt{V}} \sqrt{\left(\frac{m}{E_{\overrightarrow{p'}}}\right)} d^{\dagger}_{\overrightarrow{p'},\beta,b} v^{(\beta)}(\overrightarrow{p'}) e^{i\widetilde{p'}.\,\widetilde{x}} \tag{444}$$

are the creation parts of the fields for the constituent particle and antiparticle respectively in the macrodomain. $G^{(a,b)}(l)$ is given by

$$G^{(a,b)}(l) = \int d^4\nu \; d \in \chi^{\dagger}(\widetilde{\nu}, \in) \overline{\phi}^{(a)^T}(\widetilde{\nu}) \phi^{(b)^T}(\widetilde{\nu}) \chi(\widetilde{\nu}, \in) \tag{445}$$

[the range of integration for $\in$ is from 0 to l , a fundamental length]
Similarly, the destruction part of the meson field can be found to be

$$\phi^{(+)}_M(\widetilde{x}) = C'T\overline{\psi}^{(+)(\beta,b)}(\widetilde{x})\gamma^5 \psi^{(+)(\alpha,a)}(\widetilde{x}) G^{(a,b)}(l) \tag{446}$$

where $\psi^{(+)(\alpha,a)}(\widetilde{x})(\widetilde{x})$ and $\overline{\psi}^{(+)(\beta,b)}(\widetilde{x})$ are the destruction parts of the fields for the constituent particle and antiparticle in the macrodomain. Thus, the meson field $\phi_M(\widetilde{x})$ is given by

$$\phi_M(\widetilde{x}) = \phi^{(+)}_M(\widetilde{x}) + \phi^{(-)}_M(\widetilde{x}) \tag{447}$$

Now, the normalization factors in (442) and (446) can be specified in consistent with the following normalization condition for the meson field:

$$< 0 \mid \phi_M(\widetilde{x}) \mid \widetilde{P} >= \sqrt{Z_3}\frac{1}{\sqrt{V}}\frac{1}{\sqrt{2\omega_{\overrightarrow{P}}}}e^{-i\widetilde{P}.\ \widetilde{x}} \tag{448}$$

or

$$< \widetilde{p} \mid \phi_M(\widetilde{x}) \mid 0 >= \sqrt{Z_3}\frac{1}{\sqrt{V}}\frac{1}{\sqrt{2\omega_{\overrightarrow{P}}}}e^{i\widetilde{P}.\ \widetilde{x}} \tag{449}$$

where Z_3 (non zero) is the wave function renormalization constant. The normalization condition can also be taken as

$$< 0 \mid \phi_M(\widetilde{x}) \mid \widetilde{P} >= \frac{1}{\sqrt{V}}\frac{1}{\sqrt{2\omega_{\overrightarrow{P}}}}e^{-i\widetilde{P}.\ \widetilde{x}} \tag{450}$$

These normalization conditions give rise to the specified forms of the factors C and C' and consequently the following expressions for $\phi_M^{(-)}(\widetilde{x})$ and $\phi_M^{(+)}(\widetilde{x})$:

$$\phi_M^{(-)}(\widetilde{x}) = \frac{1}{\sqrt{V}}\frac{\sqrt{Z_3}}{\sqrt{2\omega_{\overrightarrow{P}}}}\frac{T\overline{\psi}^{(-)(\alpha,a)}(\widetilde{x})\gamma^5\psi^{(-)(\beta,b)}(\widetilde{x})G^{(a,b)}(l)}{< \widetilde{P} \mid T\overline{\psi}^{(-)(\alpha,a)}(0)\gamma^5\psi^{(-)(\beta,b)}(0)G^{(a,b)}(l) \mid 0 >} \tag{451}$$

$$\phi_M^{(+)}(\widetilde{x}) = \frac{1}{\sqrt{V}}\frac{\sqrt{Z_3}}{\sqrt{2\omega_{\overrightarrow{P}}}}\frac{T\overline{\psi}^{(+)(\beta,b)}(\widetilde{x})\gamma^5\psi^{(+)(\alpha,a)}(\widetilde{x})G^{(a,b)}(l)}{< 0 \mid T\overline{\psi}^{(+)(\beta,b)}(0)\gamma^5\psi^{(+)(\alpha,a)}(0)G^{(a,b)}(l) \mid \widetilde{P} >} \tag{452}$$

For the normalization condition (450) one has to replace Z_3 in (451) and (452) by unity. Thus, we have

$$\begin{aligned}\phi_M(\widetilde{x}) = \frac{1}{\sqrt{V}}\frac{\sqrt{Z_3}}{\sqrt{2\omega_{\overrightarrow{P}}}}[&\frac{T\overline{\psi}^{(-)(\alpha,a)}(\widetilde{x})\gamma^5\psi^{(-)(\beta,b)}(\widetilde{x})}{< \widetilde{P} \mid T\overline{\psi}^{(-)(\alpha,a)}(0)\gamma^5\psi^{(-)(\beta,b)}(0) \mid 0 >} \\ &+\frac{T\overline{\psi}^{(+)(\beta,b)}(\widetilde{x})\gamma^5\psi^{(+)(\alpha,a)}(\widetilde{x})}{< 0 \mid T\overline{\psi}^{(+)(\beta,b)}(0)\gamma^5\psi^{(+)(\alpha,a)}(0) \mid \widetilde{P} >}]\end{aligned} \tag{453}$$

This can be compared with the pseudoscalar meson field $\phi(\widetilde{x})$ obtained by Haag (1958), Zimmermann (1958) and Nishijima (1958, 1961, 1964) as

$$\phi(\widetilde{x}) = \lim_{\widetilde{\xi}^2>0 \widetilde{\xi}\longrightarrow 0} \frac{1}{\sqrt{V}}\frac{\sqrt{Z_3}}{\sqrt{2\omega_{\overrightarrow{k}}}}\frac{T\overline{\psi}\left(\widetilde{x}+\frac{\widetilde{\xi}}{2}\right)\gamma_5\psi\left(\widetilde{x}-\frac{\widetilde{\xi}}{2}\right)}{< 0 \mid T\overline{\psi}\left(\frac{\widetilde{\xi}}{2}\right)\gamma_5\psi\left(\frac{-\widetilde{\xi}}{2}\right) \mid \overrightarrow{k} >} \tag{454}$$

The one particle meson state $\mid \widetilde{P} >$ in the above expressions $\phi_M^{(+)}(\widetilde{x})$ and $\phi_M^{(-)}(\widetilde{x})$ is, in fact, given by

$$\mid \widetilde{P} > \ \equiv c^{\dagger}_{\overrightarrow{p},\alpha,a}d^{\dagger}_{\overrightarrow{p'},\beta,b} \mid 0 > \ \equiv \ \mid \overrightarrow{p},\overrightarrow{p'},p^0,p'^0,\alpha,\beta,a,b > \ = \ \mid \overrightarrow{p}+\overrightarrow{p'},p^0+p'^0,\alpha,\beta,a,b > \tag{455}$$

or

$$\mid \overrightarrow{P},P^0 > \ = \ \mid \overrightarrow{p}+\overrightarrow{p'},p^0+p'^0,\alpha,\beta,a,b > \tag{456}$$

That is, we have taken the following normalization for the state

$$< \widetilde{P} \mid \overrightarrow{p},\overrightarrow{p'},p^0,p'^0,\alpha,\beta,a,b > \ = \delta_{\overrightarrow{P},\overrightarrow{p}+\overrightarrow{p'}}\ \delta_{P^0,p^0+p'^0} \tag{457}$$

For continuous momemta case, the necessary smearing in momentum space is to be understood. Thus, the state $\mid \widetilde{P} >$ is the eigenstate of the momentum operator with eigenvalue $P^\mu = p^\mu + p'^\mu$. This specification of one particle meson state given by (456) ensures the normalization condition (449). Explicitly, this can be verified as follows:

$$< \widetilde{P} \mid \phi_M^{(-)}(\widetilde{x}) \mid 0 >= \frac{1}{\sqrt{V}}\frac{\sqrt{Z_3}}{\sqrt{2\omega_{\overrightarrow{P}}}}\frac{< \widetilde{P} \mid T\overline{\psi}^{(-)(\alpha,a)}(\widetilde{x})\gamma^5\psi^{(-)(\beta,b)}(\widetilde{x}) \mid 0 >}{< \widetilde{P} \mid T\overline{\psi}^{(-)(\alpha,a)}(0)\gamma^5\psi^{(-)(\beta,b)}(0) \mid 0 >}$$

$$= \frac{1}{\sqrt{V}}\frac{\sqrt{Z_3}}{\sqrt{2\omega_{\overrightarrow{P}}}}\frac{< \widetilde{P} \mid \sum_{\overrightarrow{p},\overrightarrow{p'}}\frac{1}{V}\sqrt{\left(\frac{m^2}{E_{\overrightarrow{p}}E_{\overrightarrow{p'}}}\right)}T\overline{u}^{(\alpha)}(\overrightarrow{p})\gamma^5 v^{(\beta)}(\overrightarrow{p'})e^{i(\widetilde{p}+\widetilde{p'}).\widetilde{x}}c^{\dagger}_{\overrightarrow{p},\alpha,a}d^{\dagger}_{\overrightarrow{p'},\beta,b} \mid 0 >}{< \widetilde{P} \mid \sum_{\overrightarrow{p},\overrightarrow{p'}}\frac{1}{V}\sqrt{\left(\frac{m^2}{E_{\overrightarrow{p}}E_{\overrightarrow{p'}}}\right)}T\overline{u}^{(\alpha)}(\overrightarrow{p})\gamma^5 v^{(\beta)}(\overrightarrow{p'})c^{\dagger}_{\overrightarrow{p},\alpha,a}d^{\dagger}_{\overrightarrow{p'},\beta,b} \mid 0 >}$$

$$= \frac{1}{\sqrt{V}}\frac{\sqrt{Z_3}}{\sqrt{2\omega_{\overrightarrow{P}}}}\frac{\sum_{\overrightarrow{p},\overrightarrow{p'}}\frac{1}{V}\sqrt{\left(\frac{m^2}{E_{\overrightarrow{p}}E_{\overrightarrow{p'}}}\right)}T\overline{u}^{(\alpha)}(\overrightarrow{p})\gamma^5 v^{(\beta)}(\overrightarrow{p'})e^{i(\widetilde{p}+\widetilde{p'}).\widetilde{x}}\delta_{\overrightarrow{P},\overrightarrow{p}+\overrightarrow{p'}}\ \delta_{P^0,p^0+p'^0}}{\sum_{\overrightarrow{p},\overrightarrow{p'}}\frac{1}{V}\sqrt{\left(\frac{m^2}{E_{\overrightarrow{p}}E_{\overrightarrow{p'}}}\right)}T\overline{u}^{(\alpha)}(\overrightarrow{p})\gamma^5 v^{(\beta)}(\overrightarrow{p'})\delta_{\overrightarrow{P},\overrightarrow{p}+\overrightarrow{p'}}\ \delta_{P^0,p^0+p'^0}}$$

$$= \frac{1}{\sqrt{V}}\frac{\sqrt{Z_3}}{\sqrt{2\omega_{\overrightarrow{P}}}}\frac{\sum_{\overrightarrow{p}}\frac{1}{V}\sqrt{\left(\frac{m^2}{E_{\overrightarrow{p}}E_{\overrightarrow{P}-\overrightarrow{p}}}\right)}T\overline{u}^{(\alpha)}(\overrightarrow{p})\gamma^5 v^{(\beta)}(\overrightarrow{P}-\overrightarrow{p})e^{i\widetilde{P}.\widetilde{x}}}{\sum_{\overrightarrow{p}}\frac{1}{V}\sqrt{\left(\frac{m^2}{E_{\overrightarrow{p}}E_{\overrightarrow{P}-\overrightarrow{p}}}\right)}T\overline{u}^{(\alpha)}(\overrightarrow{p})\gamma^5 v^{(\beta)}(\overrightarrow{P}-\overrightarrow{p})}$$

$$= \frac{1}{\sqrt{V}}\frac{\sqrt{Z_3}}{\sqrt{2\omega_{\overrightarrow{P}}}}e^{i\widetilde{P}.\widetilde{x}}$$

[We have used equations (443) and (444)]

which is the normalization condition (449). This also follows from the property of invariance under infinitesimal spacetime translations whose generators are P^μ (operators). In fact, for conservation of P^μ we have

$$\overline{\psi}^{(-)(\alpha,a)}(\widetilde{x}) = e^{i\widetilde{P}.\ \widetilde{x}}\overline{\psi}^{(-)(\alpha,a)}(0)e^{-i\widetilde{P}.\ \widetilde{x}}$$
$$\psi^{(-)(\beta,b)}(\widetilde{x}) = e^{i\widetilde{P}.\ \widetilde{x}}\psi^{(-)(\beta,b)}(0)e^{-i\widetilde{P}.\ \widetilde{x}} \tag{458}$$

and consequently,

$$< \widetilde{P} \mid T\overline{\psi}^{(-)(\alpha,a)}(\widetilde{x})\gamma^5\psi^{(-)(\beta,b)}(\widetilde{x}) \mid 0 >=< \widetilde{P} \mid Te^{i\widetilde{P}.\ \widetilde{x}}\overline{\psi}^{(-)(\alpha,a)}(0)e^{-i\widetilde{P}.\ \widetilde{x}}\gamma^5 e^{i\widetilde{P}.\ \widetilde{x}}\psi^{(-)(\beta,b)}(0)e^{-i\widetilde{P}.\ \widetilde{x}} \mid 0 >$$
$$= e^{i\widetilde{P}.\ \widetilde{x}} < \widetilde{P} \mid T\overline{\psi}^{(-)(\alpha,a)}(0)\gamma^5\psi^{(-)(\beta,b)}(0) \mid 0 > \tag{459}$$

since $\mid 0 >$ and $\mid \widetilde{P} >$ are eigenstates of P^μ (operators) with eigenvalues 0 and P^μ respectively. Then, the normalization condition (449) follows (using (453)).
Here it should be noted that the hadron fields and states as constructed above contain the indices a, b, ...representing the internal helicities. These helicities may be regarded as the manifestation of the extended hadron structure (as composite of the constituents) in an anisotropic Finslerian microdomain. The fields and state vectors also contain spin indices α, β,...... which are responsible for the spin of the hadron constructed from the constituents, whereas from the indices a, b, ... one can have the internal quantum numbers such as strangeness, baryon number, hypercharge, etc. for the hadron. This has been discussed in the previous sections for the meson field or the one particle state $\mid \overrightarrow{P} > \ = \ \mid \overrightarrow{p}+\overrightarrow{p'}, p^0+p'^0, \alpha, \beta, a, b >$, the baryon number as well as strangeness are, in fact, zero as a and b have opposite values for this case.
We have found the pseudoscalar meson field $\phi_M(\widetilde{x})$ in (453). As an example, we can find the field for a neutral scalar if we regard $\overline{\psi}^{(\pm)(\alpha,a)}(\widetilde{x})$ and $\psi^{(\pm)(\beta,b)}(\widetilde{x})$ to correspond the operators (destruction and creation) for the particles μ^+ and μ^- or those for ν_μ and $\overline{\nu}_\mu$. The destruction and creation operators

$a^M_{\vec{k}\,in}$, $a^{M\dagger}_{\vec{k}\,in}$, $a^M_{\vec{k}\,out}$, $a^{M\dagger}_{\vec{k}\,out}$ can then be found by the standard procedure. For example, the incoming meson field $\phi^{in}_M(\tilde{x})$ is given by

$$\phi^{in}_M(\tilde{x}) = \sum_{\vec{k}} \left(a^M_{\vec{k}\,in} f_{\vec{k}}(\vec{x},t) + a^{M\dagger}_{\vec{k}\,in} f^*_{\vec{k}}(\vec{x},t) \right) \tag{460}$$

where $f_{\vec{k}}$ is the solution of Klein-Gordon equation for a mass M, the mass of the meson. These in and out fields are defined with the retarded and advanced functions for this mass M. They are given by (Lurió, 1968)

$$\begin{aligned} \phi^{in}_M(\tilde{x}) &= \phi_M(\tilde{x}) + \int \triangle_R(\tilde{x} - \tilde{y}; M)(\Box_y + M^2)\phi_M(\tilde{y}) d^4y \\ \phi^{out}_M(\tilde{x}) &= \phi_M(\tilde{x}) + \int \triangle_A(\tilde{x} - \tilde{y}; M)(\Box_y + M^2)\phi_M(\tilde{y}) d^4y \end{aligned} \tag{461}$$

The retarded and advanced functions are given by

$$\begin{aligned} \triangle_R(\tilde{x}) &= -\triangle(\tilde{x}) \qquad x^0 > 0 \\ &= 0 \qquad\qquad\; x^0 < 0 \end{aligned} \tag{462}$$

$$\begin{aligned} \triangle_A(\tilde{x}) &= 0 \qquad\quad x^0 > 0 \\ &= \triangle(\tilde{x}) \quad x^0 < 0 \end{aligned} \tag{463}$$

where

$$\triangle(\tilde{x}) = \frac{1}{(2\pi)^4} \int d^3k \int_C dk^0 \frac{e^{-i\tilde{k}.\,\tilde{x}}}{\tilde{k}^2 - M^2} \tag{464}$$

The contour is being exhibited in Fig.2

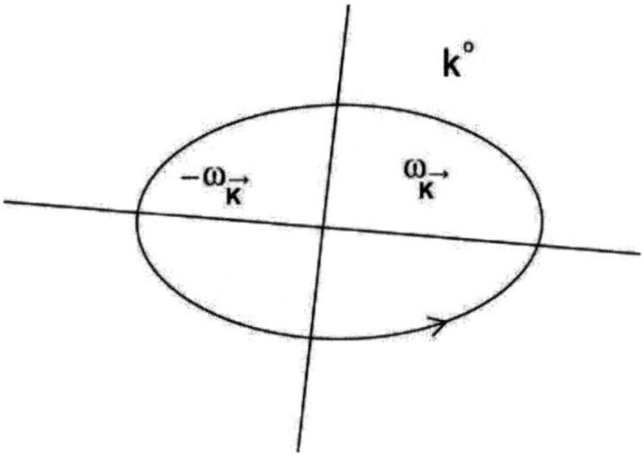

Figure 2: Contour C in k^0 complex plane.

Lurió (1968) has discussed how these incoming and outgoing fields could be identified as the correct incoming and outgoing fields for the composite particle (neutral meson). Also, as in the case of elementary particle fields the following asymptotic conditions for the composite particle field

$$\lim_{t \to -\infty} < a \mid a^M_{\vec{k}}(t) \mid b >= \sqrt{Z_3} < a \mid a^M_{\vec{k}\;in} \mid b > \tag{465}$$

$$\lim_{t\to+\infty} < a \mid a^M_{\vec{k}}(t) \mid b >= \sqrt{Z_3} < a \mid a^M_{\vec{k}\ iout} \mid b > \tag{466}$$

remain valid. Here, $\mid a >$ and $\mid b >$ are any two normalizable state vectors. In fact, as far as asymptotic conditions and the consequent reduction formulae (that expresses the S-matrix element in terms of time-ordered product of field operators) are concerned there remains no clear distinction between elementary and composite fields. Thus, the composite fields obtained here for the hadrons can be used in the reduction formulae. As the asymptotic conditions and reduction formulae have no relation with the perturbation theory these can be applied to the strong interaction of hadrons, in which case the perturbation technique does not work. Again since the reduction technique is independent of the detailed form of the Lagrangian, one can use it in strong interaction in deriving the dispersion relations from some general properties like Lorentz invariance, microscopic causality etc. In the following section, however, we shall phenomenologically consider the strong interaction of hadrons and discuss the relevance of both the postulates of S-matrix Theory and the field theory with the perturbation technique in this picture of composite hadron structure.

4.7: DYNAMICS OF STRONG INTERACTION AND S-MATRIX AXIOMS

We have investigated the non hypercharge exchange two body hadron reactions of the type $AB \longrightarrow CD$ (Bandyopadhyay and De, 1973; 1975 a, b). These reactions are basically dependent on the $\pi\pi$-interaction where the interacting pions are in the structure of the incident hadrons. In fact, we have seen in the previous section that a hadron field is constructed from those of the constituents μ^+, μ^-, ν_μ (Majorana or Dirac spinor) in its configuration as a composition of these fields in particle - antiparticle pairs. In the case of hadron with odd number of constituents one constituent, of course, remains alone in the composition of the hadron field. These pairs of particle - antiparticle are, in fact, the pseudoscalar mesons (π^+, μ^-, π^0) in the hadron configuration. Phenomenologically one may think of a hadron to be a cluster of pions, may be, with a single spinor constituent (muon or neutrino). Consequently, the strong interaction between the hadrons may be regarded as the strong interaction between the pions in the structures of incident hadrons. In order to find the amplitude for the two body hadron reaction one should also take into account the term arising out due to the rearrangement of constituents apart from the contribution of $\pi\pi$-interaction derived from the Lagrangian field theory. The amplitude is of the form

$$A(AB \longrightarrow CD) = \alpha A[\pi\pi \to \pi\pi(\omega)]T(s,t) \tag{467}$$

where α is a numerical factor depending on the number of interacting pions in the structure of A and B. $A[\pi\pi \to \pi\pi(\omega)]$ is the amplitude for $\pi\pi \to \pi\pi(\omega)$ interaction derived from the Lagrangian field theory and T(s,t) represents the rearrangement amplitude due to the rearrangement of constituents. Here, s is the square of the centre mass energy and t stands for the square of momentum transfer between A and C (or between B and D). The amplitude T(s,t) behaves like $s^{-n\gamma}$ for large s, where γ is a suitable parameter and has a correspondence with the Regge amplitude with strongly degenerate trajectories $\alpha(t)$ and residue $\beta(t)$ in the forward regions if one takes $-2\gamma + 1 = \alpha(t)$. n is the number of constituents rearranged. This gives the effective coupling $g(s) = gs^{-n\gamma}$ and inserting this factor for each vertex in the perturbation expansion, the final expression can be made convergent because the higher order terms will not be large enough to contribute and thus any inconsistency is not faced as in the naive form of the field theory. The general form of the amplitude for this case which satisfies crossing is

$$A(s,t) = \alpha g^2 \frac{s+t-\mu^2}{(s-\mu^2)(t-\mu^2)} \frac{1}{(s+t)^{n\gamma}} \frac{F(s)F(t)}{F(s+t)} + (t \to u) + (u \to s) \tag{468}$$

Here, g is the $\rho\pi\pi$ coupling constant and μ is the mass of the exchanged meson (viz. the ρ - meson) in the $\pi\pi$-interaction. F(x) is a form factor parametrizied as $F(0) = 1$ and is of the general type

$$F(x) = \frac{1}{1 + a|x| + bx^2 + ...} \tag{469}$$

This form factor arises because we have assumed that the basic unit of strong interaction is a π-meson, which in itself is not a fundamental particle but is composed of a muon-antimuon pair and it has been shown (De, 1977) that physically it corresponds to the electromagnetic form factor of the pair. One should note that for the $\pi^0\pi^0$-interaction it has to take into account the $f^0\pi^0\pi^0$ tensor coupling as the $\rho^0\pi^0\pi^0$ interaction is forbidden by the selection rule and therefore the expression (468) should be changed accordingly. If γ is not taken as a constant, it may be a function of t or s and in this case the generalized amplitude takes the following form:

$$A(s,t) = \alpha g^2 \frac{s+t-\mu^2}{(s-\mu^2)(t-\mu^2)} \frac{1}{(s+t)^{n\gamma(\frac{st}{s+t})}} \frac{F(s)F(t)}{F(s+t)} \tag{470}$$

Now, for the direct channel process $s \longrightarrow large$, $t \longrightarrow 0$, this amplitude reduces to the form:

$$A(s,t) = \alpha g^2 \frac{1}{(t-\mu^2)} \frac{1}{s^{n\gamma}} F(t) \tag{471}$$

On the other hand, for the t-channel process $t \longrightarrow large$, $s \longrightarrow 0$, it becames

$$A(s,t) = \alpha g^2 \frac{1}{(s-\mu^2)} \frac{1}{t^{n\gamma}} F(s)$$

Thus the equations (468) and (470) satisfies the requirement of crossing. It should also be noted that in this picture of strong interactions where the basic ingredient is Lagrangian field theory for $\pi\pi$-scattering, it encompasses the problem of unitarity in multiparticle scattering by a Feynman diagram series and also we have the benifit of potential scattering which ensures the requirement of analyticity. Thus, the axioms of the S-matrix theory are being satisfied in the two body non hyperchange exchange reactions. We have earlier shown that the amplitude does not violate the Froissart bound at the high energies (Bandyipadhyay and De, 1975b).

For the hyperchange exchange reactions we suggest that these processes are dominated by the direct interactions such as knock out or stripping processes which are familiar in Nuclear Physics (De, 1986). In the version of the constituent rearrangement, the so called 'line physics', these direct interactions can be thought as the conservation of constituent lines in course of the reactions. Although, in this picture, it has to introduce the annihilation of constituents (mesons, muons or neutrinos) into the vacuum or the converse process, the creation of the pair of constituent particle and antiparticle from the vacuum. In fact, in the crossed channel reactions, some constituents annihilate into the vacuum and transfer their four momenta into the other constituents of the hadrons participating in the reaction. Also, the reverse phenomenon, that is, the generation of a pair of constituents from the vacuum can occur after gaining momenta from the other constituents which are rearranged in the reaction.

Thus, the hypercharge-exchange reactions are the manifestation of direct exchange (pick up) of the constituent or annihilation (creation) of constituent lines into (from) the vacuum in contrast to the non hypercharge-exchange two body interactions which are, as already been stated, dominated by the strong interactions of the constituent mesons (pseudoscalar or vector) in the structure of hadrons, together with the rearrangement of the constituents. It is, here, conjecture that the amplitude is dominated by the pole terms due to the exchange of particles whether the exchanged particles picked up by the final particles or one being annihilated into the vacuum with some constituents of the other initial particle. Specifically, we shall consider here the hypercharge-exchange reaction $K^-p \longrightarrow \pi^-\Sigma^+$ although this mechanism is very general in character and the amplitudes of the similar reactions, such as $K^-p \longrightarrow \pi^+\Sigma^-$, $\pi^0\Sigma^0$, $\pi^0\Lambda$, $\rho^0\Lambda$, $\omega\Lambda$ etc., will follow in the similar fashion except for the last two cases where π mesons have been excited to the level of ρ or ω mesons. It might be mentioned that in this model the structures of π and ρ or ω mesons are the same but they differ only in the internal orbital angular momenta of their constituents. Considering $K^-p \longrightarrow \pi^-\Sigma^+$ as s-channel reaction the crossed channels are $K^-\pi^+ \longrightarrow \bar{p}\Sigma^+$ and $K^-\overline{\Sigma}^+ \longrightarrow \bar{p}\pi^-$. To understand the dynamics of these reactions, let us first specify the internal helicities or S_3-components of the constituents in the configurations of the initial and final particles of the reaction. We write only the internal S_3-components of the constituents of these particles in the following manner:

$$For \ \ p\ , \ \ S_3 = \left(\frac{1}{2}, -\frac{1}{2}, \frac{1}{2}, -\frac{1}{2}^*, \frac{1}{2}^*\right) \tag{472}$$

$$For\ K_-\ ,\quad S_3 = \left(-\frac{1}{2}, \frac{1}{2}, -\frac{1}{2}^*, -\frac{1}{2}^*\right) \tag{473}$$

$$For\ \pi^-\ ,\quad S_3 = \left(-\frac{1}{2}, \frac{1}{2}\right) \tag{474}$$

$$For\ \Sigma^+\ ,\quad S_3 = \left(\frac{1}{2}, -\frac{1}{2}, \frac{1}{2}, -\frac{1}{2}^*, \frac{1}{2}^*, -\frac{1}{2}^*, -\frac{1}{2}^*\right) \tag{475}$$

The asterisks on the S_3-components indicate that the corresponding constituents are Majorana spinors (neutrinos). We have stated in section 4.5 that the strangeness S is given by the sum of S_3-components (internal) of the Majorana constituents whereas the baryonic number B is twice the sum of internal S_3-components of the other constituents. It is thus, apparent from above that the strangeness and baryonic numbers of the particles p, K^-, π^- and Σ^- are

	p	K^-	π^-	Σ^-
S	0	-1	0	-1
B	1	0	0	1

(476)

Consequently, the hypercharge $Y = B + S$ of the particles p, K^-, π^- and Σ^- are 1, -1, 0, 0 respectively. Now, the s-channel reaction $K^-p \longrightarrow \pi^-\Sigma^+$, which is very similar to the nuclear pick up phenomenon is depicted in Figure 3.

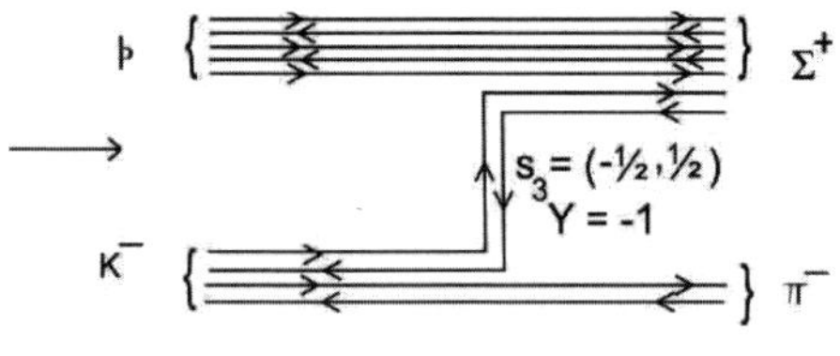

Figure 3: The s-channel $K^-p \longrightarrow \pi^-\Sigma^+$ reaction by the pick up of two constituents forming a π^0 meson π^0 carries the hypercharge - 1.

Here, particle p picks up two Majorana spinors from the configuration of K^-. These two constituents form a π^0 meson in the virtual state while being picked up. As both of these two constituents have internal S_3-components $-\frac{1}{2}$, the π^0 meson constituted by them carries strangness $S = -1$ and hence a hypercharge -1. It is, in fact, possible that π^0 has strangeness -1 only because this particle is in the virtual state. Also, while shifting from the configuration of K^- to that of the proton it maintains the same values of the third components of the internal spin angular momentum. The third component of the spin (internal), as pointed out earlier, is related to the specification of the particle and antiparticle states and also that the internal space of the constituents (that is, the extended hadron structure) is a locally anisotropic space (the flat Finslerian space). Thus the anisotropy of the space is maintained here in these pick up phenomena and this should be regarded as the special characteristic of these types of strong processes.
The amplitude of this reaction is dominated by the pole of t at μ^2 in the unphysical region of the reaction, where μ is the mass of the pion, because the situation is very similar to that of one pion-exchange-reaction. The energy dependence of this reaction is contributed by the rearrangement of the constituents as before. The clustering effect of the constituents is to be included in the form factor which, in fact, is contributed by this clustering effect as well as by the absorption of both the initial and final particles. The absorption effect is generally introduced in order to modify the usual one pion exchange amplitude and in any case,

the dip phenomenon that we encounter in some reactions, may arise from this part of the form factor. Of course, it is very difficult to ascertain the exact structure of this form factor without the knowledge of the specific model of the diffraction occurring in any reaction. However, we can write down the general form of the amplitudes of these reactions which are analytic, crossing symmetric and satisfy unitarity. The crossed channel reactions are depicted in figures 4 and 5. In the t-channel the two constituents (

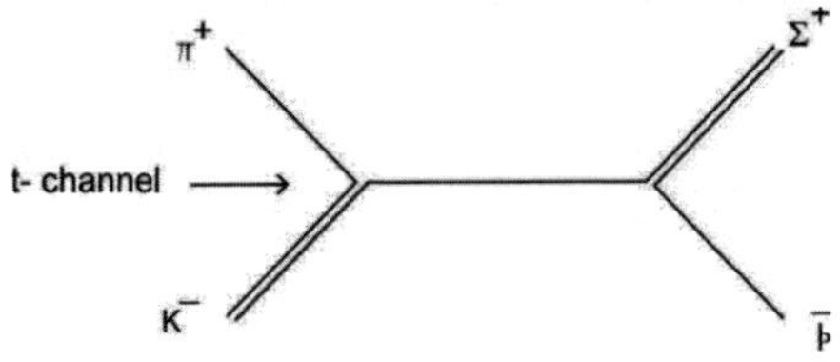

Figure 4: The t-channel reaction $K^- \, \pi^+ \longrightarrow \overline{p} \, \Sigma^+$.

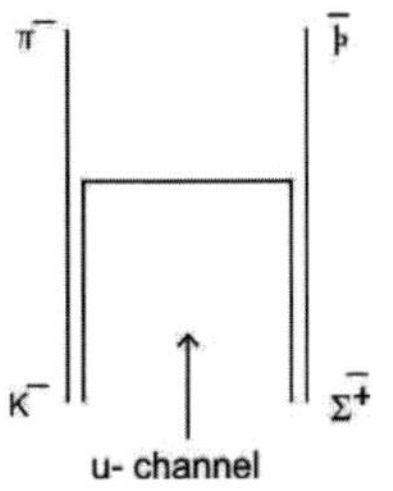

Figure 5: The u-channel reaction $K^- \, \overline{\Sigma}^{\,+} \longrightarrow \overline{p} \, \pi^-$.

collectively shown as a single line in the figure) in the structure of each initial particle annihilate into vacuum (as they have particle-antiparticle relation) transferring corresponding four momenta into the survived constituents of K^-. These survived constituents then create $\sum^+$ and p^- where some of the constituents in their structures have been generated from the vacuum receiving the four momenta from the intermediate particles (a reverse process of the annihilation of particle and antiparticle into the vacuum with transfer of four momenta). The intermediate particles (shown in the figure as a single line) originally in the structure of K^- meson from the virtual π^0 meson state and the amplitude of this reaction is dominated by s-channel exchange of π^0 meson. Similar phenomenon of annihilation also occurs in the u-channel through an exchange of π^0 meson. Here, also in the figure a line represents the constituents in π^- or in $\overline{p}$ and some of the constituents in the structure of K^- or $\overline{\Sigma}^+$ collectively. The exchanged particle π^0 being in the virtual state is composed of two constituents (Majorana particles) which are depicted in the figure by a single line, and gives rise to a pole in t in the unphysical region of that channel. The quantum numbers carried by the intermediate and exchanged particles remain the same as that for the picked up particle in the s-channel. The significant difference of the t- and u-channels from the s-channel is that in the former processes annihilation (or creation) of the constituents which have particle-antiparticle relations, into (from) the vacuum occurs transferring their momenta into other constituents (or receiving momenta from them). The amplitudes of these hypercharge-exchange rections can now be written by using the one-particle-exchange model with the modification by a distorted wave Born approximation (DWBA). In general, the one particle exchange models violate the unitarity limit at sufficiently high energy but it must be noted that in the present case where the exchanged particle is spin zero this violation does not arise. Also, the term due to the rearrangement of the constituents particles produce such a s-dependence that ensures the unitarity limit. The DWBA has been introduced to assure both the dip phenomenon where it has been absorbed experimentally as well as the unitarity. This DWBA effect might be observed into the form factor F(t). In fact, we can write $F(t) = f(t) + \widetilde{f}(t)$ where f(t) is the electromagnetic form factor of pion in the hadron configuration as discussed in De(1997) and $\widetilde{f}(t)$ is the contribution to the form factor by they rescattering of the incident and outgoing waves. In the impact parameter formalism, the amplitude (for one particle exchange) can be represented, apart from some factors (both the numerical and s-dependence), as

$$\frac{f(t)}{m_e^2 + |t|} = \int_0^{\infty} bdbJ_n(b\sqrt{|t|})G(m_e b) \tag{477}$$

where b is the impact parameter and m_e is the mass of the exchanged particle.
Again, the absorption corrected amplitude f_{abs} is proportional to

$$\int_0^{\infty} bdbJ_n(b\sqrt{|t|})G(m_e b)S_{el}(s,b) \tag{478}$$

where $S_{el}(s,b)$ is the elastic scattering. Now, for a specific model for this elastic scattering, one can write

$$f_{abs} \propto \frac{f(t)}{m_e^2 + |t|} = \frac{f(t) + \widetilde{f}(t)}{m_e^2 + |t|} \tag{479}$$

In fact, for Gaussian form of $S_{el}(s,b)$ that is,

$$S_{el}(s,b) = 1 - C\exp\left(-\frac{b^2}{R^2}\right) \tag{480}$$

where R represents the hadron radius, $\widetilde{f}(t)$ is given by

$$\frac{\widetilde{f}(t)}{m_e^2 + |t|} \propto (-C)\int_0^{\infty} bdbJ_n(b\sqrt{|t|})G(m_e b)\exp\left(-\frac{b^2}{R^2}\right) \tag{481}$$

In sharp cut off model, where

$$\begin{aligned} S_{el}(s,b) &= 0, \quad b < R \\ &= 1, \quad b > R \end{aligned} \tag{482}$$

$\widetilde{f}(t)$ is given by

$$\frac{\widetilde{f}(t)}{m_e^2 + |t|} \propto (-1)\int_0^{R} bdbJ_n(b\sqrt{|t|})G(m_e b) \tag{483}$$

$\widetilde{f}(t)$, thus, can be computed, for a specific model of the absorption, from the knowledge of the part of the form factor, f(t), which is the electromagnetic form factor of pion. Hence, the absorption introduced here can produce a change in the t-dependence in the amplitude, which can be computed analytically or by numerical method and it can be included into the form factor F(t).
The crossing symmetric amplitude (which satisfies unitarity too) can be written as follows:

$$A(s,t) = G^2 \frac{1}{(t-\mu^2)} \frac{1}{(s+t)^{n\gamma}} \frac{\widetilde{F}(s)\widetilde{F}(t)}{\widetilde{F}(s+t)} \tag{484}$$

where γ is a slowly varying suitable parameter function of s and t. It is taken to be

$$\gamma = \gamma\left(\frac{st}{s+t}\right) \tag{485}$$

n is a numerical factor which depends on the number of rearranged constituents and G is the constant which is related to be strength of the reaction (such as pick-up or stripping reactions). Here, $\widetilde{F}(t) =$

$F(t)\eta(t)$ where F(t) is the form factor mentioned above and $\eta(t)$ is the product of the t-dependence arising out from the internal quantum number of the exchanged particle and the kinematical t-dependence. The s-dependence for the exchanged particle has been absorbed into the factor $s^{-n\gamma}$.
now, the s-channel, high energy (i.e. s large) and small angle (t small < 0) phenomemnon, the amplitude becomes

$$A(s,t) \longrightarrow G^2 \frac{1}{(t-\mu^2)} \frac{1}{s^{n\gamma}} \widetilde{F}(t) \tag{486}$$

The form of this amplitude itself manifests the previously described dynamical picture of the reaction. Similarly, for t– and s- channels, the corresponding high energy amplitudes are, respectively,

$$A(s,t) \longrightarrow G^2 \frac{1}{(t_s-\mu^2)} \frac{1}{s_t^{n\gamma}} \widetilde{F}(t_t) \qquad t > 0, \; large \; ; \; s < 0 \, , \; small \tag{487}$$

$$A(s,t) \longrightarrow G^2 \frac{1}{(t_u-\mu^2)} \frac{1}{-s_u^{n\gamma}} \widetilde{F}(t_u) \qquad u > 0, \; large \; ; \; t < 0 \, , \; small \tag{488}$$

where the subscripts indicate variables for the corresponding channels. These amplitudes are also consistent with the corresponding dynamical phenomena of the reactions in the respective channels. Thus we see that for both the non hypercharge exchange and the hypercharge exchange two body reactions the postulates of the the S-matrix elements become relevant and the dynamics has been generated from the structural aspects of the hadrons. The noteworthy fact is that both the field theory in the modified form so that the perturbative expansion remains valid and the S-matrix approach for strong interactions, become united in the general premises of the constituent character of the hadron or extended hadron structure in the microlocal anisotropic spacetime.

REFERENCES

(1) Bandyopadhyay, P. (1989). Int. J. Mod. Phys. A **4**, 4449
(2) Bandyopadhyay, P. and De, S.S.(1973). Nuovo Cimento **14 A** , 285
(3) Bandyopadhyay, P. and De, S.S.(1975a). Nuovo Cimento **27 A** , 295
(4) Bandyopadhyay, P. and De, S.S.(1975b). Acta. Phys. Pol. **B6** , 241
(5) Bandyopadhyay, P. and Ghosh, P.(1989). Int. J. Mod. Phys. **A 4**, 4449
(6) Budini, P. (1979). Nuovo Cimento **53 A** , 31
(7) Budini, P. , Furlan, P. and Raczka, R. (1979). Nuovo Cimento **52 A** , 191
(8) Cartan, E. (1966). The Theory of Spinors, Paris
(9) Choudhuri, V.S.P. (1981). Bull.Cal.Math.Soc , **73**, 11
(10) De, S.S.(1977). Acta. Phys. Pol. **B8** , 521
(11) De, S.S.(1986). Hadronic Journal Supplement, **2** , 412
(12) De, S.S.(1997). Int. J.Theor.Phys. **36** , 89
(13) De, S.S.(2002a). Int. J.Theor.Phys. **41**, 1291
(14) De, S.S.(2002b). Int. J.Theor.Phys. **41**, 1307
(15) Dirac, P.M.(1936). Annals of Mathematics, **37** , 429
(16) Haag, R.(1958). Phys.Rev. **112**, 669
(17) Lurió, D. (1968). Particles and Fields, Interscience Publishers, New York, pp.442-444.
(18) Nishijima, K.(1958). Phys.Rev. **111**, 995
(19) Nishijima, K.(1961). Phys.Rev. **122**, 298
(20) Nishijima, K.(1964). Phys.Rev. **133B**, 204; 1092
(21) Perl , Martin L. (1974).High Energy Hadron Physics, John Wiley & Sons, New York, P131
(22) Zimmermann, W.(1958). Nuovo Cimnto , **10**, 597.

CHAPTER FIVE:

COSMOLOGICAL CONSEQUENCES : EARLY UNIVERSE

5.1: EARLY UNIVERSE AS A THERMODYNAMICALLY OPEN SYSTEM

In the preceding chapter we have come across the epoch dependence of the mass of the particle (lepton). From the relations (376) and (378) it is apparent that this epoch dependence of mass could have predominant effect in the early Universe, specially, in the epoch times $t \leq \alpha$. In fact, the masses of the particles were much large before this epoch time α. In particular, at the Planck-order time the masses of the particles might have been to the order of Planck mass. Presently, we shall discuss the evolution of the early universe with the incorporation of this epoch dependence of particle masses, The early universe will be regarded as an open thermodynamic system. This consideration was earlier made in De(1993).

Ordinarily, the universe is taken as a closed system, in which the laws of thermodynamics have the following form:

$$d(\rho V) = dQ - pdV \tag{489}$$

and

$$TdS = d(\rho V) + pdV \tag{490}$$

where

ρ = *energy density*, p = *themodynamic pressure*, S = *entropy of the system* , Q = *Thermal energy* , T = *temperature* , V = *comoving volume* = $R^3(t)$, $R(t)$ *being the scale factor*

.

From these two equations it follows that

$$TdS = dQ \tag{491}$$

which gives the entropy change. For adiabatic systems $dQ = 0$ and consequently we have $dS = 0$. This indicates no change in the entropy of these systems. Thus, in the standard model one has either to assume the observed value of the entropy of the universe as an initial condition or to make some dissipative mechanism in the system. On the contrary, if we consider the early universe as an open system allowing for irreversible matter creation from the gravitational field then entropy is produced in this early stage of evolution of the universe. In fact, Prigogine(1961) has modified the thermodynamical energy conservation law for an open system as follows:

$$d(\rho V) = dQ - pdV + \frac{h}{n}dN \tag{492}$$

where $h = p + \rho$ is the enthalpy per unit comoving volume and N is the total number of particles in the volume V. Here, $n = N/V$ is the particle number density. The modified form of the entropy change dS for the open thermodynamic system is given as

$$TdS = \frac{H}{N}dN - \mu dN = \frac{TS}{N}dN \tag{493}$$

where $H = hV$ is the total enthalpy and μ is the chemical potential given by

$$\mu N = H - TS \tag{494}$$

Now, for adiabatic transformation ($dQ = 0$) we have from(5.1.4)

$$d(\rho V) + pdV - \frac{h}{n}d(nV) = 0 \tag{495}$$

This equation can be reduced to

$$\dot{\rho} = \frac{\dot{n}}{n}(p + \rho) \tag{496}$$

where overdots mean time derivatives. This equation replaces the usual Einstein equation (Bianchi identities for homogeneous and isotropic universe), i.e.,

$$\dot{\rho} = -3H(t)(p + \rho) \tag{497}$$

where $H(t) = \dot{R}(t)/R(t)$ is the Hubble parameter. The other Einstein equation remains the same. For the zero-curvature three-dimensional space, this equation is

$$K\rho = 3H^2 \tag{498}$$

where $K = 8\pi G$, G being Newton gravitational constant. These equations (497) and (498), in fact, follow from the Einstein field equation

$$G_{\mu\nu} = -8\pi G T_{\mu\nu} \tag{499}$$

with the natural unit ($c = \hbar = 1$), for the homogeneous and isotropic Friedman-Robertson-Walker(FRW) line element

$$ds^2 = dt^2 - R^2(t)(\frac{dr^2}{1 - kr^2} + r^2 d\theta^2 + r^2 \sin^2\theta d\phi^2) \tag{500}$$

in the case of zero curvature($k = 0$).
In an alternative interpretation of the conservation equation(495) or (496) for the open system, we may retain the usual conservation law(Bianchi iidentities) with a phenomenological pressure $\hat{p}$ instead of the above true thermodynamic pressure p.That is,

$$d(\rho V) = -\hat{p}dV \tag{501}$$

where these two pressure $\hat{p}$ and p are related by

$$\hat{p} = p + p_c \tag{502}$$

In fact, in the context of an open system the perfect fluid stress-energy tensor $T_{\mu\nu}$ in the Einstein field equation(499) takes the following form

$$T_{\mu\nu} = (\rho + p + p_c)u_\mu u_\nu - (p + p_c)g_{\mu\nu} \tag{503}$$

where u_μ is a unit vector in the time direction.

$$u^\mu u_\nu = 0 \;\; for \; \mu \neq \nu \;\; ; \;\; u^0 u_0 = 1 \; , \;\; u^1 u_1 = u^2 u_2 = u^3 u_3 = 0 \tag{504}$$

Here, the pressure p_c is given by (using(5.1.5))

$$p_c = -\frac{p + \rho}{3H(t)}\frac{\dot{S}}{S} = -\frac{p + \rho}{3H(t)}\frac{\dot{N}}{N} \tag{505}$$

This pressure p_c, negative or zero, corresponds to the creation of particles.In fact, when $p_c = 0$, the creation of particles stops and in this case $\hat{p} = p$. consequently, the usual conservation law holds, or, other wards, the usual general relativity with the usual Einstein equations holds. Thus, for negative pressure p_c, the creation of particles occurs because of the transfer of energy from the gravitational field and it acts as a source of internal energy. Again from the equation (493) for the entropy change dS, the only admissible particle number variations are given by

$$dN \geq 0 \tag{506}$$

due to the second law of thermodynamics. This inequality implies irreversibility of the process of particle creation at the expense of gravitational energy as the reverse process is thermodynamically forbidden. Thus, the second law of thermodynamics is incorporated into the evolution of the system in a more meaningful way.
Now, let us assume the 'gamma-law' equation of state

$$p = (\gamma - 1)\rho,\ 0 \leq \gamma \leq 2 \tag{507}$$

where γ is the adiabatic index which is, here, supposed to be dependent on the Hubble parameter H or on the expansion scalar $\theta = \frac{\dot{V}}{V} = 3H$. We write

$$\gamma = f(\theta) + 1 \tag{508}$$

Therefore, the equation of state becomes

$$p = f(\theta)\rho,\ -1 \leq f(\theta) \leq 1 \tag{509}$$

Then, we have from equation (496), (502) and (505),

$$\dot{\rho} = \frac{\dot{n}}{n}[1 + f(\theta)]\rho \tag{510}$$

$$\begin{aligned} p_c = \hat{p} - p &= -\frac{d(\rho V)}{dV} - f(\theta)\rho \\ &= -\rho - V\frac{\dot{\rho}}{\dot{V}} - f(\theta)\rho \\ &= -[\rho(1 + f(\theta)) + \frac{\dot{\rho}}{\theta}] \end{aligned} \tag{511}$$

Using (5.1.10), we also have from equations(5.1.14) and (5.1.22),

$$p_c = -\rho\left[1 + f(\theta) + \frac{2\dot{\theta}}{\theta^2}\right] \tag{512}$$

$$\hat{p} = -\rho\left[1 + \frac{2\dot{\theta}}{\theta^2}\right] \tag{513}$$

Again, in the perfect fluid model, the phenomenological pressure $\hat{p}$ can be related to the density ρ by the relation

$$\hat{p} = \frac{1}{3}\rho\frac{<\nu^2>}{c^2} \tag{514}$$

where c is the velocity of light and $<\nu^2>$ is the mean square velocity of the particles. Thus, we can have, using equation(513),

$$\frac{<\nu^2>}{c^2} = -3\left(1 + \frac{2\dot{\theta}}{\theta^2}\right) \tag{515}$$

In the 'particle production' era of the very early universe the energy density $\rho = \rho_m + \rho_\gamma$ shuld be dominated by the matter density $\rho_m = mn_m$ where n_m is the number density of particles and m is the particle mass at the epoch times that we are considering. The radiation energy density ρ_γ will dominate only after the particle production stops and, in fact, when the radiation era begins. The radiation density ρ_γ is given by

$$\rho_\gamma = (\gamma - 1)mn_m + E_\gamma n_\gamma \tag{516}$$

where E_γ and n_γ are the energy per photon and photon number density, respectively. Here, of course, we are assuming for simplicity that the created particles have the samemasses. γ is related to the mean square velocity of the particles by

$$\gamma = \left(1 - \frac{<\nu^2>}{c^2}\right)^{-1/2} = \frac{1}{2}\left(1 + \frac{3\dot{\theta}^2}{2\theta^2}\right)^{-1/2} \tag{517}$$

Thus, for the radiation era to begin γ must be large and the matter density γmn_m contributes to the energy density

$$\rho = \rho_m + \rho_\gamma = \gamma mn_m + E_\gamma n_\gamma \tag{518}$$

But when the particle production occurs γ is nearly equal to one and $\rho \simeq mn_m$. In the following the particle production in the early stages of the universe from the Plank time $t_{pl} = 5.4 \times 10^{-44} sec$ to the time $t = \alpha = 0.26 \times 10^{-23} sec.$ will be considered. In fact, it will be seen that around the epoch α the particle production stops and subsequently the radiation era begins. For this particle and entropy-production era, the equations that govern the cosmology of this era follow from equations (378), (496), (498) and (510). Thcsc arc

$$\frac{\dot{\rho}}{\rho} = \frac{2\dot{H}}{H} = \frac{2\dot{\theta}}{\theta} = \frac{\dot{n}}{n}(1 + f(\theta)) \tag{519}$$

$$\frac{\dot{m}}{m} = \frac{2\alpha\dot{H}}{1 + 2\alpha H} = \frac{2\alpha\dot{\theta}}{3 + 2\alpha\theta} \tag{520}$$

(Here, 'm' is used for the epoch dependent mass in place of $\bar{m}$ in (378)).
In the above equation (519), n represents the total particle number density, i.e., $n = n_m + n_\gamma$. Again, one can write for the radiation density $\rho_\gamma = 3\hat{p}$ and consequently we have from equations(514) and (517)

$$\frac{\rho_\gamma}{\rho_m} = \gamma^2 - 1 = -\frac{3}{4}\frac{1 + 2\dot{\theta}/\theta^2}{1 + 3\dot{\theta}/2\theta^2} \tag{521}$$

This equation can also be achieved in the perfect fluid model by considering the FRW universe as the conformal Minkowski space-time. For this we write the line element of the conformal Minkowski space-time related to that of FRW universe for $k = 0$ given by (500), in the following form:

$$ds^2 = \Omega^2(\eta)\left[d\eta^2 - \Sigma_{i=1}^3(dx^i)^2\right] \tag{522}$$

where

$$d\eta = \frac{dt}{\Omega(\eta)} = \frac{dt}{R(t)}$$
$$or\ \eta = \int \frac{dt}{R(t)} \tag{523}$$

The conformal Ricci scalar and tensor $\bar{R}$ and $\bar{R}^\nu_\mu$ can be computed to be

$$\bar{R} = 6\Omega^{-3}\Omega_{;00}$$
$$\bar{R}^\nu_\mu = -2\Omega^{-1}(\Omega^{-1})_{;\mu 0}\, g^{0\nu} + \frac{1}{2}\Omega^{-4}(\Omega^2)_{;00}\, \delta^\nu_\mu \tag{524}$$

where $g^{\mu\nu}$ is the conformal metric.

In the perfect fluid model, we have

$$-2\Omega^{-1}(\Omega^{-1})_{;\nu 0}g^{0\mu} + \delta^{\mu}_{\nu}\left[\frac{1}{2}\Omega^{-4}(\Omega^{2})_{;00} - 3\Omega^{-3}\Omega_{;00}\right] = -K\left[(\rho+\hat{p})u^{\mu}u_{\nu} - \hat{p}\delta^{\mu}_{\nu}\right]$$

Then we have the following expressions for the density and pressure:

$$K\rho = 3\dot{\Omega}^2/\Omega^4$$
$$K\hat{p} = (\dot{\Omega}^2 - 2\Omega\ddot{\Omega})/\Omega^4 \tag{525}$$

where the overdots represent the differentiation with respect to the conformal time η. Writing $\rho_\gamma = 3\hat{p}$ again, we have

$$\frac{\rho_\gamma}{\rho} = 1 - \frac{2\Omega\ddot{\Omega}}{\dot{\Omega}^2}$$
$$\frac{\rho_\gamma}{\rho_m} = \frac{\dot{\Omega}^2}{2\Omega\ddot{\Omega}} - 1 = \frac{(\frac{dR}{dt})^2}{2[(\frac{dR}{dt})^2 + R\frac{d^2R}{dt^2}]} - 1 \tag{526}$$

Now, since $\theta = 3\dot{R}/R$, we can get the same expression(521) for the ratio ρ_γ/ρ_m in terms of the expansion scalar $\theta(t)$. Again, the relation between γ and $\theta(t)$ follows from the equation(517).

From the relation (518) and (521) one can arrive at the following relations:

$$\frac{n_m}{n} = \frac{E_\gamma}{E_\gamma + \gamma(\gamma-1)m}$$
$$\frac{n_\gamma}{n} = \frac{\gamma(\gamma-1)m}{E_\gamma + \gamma(\gamma-1)m} \tag{527}$$

$$\left(\frac{\dot{E}_\gamma}{E_\gamma} + \frac{\dot{n}_\gamma}{n_\gamma}\right) - \left(\frac{\dot{m}}{m} + \frac{\dot{n}_m}{n_m}\right) = \frac{\dot{\gamma}}{\gamma}\frac{2\gamma-1}{\gamma-1} \quad for\ \gamma \neq 1 \tag{528}$$

$$\frac{\dot{E}_\gamma}{E_\gamma} + \frac{\dot{n}_\gamma}{n_\gamma} = 0 \quad for\ \gamma = 1 \tag{529}$$

and

$$\frac{2\dot{H}}{H} = \frac{2\dot{\theta}}{\theta} = \frac{\dot{\rho}}{\rho} = \frac{\dot{\gamma}}{\gamma^2} + \frac{1}{\gamma}\left(\frac{\dot{m}}{m} + \frac{\dot{n}_m}{n_m}\right) + \frac{\gamma-1}{\gamma}\left(\frac{\dot{E}_\gamma}{E_\gamma} + \frac{\dot{n}_\gamma}{n_\gamma}\right) \tag{530}$$

Consequently, we obtain the following governing equations that decide,together with equation(520), the "cosmology" of the creation era in the very early universe:

$$\frac{\dot{\rho}}{\rho} = \frac{2\dot{H}}{H} = \frac{2\dot{\theta}}{\theta} = \frac{\dot{n}}{n}(1+f(\theta)) = \frac{E_\gamma}{E_\gamma + \gamma(\gamma-1)m}\left[\frac{\dot{m}}{m} + \frac{\dot{E}_\gamma}{E_\gamma}(\gamma-1)\frac{\gamma m}{E_\gamma} + \frac{\dot{\gamma}}{\gamma}\left(2 - \frac{\gamma m}{E_\gamma}\right)\right](1+f(\theta)) \tag{531}$$

5.2: SPECIFIC ENTROPY PER BARYON

It is apparent from equations(519) and (520) [also from the equations (527) to (530)] that they possess the trivial solution

$$\dot{\rho} = \dot{\theta} = \dot{n} = \dot{m} = 0 \tag{532}$$

This solution corresponds to the inflationary stage of the very early universe. It should be noted that during this period the particle mass remains constant and after that period the mass decreases. The particle mass achieves a value of a few multiples of its present value as soon as the age of the universe is only $\alpha = 0.26 \times 10^{-23} sec$. For the scale factor $R(t) \propto exp(Dt)$ in the inflationary stage of the universe, the mass of the particle is given by

$$m = \bar{m}(1 + 2\alpha D) \simeq 2\alpha\bar{m}D \tag{533}$$

since $\alpha D >> 1$, $\bar{m}$ being the inherent mass of the particle. In fact, for the typical inflationary parameter $D \sim 10^{34} sec^{-1}$ the value of αD is of the order of 10^{11}.

For the FRW universe with Hubble parameter $H(t) = \beta/t$ the situation is quite different. In this case, the masses of the elementary particles might have been of the order of the Planck mass (or even more) around the Planck epoch so that they can achieve their present values at the present epoch of the universe or be only a few multiples of the present values at the epoch time $t = \alpha$. Thus, the creation of of such (Planck scale) massive particles in the vary early universe (say, from the Planck epoch time) ensures that the energy density ρ must be dominated by the matter density and consequently the value of γ is almost one. In other wards, these created massive particles of the "creation era" are nonrelativistic in nature and in the transition from this creation era to the radiation era around the epoch time $t = \alpha$ they become relativistic with largeγ, the value of which will be found later. Consequently, the energy density is dominated by the radiation density. It is to be noted that the pressure p_c which is responsible for the particle creation must be zero at this transition around $t = \alpha$ at which epoch time the masses of elementary particles achieve the values of the order(only few multiples) of their present values. Around this epoch, therefore, we must have $\hat{p} = p = \rho/3$ and $\dot{N}/N = \dot{S}/S = 0$, giving $\frac{\dot{n}}{n} = -\theta$ [cf. equation(496)] since $\frac{\dot{N}}{N} = \frac{\dot{n}}{n} + \theta$. Thus, the entropy production accompanying the particle creation stops and remains constant afterwards. The modified Einstein equation (496) then becomes the usual Einstein equation (497) and the usual cosmology of the radiation era follows subsequently.

For $\gamma \simeq 1$, $\rho_\gamma << \rho_m$, we have from (530) or from (519) and (520) (noting that $\frac{\dot{\gamma}}{\gamma} << \frac{\dot{m}}{m}$) the following governing equations for the creation era:

$$\frac{\dot{\rho}}{\rho} = \frac{2\dot{\theta}}{\theta} = \frac{\dot{n}}{n}(1 + f(\theta)) = \frac{\dot{m}}{m}\frac{1 + f(\theta)}{f(\theta)} \tag{534}$$

Apart from the trivial solution (532), these equations possess the following nontrivial solution:

$$f(\theta) = \frac{\alpha\theta}{3 + \alpha\theta}$$
$$n = \frac{\theta^2}{K\bar{m}(3 + 2\alpha\theta)} \tag{535}$$

For the FRW universe $H = \frac{\theta}{3} = \frac{\beta}{t}$, we have, therefore, $f(\theta) = \frac{\alpha\beta}{t+\alpha\beta}$ and for $t << \alpha$ the equation of state becomes $p = f(\theta)\rho \simeq \rho$. This corresponds to the maximally stiff equation of state (Zel'dovich,1962). The value of β in this case is 2/3 since this creation era is matter-dominated. In fact, by taking the phenomenological pressure $\hat{p} = 0$ in equation (513) we find $\theta = \frac{2}{t}$. On the contrary, when the particle creation stops, then the value of β must be close to 1/2. This value of β corresponds to the large value of γ. In fact, the relation between them is $\frac{1}{\gamma^2} = 4 - \frac{2}{\beta} + \frac{2\dot{\beta}t}{\beta}$. Thus, $f(\theta) = 1/3$ at $t = \alpha$ and $p_c = 0$, as $\hat{p} = \rho/3$ for the radiation era.

When in the transition β changes from its value 2/3 of the creation era to 1/2, then γ increases from its 'near one' value to a large value and we have the following expression of the function $f(\theta)$ in the equation of state:

$$f(\theta) = \left[\frac{E_\gamma}{E_\gamma + \gamma(\gamma-1)m}\left\{\frac{\dot{m}}{m} + \frac{\gamma(\gamma-1)m}{E_\gamma}\frac{\dot{E}_\gamma}{E_\gamma} + \frac{\dot{\gamma}}{\gamma}\left(2 - \frac{\gamma m}{E_\gamma}\right)\right\}\right] \times$$
$$\left[\frac{2\dot{\theta}}{\theta} - \frac{E_\gamma}{E_\gamma + \gamma(\gamma-1)m}\left\{\frac{\dot{m}}{m} + \frac{\gamma(\gamma-1)m}{E_\gamma}\frac{\dot{E}_\gamma}{E_\gamma} + \frac{\dot{\gamma}}{\gamma}\left(2 - \frac{\gamma m}{E_\gamma}\right)\right\}\right]^{-1} \quad (536)$$

The value of γ increases in the transition to the radiation era when the particles become relativistic. It is, in fact, our conjecture that the part of the energy density $\gamma m n_m$ due to the massive particles contributes to the radiation energy density ρ_γ and it is due to the fact that γm for large γ at the transition period around the epoch time $t = \alpha$ becomes the relativistic mass-energy. Specifically, it increases to the order of E_γ at that epoch, i.e.,

$$E_\gamma \simeq \gamma_\alpha m \quad (537)$$

where γ_α is the value of γ at $t = \alpha$. It can also be seen from (536) for large value of γ, $f(\theta)$ approaches to the value 1/3 for the radiation era at this transition epoch and epochs afterwards. The above relation (537) determines the value of γ_α from the known value of the particle mass m and E_γ at the epoch $t = \alpha$ since the usual cosmology follows subsequently. The value of E_γ at $t = \alpha$ is $10^{22} cm^{-1}$ and $m \simeq 10^{13} cm^{-1}$ for muons (in the unit $\hbar = c = k = 1$), which are taken as the "representative" particles [and, in fact, these are taken as the fundamental particles, being constituents in the hadron configuration (De,1986,1989,1997)]. These values of mass and energy per photon at the transition epoch give rise the value of γ_α as

$$\gamma_\alpha = 10^9 \quad (538)$$

Again, from the previous expressions of ρ_m and ρ_γ as well as from (5.1.33) it follows that

$$\frac{E_\gamma n_\gamma}{m n_m} = \gamma(\gamma - 1) \quad (539)$$

Now, by using equations(537) and (539) one can find the ratio of the photon to particle numbers at this transition epoch. this ratio remains constant after this transition epoch from the creation to the radiation era because the particle creation has ceased afterwards. This ratio is given by

$$\frac{n_\gamma}{n_m} = \gamma_\alpha - 1 \simeq 10^9 \quad (540)$$

The present value of the specific entropy σ per baryon can be computed from the ratio ρ_γ/ρ_m at the transition epoch $t = \alpha$. This ratio at $t = \alpha$ is

$$\frac{\rho_\gamma}{\rho_m} = \gamma_\alpha^2 - 1 \simeq 10^{18} \quad (541)$$

The expression for the specific entropy per baryon is

$$\sigma = 3.7 \frac{\rho_{\gamma_0}}{\rho_{m_0}} \frac{m_b}{T_0} \quad (542)$$

where ρ_{γ_0} and ρ_{m0} refer to the present values of the corresponding quantities, m_b and T_0 are, respectively, the baryon mass and the present universe temperature. Now, since, after the epoch $t = \alpha$, the standard cosmology follows we can use the following adiabatic constant:

$$\frac{\rho_\gamma}{\rho_m T} = constant = \frac{\rho_{\gamma 0}}{\rho_{m0} T_0} \quad (543)$$

Therefore, at $t = \alpha$,

$$\frac{\rho_\gamma}{\rho_m} = \frac{T\rho_{\gamma 0}}{\rho_{m0}T_0} = \gamma_\alpha^2 - 1 \simeq 10^{18} \tag{544}$$

Then, we have from (542) and (544) the following expression for σ:

$$\sigma = 3.7 \times 10^{18}\frac{m_b}{T} \tag{545}$$

The temperature T at the epoch can be computed by using the Einstein equation (541) and the standard relation

$$\rho_\gamma = \frac{\pi^2}{30}N_{eff}T^4 \tag{546}$$

where N_{eff} is the effective number of relativistic particle spin degrees of freedom (Blau and Guth, 1987). It can be found to be $T \simeq 10^{22}cm^{-1}$ and consequently the present specific entropy σ per baryon is computed as to be

$$\sigma = 1.76 \times 10^{10} \tag{547}$$

where we have used the proton mass as the representative mass for the baryons. This result is in good agreement with the observational facts.

Here, we point out that in the transition epoch just before $t = \alpha$ the term $\dot{\gamma}/\gamma$ begins to increase (with the increase of γ) from its very small value (which is, in fact, zero at epoch times $t << \alpha$) of the pre-transition creation era. Consequently, it is evident from equation (536) that $f(\theta)$ decreases from its value one and makes $p_0 = 0$ just before $t = \alpha$ because $\frac{\hat{p}}{\rho} = \frac{1}{3}(1 - \frac{1}{\gamma^2})$ [which follows from the equation (514) and (517)] increases from its value zero to a positive value that is equal to $f(\theta)$ at some α'(say, $\alpha' \leq \alpha$) just before α. Thus, the matter creation stops and the usual cosmology follows with increasing γ, which culminates in the values 10^9 *at* $t = \alpha$. In this very small time interval between α' *and* α, $f(\theta)$ and $\hat{p}$ may become negative since $\dot{\gamma}/\gamma$ might becomes large compared to $\dot{m}/m$ and $\dot{E}_\gamma/E_\gamma$ in this period, stimulating creation of radiation within the framework of standard cosmology. Of course, by the decays of particles or by the phase transitions the radiation energy might also be increased in this transition era. But, at $t = \alpha$ as well as in subsequent epochs, $\dot{\gamma}/\gamma$ becomes small again compared to $\dot{E}_\gamma/E_\gamma$ and $\dot{m}/m$.Then we have from equation (530) and (537), since γ is very large,

$$\frac{\dot{\rho}}{\rho} = \frac{2\dot{\theta}}{\theta} = \frac{\dot{n}}{n}(1 + f(\theta)) = \frac{\dot{E}_\gamma}{E_\gamma}\left(1 + \frac{1}{f(\theta)}\right) \tag{548}$$

with Pascal's equation of state, given by $\hat{p} = p = f(\theta)\rho = \rho/3$ for the radiation era. These equations give $n \propto t^{-3/2}$ and $E_\gamma \propto t^{-1/2}$ or, since $R(t) \propto t^{1/2}, E_\gamma \propto \frac{1}{R(t)}$.

These are the results of the standard cosmology. Thus, the maximally stiff equation of state remains valid in almost the whole particle creation era from the Planck order time to the epoch α except for the very short period of time between α' *and* α, which characterizes the "sharp" transition from this creation era to the standard FRW radiation era.

So far consequences of epoch dependence of particle mass in the early universe have been discussed. Following a phenomenological macroscopic approach proposed by Prigogine et al (1988,1989) with the introduction of thermodynamics of the open system for the early universe it is shown that the production of both particles and entropy prevalent in our universe mostly in the form of blackbody radiation is possible in this early stage of the universe. Several authors (Brout et al., 1978,1979a,b, 1980) considered earlier the particle production quantum mechanically in the framework of Einstein's equations. But because of the adiabatic and reversible nature of these equations, a huge entropy production accompanying the production of matter is not possible. The quantum production of matter (particle-creation), particularly for the Friedmann metric, was considered long ago by, e.g., Parker(1968), Grib and Mamayev(1969), and Zel'dovich and Starobinsky (1972). Similar considerations were also made by several other authors and the review by Gibbons (1979) gives a good account of these (see also Birrell and Davies, 1982). But the problem of large-scale entropy production persisted until the work of Prigogine et al. By incorporating this approach of Prigogine et al with the epoch dependence of particle masses we have arrived to the governing equations for the evolution of the early universe. The usual inflationary evolution has appeared here as the trivial solution of these equations. On the other hand, the nontrivial solution is found to be a natural possibility. This is, in fact, a FRW matter-dominated expansion phase together with an equation of state for this creation era of early universe. Due to epoch dependence of masses of subatomic particles, the particles could have masses of the order of Planck mass at the Planck order epoch times. In the following section we shall discuss how, by the quantum creation of matter in the perturbed "anisotropic" Minkowski space-time, the creation of such heavy massive matter constituents as well as the avoidance of the initial singularity are possible. More specifically, the anisotropy in the Minskowski space time occurs in a very short period of time in comparison with the Planck scale time, which signifies the "origin" of the expanding universe without a singularity (initial). In fact, Nardone(1989) has discussed the particle production quantum mechanically and by using a perturbation approach he has shown that the Minkowski space time is unstable with the created particles of heavy masses (more than 50 times of Planck mass m_{pl}), which have been regarded as black holes. Gunzig et al. (1987) and Gunzing and Nardone (1989) proposed that due to the fluctuations of conformal degrees of freedom of this initial Minkowski vacuum, de Sitter space-time might appear with the creation of matter. The black holes evaporate during this inflationary stage. The model can also account for the specific entropy per baryon of the present universe. This consideration, of course, is based on the production of heavy massive particles (masses 50 m_{pl}) only in the earlu universe.

5.3: QUANTUM CREATION OF MATTER

First we shall obtain the field equations for the isotropic space-time which is the conformal Minkowski space-time related to FRW space (in four dimensions) with flat spatial sections. The line element of that space-time has been given in (522). The mode decomposition for a scalar field ϕ in this space-time is given by

$$\phi(\tilde{x}) = \int d^3k[a_k u_k(\tilde{x}) + a_k^t u_k^*(\tilde{x})] \tag{549}$$

where the modes u_k can be written in the following separated form (Birrell and Davice, 1982):

$$u_k(\tilde{x}) = (2\pi)^{-3/2} e^{i\vec{k}.\vec{x}} \Omega^{-1}(\eta)\chi_k(\eta) \tag{550}$$

Here,η is the conformal time parameter which is related to the cosmological proper time t by the relation (523) and $k = |\vec{k}|$. Also, χ_k satisfies the following equation

$$\frac{d^2\chi_k}{d\eta^2} + (k^2 + \Omega^2(\eta)[m^2 + (\xi - \frac{1}{6})R(\eta)])\chi_k = 0 \tag{551}$$

with the normalization condition

$$\chi_k \partial_\eta \chi_k^* - \chi_k^* \partial_\eta \chi_k = i \tag{552}$$

The above equations are in fact, derived from the following Lagrangian density

$$\text{Ł}(\tilde{x}) = \frac{1}{2}[-g(\tilde{x})]^{1/2}(g^{\mu\nu}(\tilde{x})\phi(\tilde{x})_{,\mu}\phi(\tilde{x})_{,\nu} - [m^2 + \xi R(\tilde{x})]\phi^2(\tilde{x}) \tag{553}$$

with the resulting action

$$S = \int \text{Ł}(\tilde{x}) d^4x \tag{554}$$

where m is the mass of the scalar. Here, $R(\tilde{x})$ represents the Ricci scalar and ξ is a numerical factor that represents the nature of the coupling between the scalar and the gravitational fields. The variation of the action S with respect to the scalar field ϕ is demanded by the action principle and this can give rise to the following field equation for the scalar field ϕ :

$$[\,\Box\; + m^2 + \xi R(\tilde{x})\,]\phi(\tilde{x}) = 0 \tag{555}$$

where

$$\Box\,\phi = (-g)^{-\frac{1}{2}}\partial_\mu[(-g)^{\frac{1}{2}} g^{\mu\nu}\partial_\nu\phi] \tag{556}$$

with $\eta = x^0$.

We shall consider the case with $\xi = 1/6$, which is the particularly interesting case of conformal coupling. In this case, we have the equation for χ_k to be the following form:

$$\frac{d^2\chi_k}{d\eta^2} + \omega_k^2(\eta)\chi_k = 0 \tag{557}$$

where

$$\omega_k^2(\eta) = k^2 + m^2\Omega^2(\eta) \tag{558}$$

Now, in the following we shall consider an anisotropic perturbation (of extremely short duration) in the Minkowski space-time and find out the energy density of the matter created quantum mechanically with the very massive particles (of masses $\geq 50m_{pl}$) around the Planck time. As pointed out above, due to epoch dependence of particle mass, the masses of subatomic particles could have been of the order of more than 50 times the Planck mass around the Planck order epoch times. In fact, the mass of a particle could have been 53.3 times m_{pl} at the epoch time $\hat{t} = 0.05\ t_{pl}$. This result is obtained from the mass relation (378) with the Hubble parameter $H = \frac{2}{3t}$ for the FRW matter- dominated expansion phase which begins at that Planck order time due to the instability caused by these heavy massive particles, as shown by Nardone (1989). The conformal factor $\Omega(\eta)$ for this period of matter-dominated FRW expansion phase after the Planck order time is given by

$\Omega(\eta) = A^3(\frac{\eta}{\tau})^2$

where A is an appropriate constant and $\tau = 2H_0^{-1}$, H_0 being the present value of H. Now, let us specify the conformal factor $\Omega(\xi)$ before a conformal time parameter $\hat{\xi}$ corresponding to the Planck order time $\hat{t} = 0.05t_{pl}$. Since the space-time is an anisotropically perturbed Minkowski space-time, we have

$$\Omega(\xi) == B^3(\frac{\hat{\xi}}{\tau})^2\ for\ \xi \leq \hat{\xi} \tag{559}$$

After the epoch time $\hat{t}$, the conformal factor is

$$\Omega(\eta) == A^3(\frac{\eta}{\tau})^2\ for\ \eta \geq \hat{\eta} \tag{560}$$

where $\hat{\eta}$ is the conformal time parameter corresponding to the epoch time $\hat{t}$. The relation between $\hat{\eta}$ and $\hat{\xi}$ as well as that between the two constants A and B follow from the continuity of the conformal factors (also of the corresponding FRW scale factor) at the epoch $\hat{t}$. These are given by

$$\hat{\eta} = 3\hat{\xi} \tag{561}$$

and

$$3^{2/3}A = B \tag{562}$$

The epoch time and the conformal time parameters are related by

$$A\eta = (3\tau^2 t)^{1/3} \tag{563}$$

and

$$\xi = \frac{t}{B}(\frac{\tau}{\hat{t}})^{2/3} = \frac{t}{A}(\frac{\tau}{3\hat{t}})^{2/3} \tag{564}$$

Now, the line element for the anisotropically perturbed Minkowski space-time before the epoch time $\hat{t}$ can be written as

$$ds^2 = \Omega^2(\xi)[d\xi^2 - \Sigma_{i=1}^3(1 + h_i(\xi))(dx^i)^2] \tag{565}$$

with $\Omega(\xi)$ given by (559) and the perturbation functions $h_i(\xi)$ $(i = 1, 2, 3)$ are specified as follows:

$$h_i(\xi) = e^{-\alpha\xi^2} cos(\beta\xi^2 + \delta_i) \tag{566}$$

where α, β *and* δ_i are all constants. These constants are related with the duration of the anisotropic perturbation and will be specified later. Now, if we consider a scalar field $\phi(\tilde{x})$ in this anisotropically perturbed Minkowski space time, the mode decomposition for this field will be the same as that for the case of isotropic space-time. In fact, it is given by (549) with the separated modes u_k given as in (550). But the equation for χ_k given by (551) is to be modified for this anisotropically perturbed Minkowski space-time. For conformal coupling between the scalar and the gravitational fields the equation for χ_k has the following form:

$$\frac{d^2\chi_k}{d\xi^2} + [k^2 + m^2\Omega^2(\xi) - \Sigma_{i=1}^3 h_i(\xi)k_i^2]\chi_k = 0 \tag{567}$$

retaining only the first order terms in h_i.
We shall follow the method of solution for this equation as given in Birrell and Davies (1980) who developed the original method of Zel'dovich and Starobinsky (1972,1977) for the case of small perturbations about a FRW space time. Apart from taking $h_i (i = 1, 2, 3)$ to be very small compared to unity, we can impose, for simplicity, the following condition

$$\Sigma_{i=1}^3 h_i(\xi) = 0 \tag{568}$$

which gives rise to the fact that δ_i must differ from one another by $2\pi/3$. Also, the following conditions of this approximation method, i.e.,

$$\begin{aligned} h_i(\xi) &\longrightarrow 0 \; as \; \xi \longrightarrow \pm\infty \\ and \; \Omega^2(\xi) &\longrightarrow \Omega^2(\infty) = \Omega^2(-\infty) < \infty \\ & as \; \xi \longrightarrow \pm\infty \end{aligned} \tag{569}$$

are clearly satisfied. Then the normalized positive frequency solution of (567) as $\xi \longrightarrow -\infty$ is clearly given by

$$\chi_k^{in}(\xi) = (2\omega)^{-\frac{1}{2}} e^{-i\omega\xi} \tag{570}$$

where

$$\omega^2 = k^2 + m^2\Omega^2(\infty) \tag{571}$$

Now, with the initial condition the equation (567) may be transformed into the following integral equation:

$$\chi_k(\xi) = \chi_k^{in}(\xi) + \omega^{-1} \int_{-\infty}^{\xi} V_k(\acute{\xi}) sin[\omega(\xi - \acute{\xi})]\chi_k(\acute{\xi})d\acute{\xi} \tag{572}$$

where

$$V_k(\xi) = \Sigma_{i=1}^3 h_i(\xi)k_i^2, \; since \; \Omega(\xi) = \Omega(\pm\infty) = B^3(\frac{\hat{\xi}}{\tau})^2 \tag{573}$$

The equation (572) possesses the following solution in the late time region:

$$\chi_k^{out}(\xi) = \alpha_k\chi_k^{in}(\xi) + \beta_k\chi_k^{in*}(\xi) \tag{574}$$

where α_k *and* β_k are the Bogolubov coefficients. For the present case these coefficients are given by

$$\alpha_k = 1 + i \int_{-\infty}^{\infty} \chi_k^{in*}(\xi)V_k(\xi)\chi_k(\xi)d\xi \tag{575}$$

$$\beta_k = -i\int_{-\infty}^{\infty} \chi_k^{in}(\xi)V_k(\xi)\chi_k(\xi)d\xi \tag{576}$$

For small $V_k(\xi)$ one can solve (572) by iteration. The Bogolubov coefficients to first order in $V_k(\xi)$ can be found to be

$$\alpha_k = 1 + \frac{i}{2\omega}\int_{-\infty}^{\infty} V_k(\xi)d\xi \tag{577}$$

$$\beta_k = -\frac{i}{2\omega}\int_{-\infty}^{\infty} e^{-2i\omega\xi}V_k(\xi)d\xi \tag{578}$$

since $\chi_k(\xi) = \chi_k^{in}(\xi)$ to lowest order in $V_k(\xi)$. The energy density per unit proper volume is related to the Bogolubov coefficient β_k. It is given by

$$\rho = \frac{1}{(2\pi)^3\Omega^4}\int |\beta_k|^2\omega d^3k \tag{579}$$

From the expression of $V_k(\xi)$ given in (5573), it is evident that in the present case the contribution to the energy density arises out only from the anisotropy of the space-time, that is, from the functions $h_i(\xi)$. This energy density for the functions h_i given by (566) is (Birrell and Davies, 1982)

$$\rho = \frac{\tilde{m}^2}{1536\pi^{\frac{1}{2}}\Omega^4}\frac{(\alpha^2+\beta^2)^{\frac{3}{2}}}{\alpha^2}e^{-3\alpha\frac{\tilde{m}^2}{\alpha^2+\beta^2}}W_{-\frac{3}{2},\frac{3}{2}}\left(\frac{2\alpha\tilde{m}^2}{\alpha^2+\beta^2}\right) \tag{580}$$

where W is a Whittaker function and

$$\tilde{m} = \Omega(\pm\infty)m = B^3(\frac{\hat{\xi}}{\tau})^2 m = A(\frac{3\hat{t}}{\tau})^{2/3}m \tag{581}$$

Here m is the mass of the particle at the Planck order epoch time $\hat{t}$ and according to the mass formula (378), it is given by

$$m \simeq \lambda\tau\bar{m}H(\hat{t}) = 2\alpha\bar{m}H(\hat{t}) \tag{582}$$

where $\bar{m}$ is the present mass of the particle and $H(t) = \frac{2}{3t}$, which is the Hubble parameter for the matter dominated FRW universe corresponding to the conformal factor (560).
We can, now, use an integral representation of the Whittaker function (Erde'lyi, 1953) to find the following expression for the energy density:

$$\rho = \frac{\tilde{m}^6}{384\pi^{\frac{1}{2}}\Omega^4}\frac{1}{(\alpha^2+\beta^2)^{\frac{1}{2}}}\left(\Gamma(\frac{7}{2})\right)^{-1}e^{-\frac{4\alpha\tilde{m}^2}{\alpha^2+\beta^2}}\int_0^{\infty} e^{-\frac{2\alpha\tilde{m}^2 t}{\alpha^2+\beta^2}}t^{\frac{5}{2}}(1+t)^{-\frac{1}{2}}dt \tag{583}$$

We shall here consider the creation of known elementary particles in that 'era' and the muons are taken as the 'representative' particles. In fact, in earlier consideration (De, 1986) muon has been taken as one of the fundamental constituents of hadrons. The mass m of a muon at the epoch time $\hat{t}$ can be calculated by using the formula (582). In the unit $\hbar = c = k = 1$, by setting $G = \frac{1}{8\pi m_{pl}^2}$ and $t_{pl}^2 = G$, we find the present mass of a muon to be $5.37 \times 10^{12}cm^{-1}$ and $m \simeq 55.5m_{pl}$ where $m_{pl} = 1.22 \times 10^{32}cm^{-1}$.
Then from (581) we have

$$\tilde{m} = 1.978 \times 10^{-40}m_{pl}.A \tag{584}$$

Also,

$$\Omega(\hat{\xi}) = A(\frac{3\hat{t}}{\tau})^{2/3} = 3.569 \times 10^{-42}A \tag{585}$$

Let us set

$$\alpha = A^2 m_{pl}^2\delta^2$$
$$and\ \beta = p\alpha \tag{586}$$

where δ and p are constants yet to be determined. Then, we have

$$\rho = \frac{5.05\times10^{-76}m_{pl}^4}{\pi\delta^2(1+p^2)^{\frac{1}{2}}}\, e^{-\frac{1.56\times10^{-79}}{(1+p^2)\delta^2}} \int_0^\infty e^{-\frac{7.8\times10^{-80}t}{(1+p^2)\delta^2}}\, t^{\frac{5}{2}}(1+t)^{-\frac{1}{2}}dt \tag{587}$$

Let us put

$$\delta^2 = 7.8\times10^{-80}\frac{a^2}{1+p^2} \tag{588}$$

where a is a constant. With this, the matter energy density becomes

$$\rho = 2.06\times10^3(1+p^2)^{1/2}a^5 m_{pl}^4 e^{-\frac{2}{a^2}} \int_0^\infty \frac{e^{-x}x^{5/2}}{\sqrt{1+a^2x}}dx \tag{589}$$

It should be noted that the values of the constants a and p determine those of α and β; that is, the 'duration' and 'frequency' of the anisotropic fluctuation are determined by these constants a and p. One possible choice is $a = 2, p = 1$.
With these values of the constants and by numerical evaluation of the integral on the R.H.S. of (589) one can find the energy density ρ at the epoch time $\hat{t}$. It is given as

$$\rho = 2.8\times10^4 m_{pl}^4 \tag{590}$$

After the epoch $\hat{t}$ the expansion phase of the universe begins due to the instability caused by the very massive particles In fact, Nordone (1989) has shown that instability of Minskowski space-time corresponding to a global fluctuation $\delta(t)$, that is, for the scale factor $a(t)$ given by

$$a(t) = 1+\delta(t), \tag{591}$$

shows up as soon as

$$km^2 \geq 288\pi^2 \tag{592}$$

$$(k = 8\pi G)$$

$$or\ m \geq 53.3 m_{pl} \tag{593}$$

As discussed in the previous section, after the Planck order time the universe goes through the matter dominated FRW expansion upto the transition epoch $\alpha = 0.26\times10^{-23}sec.$when the particles become relativistic and contribute to the radiation density. After this transition epoch the radiation era with the usual cosmology begins.
Also, the energy density is given by

$$\rho = \left(\frac{\hat{t}}{t}\right)^2 \rho(\hat{t}) \tag{594}$$

and this energy density at the transition epoch α is fully contributed from the radiation energy density at that epoch.
Thus, $\rho(\alpha) = (\frac{\hat{t}}{t})^2\rho(\hat{t})$ where $\rho(\hat{t})$ is given by (590). By using the value of m_{pl}, one can find

$$\rho(\alpha) = 6.14\times10^{90}cm^{-4} \tag{595}$$

Also, the energy density at the epoch time $t = 10^{-23}sec$ is found to be

$$\rho = 3.84\times10^{89}cm^{-4} \tag{596}$$

Using the standard relation (546) one can find from (595) the universe temperature at the epoch time $t = 10^{-23}sec$. It is given by

$$T = 2.75\times10^{22}cm^{-1} \tag{597}$$

These results are in good agreement with the results that are calculated from the standard cosmology. Now,the perturbation functions $h_i(\xi)$ are specified for the values of parameters given in (586) and (588) with $a = 2$ and $p = 1$, as given above. They are given by

$$h_i(\xi) = e^{-\left(\frac{t}{0.045t_{pl}}\right)^2} \cos\left[\left(\frac{t}{0.045t_{pl}}\right)^2 + \delta_i\right] = e^{-\left(\frac{t}{0.9\hat{t}}\right)^2} \cos\left[\left(\frac{t}{0.9\hat{t}}\right)^2 + \delta_i\right] \tag{598}$$

The amplitudes of the perturbation functions are damped for $t >> 0.9\hat{t}$. That is, the anisotropic fluctuation of the Minkowski space-time dies out after the epoch time $\hat{t}$. The fluctuation is dominant for $|t| < 0.9\hat{t}$. This dominant 'duration' of the 'fluctuation era' depends on the values of the parameters a and p which also determine the energy density. Thus, the duration is related with the energy density. The duration of the anisotropically perturbed Minkowski space-time for such a short time, as described above, can produce particles (with masses $55.5m_{pl}$) that give rise matter energy density at the planck order epoch time $\hat{t}$. This energy density at $\hat{t}$ determines the radiation energy density at the transition epoch α or at $t = 10^{-23}sec.$ (and consequently at later epochs of the universe) correctly. It is apparent that the 'zero' epoch (the cosmological time- origin) of the present universe, given by $\xi = 0 = t$ corresponds to the maximum amplitudes of the perturbation functions and the universe is free from any initial singularity.

5.4: PARTICLE CREATION AFTER THE PLANCK ORDER TIME

In sections 5.1 and 5.2, we have considered matter creation phenomenological for the period after the Planck order time up to the transition epoch α. Now, the quantum creation of particles for this era will be considered [se also De (2001)]. It was shown that this era is a matter dominated FRW expansion state of the universe with the conformal factor given by (560). The scale factor for this era is

$$R(t) = A\left(\frac{3t}{\tau}\right)^{2/3} \tag{599}$$

We use the mass formula (378), which is given as

$$m \simeq 2\alpha\, \bar{m}\, H(t) \qquad for \quad t \leq \alpha \tag{600}$$

In fact, (600) is an approximation of (378) and is valid up to the epoch time $\leq \alpha$ In terms of the conformal time η , (5.4.2) can be written as

$$m \simeq \frac{4\alpha\bar{m}\tau^2}{A^3\eta^3} \qquad for \quad \eta \leq \eta_\alpha \tag{601}$$

where η_α is the conformal time parameter corresponding to the epoch time α. Then we can find

$$m\Omega(\eta) \simeq \frac{4\alpha\bar{m}}{\eta} \qquad for \quad \hat{\eta} < \eta < \eta_\alpha \tag{602}$$

Now, the equation (557) becomes

$$\frac{d^2\chi_k}{d\eta^2} + \omega_k^2(\eta)\chi_k = 0 \tag{603}$$

where

$$\omega_k^2(\eta) = k^2 + \frac{16\alpha^2\bar{m}^2}{\eta^2} \qquad for \quad \eta \geq \hat{\eta} \tag{604}$$

with the normalization condition for χ_k given by (552). For nonconformal particle creation, that is, for the quantized scalar field with arbitrary coupling in a nonstationary isotropic gravitational field, the equation for χ_k follows from (551) which can be written as follows:

$$\frac{d^2\chi_k}{d\eta^2} + Q^2(\eta)\chi_k = 0 \tag{605}$$

where

$$Q^2(\eta) = \omega_k^2(\eta) + q(\eta) \ for \ \eta \geq \hat{\eta} \tag{606}$$

with

$$q(\eta) = 6\left(\xi - \frac{1}{6}\right)\frac{\Omega''(\eta)}{\Omega(\eta)} \tag{607}$$

For both the cases the nonmalized positive frequency solution for χ_k as $\eta \longrightarrow -\infty$ is given by

$$\chi_k^{in}(\eta) = (2\omega)^{-1/2}e^{-i\omega\eta} \ where \ \omega^2 = k^2 + \frac{16\alpha^2\bar{m}^2}{\hat{\eta}^2} \tag{608}$$

[In fact, $\omega_k^2(\eta) = k^2 + \frac{16\alpha^2\bar{m}^2}{\hat{\eta}^2} = \omega^2 \ for \ \eta \leq \hat{\eta}$, since the space-time is Minkowskian before the conformal time $\hat{\eta}$ (c.f. equation (5.3.11)). In this case $q(\eta) = 0$ and consequently $Q^2(\eta) = \omega_k^2(\eta) = \omega^2$]
The equations (603) and (605) for the conformal and nonconformal cases respectively can be written as the following integral equation:

$$\chi_k(\eta) = \chi_k^{in}(\eta) + \frac{1}{\omega}\int_{\hat{\eta}}^{\eta}[sin\omega(\eta - \xi)]V(\xi)\chi_k(\xi)d\xi \tag{609}$$

Here, $V(\eta)$ for the two cases is given, respectively, as

$$V(\eta) = m^2[\Omega^2(\hat{\eta}) - \Omega^2(\eta)] = 16\alpha^2\bar{m}^2\left(\frac{1}{\hat{\eta}^2} - \frac{1}{\eta^2}\right) \tag{610}$$

and

$$V(\eta) = 16\alpha^2\bar{m}^2\left(\frac{1}{\hat{\eta}^2} - \frac{1}{\eta^2}\right) - 12\left(\xi - \frac{1}{6}\right)\frac{1}{\eta^2} \tag{611}$$

In the late time region the above equation (609) possesses the solution (Birrell and Davies, 1982)

$$\chi_k^{out}(\eta) = \alpha_k\chi_k^{in}(\eta) + \beta_k\chi_k^{in*}(\eta) \tag{612}$$

where the Bogolubov coefficients α_k and β_k are given by

$$\begin{aligned}\alpha_k &= 1 + i\int_{\hat{\eta}}^{\infty}\chi_k^{in*}(\eta)V(\eta)\chi_k(\eta)d\eta \\ \beta_k &= -i\int_{\hat{\eta}}^{\infty}\chi_k^{in}(\eta)V(\eta)\chi_k(\eta)d\eta\end{aligned} \tag{613}$$

From (612) we can find, by using (608),

$$\frac{i}{\omega}\frac{d\chi^{out}(\eta)}{d\eta} = \alpha_k\frac{e^{-i\omega\eta}}{(2\omega)^{1/2}} - \beta_k\frac{e^{+i\omega\eta}}{(2\omega)^{1/2}}$$

Consequently, it follows that

$$\chi_k^{out}(\eta) - \frac{i}{\omega}\frac{d\chi^{out}(\eta)}{d\eta} = 2\beta_k\frac{e^{+i\omega\eta}}{(2\omega)^{1/2}} \tag{614}$$

Therefore, we get

$$\frac{2}{\omega}|\beta_k|^2 = |\chi_k^{out}(\eta)|^2+\frac{1}{\omega^2}|\frac{d\chi_k^{out}(\eta)}{d\eta}|^2+\frac{i}{\omega}[\chi_k^{out}(\eta)\frac{d\chi_k^{out*}(\eta)}{d\eta}-\chi_k^{out*}(\eta)\frac{d\chi_k^{out}(\eta)}{d\eta}] = |\chi_k^{out}(\eta)|^2+\frac{1}{\omega^2}|\frac{d\chi_k^{out}(\eta)}{d\eta}|^2+\frac{i^2}{\omega}$$

[because of the normalization condition (552)]
Thus, we arrive at the following expression for $|\beta_k|^2$:

$$|\beta_k|^2 = \frac{1}{2\omega}[|\frac{d\chi_k^{out}(\eta)}{d\eta}|^2 + \omega^2|\chi_k^{out}(\eta)|^2] - \frac{1}{2} \tag{615}$$

This formula may be compared with the spectral quasiparticles density, related to unit volume, given by the squared modulus of the Bogoliubov transformation coefficient in the diagonalization procedure of the quantized field Hamiltonian in creation- annihilation operators (Grib et al. 1994).
Now, the correctly normalized exact solutions of the equations (603) and (605) for χ_k can be written as

$$\chi_k(\eta) = \frac{1}{2}e^{-i\nu\pi/2}(\pi\eta)^{1/2}H_\nu^{(2)}(k\eta) \tag{616}$$

where

$$\nu^2 = \frac{1}{4} - 16\alpha^2\bar{m}^2 \tag{617}$$

and

$$\nu^2 = \frac{1}{4} - 16\alpha^2\bar{m}^2 - 12(\xi - \frac{1}{6}) \tag{618}$$

respectively, that is, for the conformal and nonconformal cases. Here, $H_\nu^{(2)}$ is a Hankel function of the second kind. That this solution (616) is normalized according to (552)can be verified by using the following formula

$$H_\nu^{(2)}(x)\frac{d}{dx}H_\nu^{(2)*}(x) - H_\nu^{(2)*}(x)\frac{d}{dx}H_\nu^{(2)}(x) = \frac{4ie^{i\nu\pi}}{\pi x}$$

where $\nu = ib$ is purely imaginary.This formula, in fact, can be deduced from the Wronskian of the pair of Hankel functions $H_\nu^{(1)}(x)$ and $H_\nu^{(2)}(x)$, which is equal to $-4i/\pi x$ and by using the following integral representations of $H_\nu^{(1)}(x)$ and $H_\nu^{(2)}(x)$:

$$H_\nu^{(1)}(x) = \frac{2e^{-\nu\pi i/2}}{\pi i}\int_0^\infty e^{ixcht}ch\ \nu t dt \tag{619}$$

$$H_\nu^{(2)}(x) = -\frac{2e^{\nu\pi i/2}}{\pi i}\int_0^\infty e^{-ixcht}ch\ \nu t dt \tag{620}$$

for $-1 < Re\ \nu < 1\ ,\ \ x > 0$ (Gradshteyn and Ryzhik, 1980).
It is to be noted that the solution (616) is consistent with the condition that is to be satisfied for an adiabatic vacuum to be a reasonable definition of a no particle state as $\eta \to \pm\infty$. This condition is, in fact, that the A th order adiabatic approximation $\chi_k^{(A)}$ to χ_k should satisfy the following condition:

$$\chi_k^{(A)} \longrightarrow \frac{1}{(2k)^{\frac{1}{2}}}e^{-ik\eta} \tag{621}$$

for large k or η (for details see Birrell and Davies, 1982). It is pointed out above that $\nu = ib$ is purely imaginary. In fact, by using the present mass of a muon we find $\alpha\bar{m} = 0.416$ (in the unit $c = \hbar = 1$) and consequently from (617) and (618), it follows that for conformal and nonconformal cases respectively

$$\nu^2 = -2.519 \tag{622}$$

and

$$\nu^2 = -0.519 - 12\xi < 0\ for\ \xi \geq 0 \tag{623}$$

From the solution (616) for $\chi_k(\eta)(\eta \geq \hat{\eta} > 0)$ one can find $\chi_k^{out}(\eta)$ which is given as

$$\chi_k(\eta) \longrightarrow \chi_k^{out}(\eta) \ \ as \ \ \eta \ \longrightarrow \infty \tag{624}$$

by using asymptotic formula for the Hankel function $H_\nu^{(2)}(x)$,which is

$$H_\nu^{(2)}(k\eta) \sim \sqrt{\frac{2}{\pi k\eta}} e^{-i(k\eta - \frac{\nu\pi}{2} - \frac{\pi}{4})} \qquad for \ \ large \ \ |k\eta| \tag{625}$$

By using this solution for $\chi_k^{out}(\eta)$ into (615) we can find out for the conformal case the following expression for $|\beta_k|^2$:

$$|\beta_k|^2 = \frac{(\omega - k)^2}{4\omega k} \tag{626}$$

It is interesting to note that $|\beta_k|^2$ is independent of the value of ν or b and thus for nonconformal case we have the same expression (626) for $|\beta_k|^2$. Also, we see that $|\beta_k|^2$ vanishes as k becomes large because $\omega \longrightarrow k$ for large k.

Now, the number of particles created in the Lagrange volume by the gravitational field until the moment η is given by

$$N(\eta) = \Omega^3(\eta)n(\eta) \tag{627}$$

where the number density $n(\eta)$ is given by

$$n(\eta) = \frac{1}{2\pi^2\Omega^3(\eta)} \int_0^\infty k^2 dk |\beta_k|^2 \tag{628}$$

Here, the conformal factor $\Omega(\eta) \equiv R(t)$ is given by (560) and (599). We can now calculate the total particle number N for large η, which is given by

$$N = \frac{1}{2\pi^2} \int_0^\infty k^2 dk |\beta_k|^2 \ for \ \ large \ \eta \tag{629}$$

As, for large η, $|\beta_k|^2$ is given by (626) we have

$$N = \frac{1}{2\pi^2} \int_0^\infty \frac{k}{4\omega}(\omega - k)^2 dk \tag{630}$$

where ω is given in (608). The integral can be computed with the substitution $k = \frac{4\alpha\bar{m}}{\hat{\eta}} shx$. Then

$$\begin{aligned} N = \frac{1}{16\pi^2}\left(\frac{4\alpha\bar{m}}{\hat{\eta}}\right)^3 \int_0^\infty (e^{-x} - e^{-3x})dx \\ = \frac{1}{24\pi^2}\left(\frac{4\alpha\bar{m}}{\hat{\eta}}\right)^3 \\ = \frac{8}{3\pi^2}\left(\frac{\alpha\bar{m}}{\hat{\eta}}\right)^3 \end{aligned} \tag{631}$$

Now, from (563), it follows that

$$A\hat{\eta} = (3\tau^2\hat{t})^{1/3} \ and \ A\eta_\alpha = (3\tau^2\alpha)^{1/3} \tag{632}$$

Consequently,

$$\frac{\hat{\eta}}{\eta} = \left(\frac{\hat{t}}{\alpha}\right)^{1/3} = 10^{-7}(since \ \hat{t} = 0.05t_{bl} \tag{633}$$

Also,

$$R(\alpha)=\Omega(\eta_\alpha)=A^3\left(\frac{\eta_\alpha}{\tau}\right)^2=\frac{3\alpha}{\eta_\alpha}=\frac{3\alpha}{\hat{\eta}}\frac{\hat{\eta}}{\eta_\alpha}=\frac{3\alpha}{\hat{\eta}}\left(\frac{\hat{t}}{\alpha}\right)^{1/3} \qquad (using\ (632))$$

Therefore,

$$\frac{1}{\hat{\eta}^3}=R^3(\alpha)(\frac{\alpha}{\hat{t}})\frac{1}{27\alpha^3} \tag{634}$$

Using this relation we have from (631) the total particle number N as

$$N=\frac{8}{81\pi^2}(\frac{\alpha}{\hat{t}})\frac{(\alpha\bar{m})^3}{\alpha^3}R^3(\alpha) \tag{635}$$

We are considering here the expansion phase of the early universe (after the epoch time $\hat{t}$) governed by the conformal factor $\Omega(\eta)$ as given in (560), which corresponds to a matter-dominated FRW universe with the scale factor $R(t)$ given by (599).It was discussed in section 5.2 that after the epoch $t=\alpha$ the universe transits to a radiation-dominated FRW expansion with no particle production. That is, the standard cosmology follows after this transition epoch α. In a subsequent section we shall consider the early universe in the framework of modified general relativity originally proposed by Rastall (1972) and later by Al-Rawaf and Taha (1996a,b).There it will be shown that the evolution in the era $(\hat{t},\alpha)$ is a mild inflation with the scale factor proportional to the epoch time. This mild inflation is also shown to be turned off automatically and transited into the radiation era with the usual cosmology after the epoch time α. In this connection it is also to be noted that the expression for $m\Omega(\eta)$ given in (602) is valid upto the epoch time α which corresponds to the conformal time parameter η_α. Again, it will now be shown that after the epoch time α, when the universe is radiation-dominated FRW universe with the conformal factor

$$\Omega(\eta)=b\eta \tag{636}$$

(This corresponds to the scale factor $R(t)=(2bt)^{1/2}, t$ being the cosmological time) there is no quantum creation of particles by the gravitational field. For this it will be convenient to employ the Hamiltonian diagonalization procedure. If we consider the conformal particle production then the formula for $|\beta_k|^2$ is given by

$$|\beta_k|^2=\frac{1}{2\omega_k(\eta)}\left(|\partial_\eta\chi_k|^2+\omega_k^2(\eta)|\chi_k|^2\right)-\frac{1}{2} \tag{637}$$

On the other hand, if nonconformal particle production is considered then the same formula (637) for the spectral quasiparticle density given by $|\beta_k|^2$ remains valid with a 'changed' concept of particles in this nonconformal case (Bezerra et al., 1997). There, in fact, $|\beta_k|^2$ is equated to R.H.S. of (637) by definition with the "switched off" external field in the general expression for it in the case $\xi\neq\frac{1}{6}$ and, of course, with the replacement of $\omega_k(\eta)$ by $Q(\eta)$. Thus, for nonconformal case

$$|\beta_k|^2=\frac{1}{2Q(\eta)}\left(|\partial_\eta\chi_k|^2+Q^2(\eta)|\chi_k|^2\right)-\frac{1}{2} \tag{638}$$

For the conformal (636), we have $q(\eta)=0$ and hence $Q^2(\eta)=\omega_k^2(\eta)=k^2+\bar{m}^2(b\eta+\frac{2\alpha}{\eta})^2$ (by using the mass formula(378)). Consequently, for $\eta\geq\eta_\alpha$ we have

$$Q^2(\eta)=\omega_k^2(\eta)\simeq k^2+\bar{m}^2b^2\eta^2 \tag{639}$$

(In fact, at $\eta=\eta_\alpha,\ b\eta_\alpha=\frac{2\alpha}{\eta_\alpha}$ *and as* $\eta>\eta_\alpha, b\eta>\frac{2\alpha}{\eta_\alpha}$) Thus we get the same equation for χ_k for both the conformal and nonconformal cases. Also, the expression for $|\beta_k|^2$ is the same for these cases. Now, in Birrell and Davies (1982) the exact solution for χ_k for the case with $\omega_k(\eta)$ given by (639) has been given. Also, for large $\eta, \chi_k(\eta)$ has been found to be

$$\chi_k(\eta)=(2\bar{m}b|\eta|)^{-\frac{1}{2}}e^{-\frac{imb\eta^2}{2}} \tag{640}$$

$$(\eta > 0, \ therefore \ |\eta| = \eta \ , \ \eta \ large \ )$$

$$\partial_\eta \chi_k(\eta) = \frac{-1}{(2\bar{m}b|\eta|)^{1/2}} \left[i\bar{m}b\eta + \frac{1}{2\eta} \right] e^{-\frac{i \bar{m} b\eta^2}{2}}$$

Therefore, for large η,

$$|\beta_k|^2 = \frac{1}{16(\bar{m}b)^2\eta^4} \longrightarrow 0$$

That is, there is no massive particle creation in this radiation era after the conformal time paramerer η_α or equivalently after the epoch time α. Thus, the particle creation occurs only in $(\hat{t}, \alpha)$. In the consideration of particle creation in the period $(\hat{t}, \alpha), \chi(\eta)$ for large η has been regarded as $\chi_k^{out}(\eta)$. In fact, around the conformal time parameter η_α (that is, around the cosmological time α), the asymptotic formula for the Hankel function can be applicable because of the fact that

$$k\eta_\alpha = k\hat{\eta} \left(\frac{\eta_\alpha}{\hat{\eta}} \right) \simeq \bar{m}\alpha \left(\frac{\eta_\alpha}{\hat{\eta}} \right) = 0(10^7)$$

Now, the scale factor in (635) can be obtained by using the cosmological invariant

$$RT = R_0T_0 = 1.18 \times 10^{29}u, \ 1 < u < 1.8 \tag{641}$$

where R_0 *and* T_0 are the present scale factor and the temperature of the universe respectively. They are given by

$$\begin{aligned} R_0 &= 10^{28}u \ cm, \\ T_0 &= 11.8cm^{-1} \end{aligned} \tag{642}$$

This cosmological invariant is valid after the epoch time α since after that epoch the standard cosmology follows. Now, as the universe temperature T around the epoch time $10^{-23}sec$ is $T \simeq 10^{22} \ cm^{-1}$ (see also the calculated value of the previous section), $R(10^{-23}sec)$ is given by $R(10^{-23}sec) \simeq 1.18 \times 10^7u \ cm$ and consequently

$$R(\alpha) \simeq 5.9 \times 10^6u \ cm \tag{643}$$

Using this value of $R(\alpha)$ and those of $\alpha\bar{m}, \frac{\hat{t}}{\alpha}$(from (633)) one can easily find, from (635), the total particle number N of the particles created from the gravitational field. It is given by

$$3.12 \times 10^{77} < N < 1.82 \times 10^{78} \tag{644}$$

This particle number at the epoch α is, in fact, the total particle number at the present epoch of the universe as it remains constant after that epoch α. This number is comparable with the baryon number of the present universe, as obtained from the observational facts. Interestingly, one can compute the photon-to-baryon ratio at the epoch α by using the radiation density at that epoch from (594). In fact, by using the following standard relations

$$\begin{aligned} \rho_\gamma &= \frac{\pi^2}{15}T^4 \\ n_\gamma &= \frac{\pi}{13}T^3 \end{aligned} \tag{645}$$

one can compute the universe temperature and photon number density at the epoch α. These are given by

$$\begin{aligned} T &= 5.52 \times 10^{22}cm^{-1} \\ n_\gamma &= 4.08 \times 10^{67}cm^{-1} \end{aligned} \tag{646}$$

From (643) and (644) we can find particle number density n_m to be

$$n_m = 1.52 \times 10^{57}cm^{-3} \tag{647}$$

Consequently, we have at the epoch time α

$$\frac{n_\gamma}{n_m} = 2.68 \times 10^{10} \tag{648}$$

As standard cosmology follows after the epoch α, this ratio remains constant afterwards and gives its present value.

5.5: PARTICLES WITH VERY LARGE MASS

We have discussed in previous sections the creation of heavy particles in the classical nonstationary space-time (the anisotropic fluctuations of Minkowski space time, and RW space-time).The quantum creation of particles from vacuum in the gravitational field is regarded as the 'creation' of the universe at the Planck order epoch after which the matter-dominated 'creation era' and subsequently (after a transition epoch) the usual radiation-dominated era follow. The created particles with very large mass may be known elementary particles such as muons, electrons, and massive neutrinos or the primordial black holes. The masses (of the order of Planck mass) of the elementary particles reduce to their present values at the present epoch of the universe owing to their epoch-dependence. Grib (1989) has discussed the quantum effects of vacuum polarization in early FRW space-time, which give rise to an effective change of the gravitational constant. Such a change in gravitational constant leads to the possibility of creation of particles with macroscopic masses to the order of the mass of the observable universe or some effective mass equivalent to its entropy. Subsequent change of the gravitational constant compels these particles to explode as black holes and his assertion is that only particles with microscopic masses can be created from the vacuum in the present era because of the present value of the gravitational constant. Thus, no big bang is possible now. Also, *Schrödinger* cats observers and other macroscopic bodies cannot be created from the vacuum quantum mechanically at the present era. Only in the quantum era of the universe, when the gravitational constant is small enough *Schrödinger* cats may be observable. With the change of the value of G, the gravitational constant, the universe becomes macroscopic and classical. In Grib et al. (1994) the quantum creation of massive particles in strong field has also been discussed. In there finite expressions for the density of created particles were obtained with the use of the method of diagonalization of the instantaneous Hamiltonian. The created particles are real and not virtual as the gravitational field acts as the energetic reservoir.
The interesting fact is that Grib (1989) obtained the following relation that connects the mass m of the created particles with the effective constant $\tilde{G}$:

$$\frac{1}{8\pi\tilde{G}} = \frac{1}{8\pi G} + \frac{m^2}{288\pi^2} \tag{649}$$

where G is the modern value of this constant. It is apparent that small value of $\tilde{G}$ makes the large mass m of the created particles possible. It is also clear from the uncertainty relation $\Delta E \Delta t \geq h$ that a large mass can be created from vacuum in case Δt is much less than the 'modern' value Planck time. In fact, for small $\tilde{G}$, one can have small of Planck time $\tilde{t}_{pl} \sim \tilde{G}^{1/2}$. Thus, is possible for gravity to remain classical even for $t << t_{pl} \sim G^{1/2}$. This fact justifies our considerations (in previous sections) of quantum creation of particles of Planck order masses from classical gravity around the Planck order time. Even heavier particles may be created at an earlier epoch of the universe, when the gravity is still classical. If we substitute for m in (649) from the relation (378) for epoch-dependence of mass for $t << \alpha$, that is,

$$m = 2\alpha\bar{m}H(t) = 0.832H(t) \tag{650}$$

we get

$$\frac{1}{8\pi\tilde{G}} = \frac{1}{8\pi G} + \frac{(0.832)^2(H(t))^2}{288\pi^2}$$

For the matter-dominated era of 'creation', $H(t) = \frac{2}{3t}$ and consequently we find

$$\frac{1}{8\pi\tilde{G}} \simeq 2\left(\frac{0.832}{36\pi t}\right)^2$$

or,

$$\frac{1}{4\tilde{t}_{pl}} \simeq 2\frac{0.832}{36\sqrt{\pi t}}$$

or,

$$t \simeq 0.05\tilde{t}_{pl} \tag{651}$$

Thus, we see that the epoch of the creation of particles of large masses is of the order of the 'effective', that is, changed Planck time, (for the change in G due to vacuum polarization effect) when the gravity remains classical. Thus, however large may be the mass of the created particles the gravity remains classical at the creation epoch of these particles. These particles, as pointed out earlier, may be primordial black holes or even the usual elementary particles (muons, neutrions).
The epoch-dependence of particle mass seems to be very encouraging for the question raised by Dicke and Peebles (1979) with respect to the beginning of the universe as a 'quantum fluctuation' with $K^{-1} \sim 1$ where $K = \frac{Gm_p^2}{\hbar c}$ (m_p being the mass of the proton). In fact, in our framework $K^{-1} \sim 1$ (in the unit $\hbar = c = k = 1$, $Gm_{pl}^2 \sim 1$ and since at the Planck time t_{pl}, the proton mass $m_p \sim m_{pl}$ if one consider proton to be a 'cluster' of muons, neutrinos and their antiparticles; consequently we find $K \sim 1$).One may instead consider the field of proton to satisfy the field equation in the Finsler space and can arrive at the epoch-dependence of its mass as in the case of the particles (leptons) like muons and neutrinos. In this case also, we should have $m_p \sim m_{pl}$ at the Planck time epoch. Now, the generation of the present large value of K^{-1} is possible (together with the generation of matter and entropy as considered earlier), since the proton mass decreases from its Planck-scale value to its modern value. It is also mentioned here that the GUT and SUSY theories require particles with such large masses (neutrinos) at the early universe. It is, in fact, the epoch dependence of mass admits all such Planck-scale massive particles.
Parker (1989) pointed out the possible existence of particles of masses of the order of the Planck mass (mass of $0.28m_{pl}$) in the very early universe (at the Planck time t_{pl}) in his discussion of possible anomalous decay of the neutral pion into gravitons. There it is supposed that a massive neutral particle should appear as an interpolating field in the divergence of an axial current, which contains gravitational and electromagnetic anomalies. Then if the mass of the decaying particle is of planck- scale ($0.28m_{pl}$) then the gravitational decay rate Γ (that is, for the decay of π^0 at rest into a pair of gravitons) becomes the same order in magnitude of the electromagnetic decay rate $\Gamma_{em}\ of\ \pi^0$ (that is, for $\pi^0 \longrightarrow 2\gamma$). In fact, the ratio of these decay rates is given by

$$\frac{\Gamma}{\Gamma_{em}} = \frac{1}{36\pi\alpha^2}\left(\frac{m}{m_{pl}}\right)^4 \tag{652}$$

where α is the fine structure constant. We have seen in chapter IV that π^0 is a composite of muon-antimuon or neutrino-antineutrino pair and as the constituents of hadrons, muons and neutrions have no masses due to the anisotropy of the microdomain, that is, the masses of these constitutionts have no epoch-dependent parts. On the other hand if π^0 goes through a state of interpolating π^0 field, one may regard it to be a 'cluster' of muon-antimuon or neutrino-antineutrino pair having epoch-dependent parts of masses for these constituent particles. Then we can have very large mass $m = 0.555m_{pl}$ at the Planck time t_{pl} , as follows from (650). Thus, the interpolating cluster field (a neutral particle field) having such very large mass in the very early period of the universe gives rise to a significant decay rate for π^0 decaying into a pair of gravitons. This gravitational decay of π^0 via the anomaly as discussed by Parker (1989), should result in a nonthermal cosmic gravitational wave background at a frequency characteristic of the rest energy of the decaying particle. It is expected that in the future progress of gravitational wave detection one can have a possible observational test for the large mass of the particle due to its epoch-dependence by studying the nature of the resulting gravitational wave from such decays, if they exist, in the the early universe.

REFERENCES

1. AL-Rawaf, A.S., and Taha, M.O. (1996a). Physics Letters B, **366**, 69.
2. AL-Rawaf, A.S., and Taha, M.O. (1996b).General Relativity and Gravitation, **28**, 935.
3. Bezerra, V.B., Mostepanenko, V.M., and Romero, C. (1997). Modern Physics Letters A, **12**, 145.
4. Birrell,N.D., and Davies, P.C.W. (1980). Journel of Physics, A: General Physics, **13**, 2109.
5. Birrell,N.D., and Davies, P.C.W. (1982).Quantum Fields in Curved space, Cambridge University Press, Cambridge.
6. Blau, S.K., and Guth, A.H. (1987). In Three Hundread Years of Gravitation, S.W. Hawking and W. Israel, eds., Cambridge University Press, Cambridge, p.524.
7. Brout, R., Englert, F., and Gunzing, E. (1978). Annals of Physics, **115**, 78-106.
8. Brout, R., Englert, F., and Gunzing, E. (1979a). General Relativity and Gravitation, **1**, 1-5.
9. Brout, R., Englert, F., and Spindel, P. (1979b). Physical Review Letters, **43**, 417.
10. Brout, R., et al. (1980). Nuclear Physics B, **170**,228.
11. De, S.S. (1986). International Journal of Theoretical Physics, **25**, 1125.
12. De, S.S. (1989). In Hadronic Mechanics and Nonpotential Interactions, M.Mijatovic, ed., Nova Science Publishers, New York, p.37.
13. De, S.S. (1993). International Journal of Theoretical Physics, **32**, 1603.
14. De, S.S. (1997). International Journal of Theoretical Physics, **36**, 89.
15. De, S.S. (2001). International Journal of Theoretical Physics, **40**, 2067.
16. Dicke, R.H., and Peebles, P.J.E. (1979). In General Relativity-An Einstein Centenary Survey, S.W.Hawking and W.Israel, eds., Cambridge University Press, Cambridge, p.504.
17. Erde'lyi, A., ed. (1953). Higher Transcendental. Functions, McGraw- Hill, New York.
18. Gibbons, G.W.(1979). In General Relativity- An Einstein Centenary Survey, S.W. Hawking and W. Israel, eds., Cambridge University Press, Cambridge, p.639.
19. Gradshteyn, I.S., and Ryzhiok, L.M. (1980). Table of Integrals, Series, and Products, Academic Press, New York.
20. Grib, A.A. (1989). International Journal of Therotical Physics, **28**,1099.
21. Grib, A.A., and Mamayev, S.G. (1969). Yadernaya Fizika, **10**, 1276 [in Russian].
22. Grib, A.A., Mamayev, S.G., and Mostepanenko, V.M. (1994). Vacuum Quantum Effects in Strong Fields, Friedmann Laboratory Publishing.
23. Gunzing, E. Geheniau, J. , Prigogine, I (1987) Nature, **330** , 621-624
24. Gunzing, E., and Nardone, P. (1989). International Journal of Theoretical Physics, **28**, 943.
25. Nardone, P. (1989). International Journal of Theoretical Physics, **28**, 935.
26. Parker, L. (1968). Physics Review Letters, **21**,562.
27. Parker, L. (1969). Physics Review, **183**, 1057.
28. Parker, L. (1989).International Journal of Theoretical Physics, **28**, 1163.
29. Prigogine, I. (1961). Thermodynamics of Irreversible Processes, Wikely, New York.
30. Prigogine, I., Geheniau, J., Gunzig, E., and Nardone, E. (1988). Proceedings of the National Academy of Sciences of the USA, **85**, 7428.
31. Prigogine, I., Geheniau, J., Gunzig, E., and Nardone, P. (1989). General Relativity and Gravitation, **21**,767.
32. Rastall, P. (1972). Physical Review D, **6**,3357.
33. Zel'dovich, Ya.B. (1962). Soviet Physics-JEPT, **16**,1163.
34. Zel'dovich, Ya.B., and Starobinsky,A.A. (1972). Soviet Physics-JETP, **34**,1159
35. Zel'dovich, Ya.B., and Starobinsky,A.A. (1977). JETP Letters, **26**, 252.

CHAPTER SIX:

MODIFIED GENERAL RELATIVITY AND BULK VISCOSITY

6.1: EARLY UNIVERSE WITH BULK VISCOSITY

In the previous chapter particle production in the early universe was considered in the context of the thermodynamics of open systems. From a re-interpretation of energy momentum tensors a negative pressure responsible for matter creation appears. This effect is irreversible in the sense that space-time can create matter as well as entropy. On the other hand, the reverse process is thermodynamically forbidden. Presently, we shall consider the effect of viscous stress in the early evolutionary stage of the universe regarded as a thermodynamically open system. In fact, it has been pointed out in Calvao et al. (1992) that the production of specific entropy per baryon is possible if there is a nonzero bulk viscous stress. In the cosmological evolution of closed system the effect of bulk viscosity has been investigated by several authors(Wienberg(1971), Waga et al. (1986), Padmanabhan and Chitre(1987), Johri and Sudharsan(1988)). It is supposed to be the only dissipative mechanism that might be incorporated in an isotropic cosmological model of the universe. Also, in decaying vacuum cosmologies where decaying vacuum acts as the source of created particles it can be shown that the time derivative of the time dependent cosmological 'constant' has a correspondence with the scalar viscous pressure of the imperfect fluid cosmology. In fact, following Gunzig et al(1998) we have for FRW universe the energy momentum tensor of matter as

$$T^M_{\mu\nu} = (\rho + p)u_\mu u_\nu - pg_{\mu\nu} \quad , \quad u_\mu = \delta^0_\mu \tag{653}$$

and that corresponding to the quantum vacuum energy is

$$T^Q_{\mu\nu} =< \widehat{T}^Q_{\mu\nu} >= \Lambda(t)g_{\mu\nu} \tag{654}$$

Then the conservation equations$\nabla^\nu(T^M_{\mu\nu} + T^Q_{\mu\nu}) = 0$ Leads to

$$\dot{\rho} + 3H(p + \rho) = -\dot{\Lambda} \tag{655}$$

The corresponding equation for energy balance in an imperfect fluid cosmology is

$$\dot{\rho} + 3H(p + \rho) = -3H\Pi \tag{656}$$

These two equations make a correspondence between the viscous pressure Π and the cosmological constant $\Lambda(t)$ as

$$\Pi = \frac{\dot{\Lambda}}{3H} \tag{657}$$

which shows the role of bulk viscous pressure as source of created particles.

In the context of open system, the evolution of FRW universe with bulk viscosity was investigated in Sudharsan and Johri(1994),Desikan(1997), Abramo and Lima(1996). In these works, the source function of particle creation has been chosen suitably so as to obtain inflationary expansion as a solution and also to have a non-singular beginning. On the contrary, the mass relation obtained earlier will be incorporated into this phenomenological macroscopic approach with the bulk viscous pressure, specifically for the transition period when the creation era of FRW universe turned into the radiation dominated FRW expansion phase with relativistic particles. As in Sudarsan and Johri(1994), bulk viscosity under equilibrium thermodynamics will be considered. Later we shall justify it in the present matter creation model. In the context of equilibrium thermodynamics the bulk viscous pressure and the stress-energy tensor are, respectively, given by

$$\Pi = -\zeta\theta \tag{658}$$

$$T_{\mu\nu} = (\rho + p + p_c + \Pi)u_\mu u_\nu - (p + p_c + \Pi)g_{\mu\nu} \tag{659}$$

Where the negative pressure is given by (505) and ζ , the coefficient of bulk viscosity, is in general a function of time. Here, $\theta = 3H = 3\frac{\dot{R}}{R}$ is the expansion scalar. For the FRW metric, with k=0, given by

$$ds^2 = dt^2 - R^2(t)[dr^2 + r^2(d\theta^2 + \sin^2\theta d\phi^2] \tag{660}$$

The Einstein field equations in the presence of bulk viscosity for open systems lead to the following equations:

$$\theta^2 = 3k\rho \tag{661}$$

$$\dot{\rho} = \zeta\theta^2 + \frac{h}{n}\dot{n} \tag{662}$$

$$\dot{n} + n\theta = \psi(t) \;\geq\; 0 \tag{663}$$

where $k = 8\pi G$ and $h = p + \rho$ is the enthalpy per unit volume. $\psi(t)$ is the source function for particle creation. These equations are supplemented by the equation (520) derived from the mass relation and it is given by

$$\frac{\dot{m}}{m} = \frac{2\alpha\dot{\theta}}{3 + 2\alpha\theta} \tag{664}$$

Also, the entropy equation in this case becomes

$$\frac{T\dot{S}}{V} = \zeta\theta^2 + \left(\frac{TS}{V}\right)\frac{\dot{N}}{N} \tag{665}$$

where $V = R^3(t)$ is the comoving volume. For the creation era in the very early universe the equation of state can be taken to be

$$p = f(\theta)\rho \tag{666}$$

with

$$\rho = mn \tag{667}$$

Because in this case the energy density is dominated by the matter density. Here, $0 \leq f(\theta) \leq 1$ and $f(\theta) = 1$ corresponds to maximally stiff equation of state (Zeldovich, 1962). Here again, in an alternative interpretation the conservation equation takes the following form:

$$d(\rho V) = -\widehat{p}\, dV \tag{668}$$

with a phenological pressure

$$\widehat{p} = p + p_c - \zeta\theta \tag{669}$$

When the entropy equation (665) is written in terms of specific entropy per particle, σ, we have

$$\dot{\sigma} = \frac{\zeta\theta^2}{nT} \tag{670}$$

Obviously, when $\zeta = 0$, σ= constant. Now, using the equation of state given by (666) and (667) we have from (661) and (662)

$$\dot{\rho} = \zeta\theta^2 + \frac{\dot{n}}{n}(1+f(\theta))\rho = 3k\rho\zeta + \frac{\dot{n}}{n}(1+f(\theta))\rho$$

or

$$\frac{\dot{\rho}}{\rho} = 3k\zeta + \frac{\dot{n}}{n}(1+f(\theta))$$

Also,

$$\frac{\dot{\rho}}{\rho} = \frac{2\dot{\theta}}{\theta} = \frac{\dot{m}}{m} + \frac{\dot{n}}{n}$$

[from (661 and 667]
Therefore,

$$\frac{2\dot{\theta}}{\theta} = 3k\zeta + (1+f(\theta))\left(\frac{2\dot{\theta}}{\theta} - \frac{\dot{m}}{m}\right)$$

which leads to

$$\zeta(t) = \frac{1}{3k}\left(\frac{\dot{m}}{m}(1+f(\theta)) - \frac{2\dot{\theta}}{\theta}f(\theta)\right)$$

By using (664), we have finally,

$$\zeta(t) = \frac{2\dot{\theta}}{3k\theta(3+2\alpha\theta)}\left[\alpha\theta - (3+\alpha\theta)f(\theta)\right] \tag{671}$$

It is evident from this equation that an exponential inflationary stage of evolution given by $\dot{\theta} = 0$ is not possible for a nonzero coefficient of bulk viscosity. Again, a non viscous state of evolution can not give rise to production of specific entropy per particle, as is evident from (670). Consequently, it is clear that an exponential inflation of the early universe can not produce specific entropy per particle. It may be recalled that the governing equations for the early evolution of the universe considered as a perfect fluid have been given in (530) of the previous chapter. Also, the energy density $\rho = \rho_m + \rho_\gamma$ is given in (518). In fact, ρ_m and ρ_γ are given by

$$\begin{aligned} \rho_m &= mn_m \\ \rho_\gamma &= (\gamma - 1)mn_m + E_\gamma n_\gamma \\ \frac{\rho_\gamma}{\rho_m} &= \gamma^2 - 1 \end{aligned} \tag{672}$$

where γ is given in (517) which is

$$\gamma = \left(1 - \frac{<\nu^2>}{c^2}\right)^{-\frac{1}{2}} = \left(4 + \frac{6\dot{\theta}}{\theta^2}\right)^{-\frac{1}{2}} \tag{673}$$

$n = n_m + n_\gamma$ is the total number density. The particle and photon number densities, n_m and n_γ respectively, are given in (527). For the matter-dominated 'particle creation era' of the early universe we have

$$\gamma \simeq 1\ , \quad \rho_\gamma << \rho_m\ , \quad n \simeq n_m$$

And consequently, we find

$$\frac{\dot{\theta}}{\theta^2} = -\frac{1}{2}$$

This gives the scale factor $R(t) \propto t^{\frac{3}{2}}$. Also, from (530) it follows that (using (664))

$$f(\theta) = \frac{\alpha\theta}{3+\alpha\theta} \qquad\qquad for\ \dot{\theta} \neq 0 \tag{674}$$

From (671) it is evident that the equation of state $p = f(\theta)\rho$ where $f(\theta)$ is given by (674) gives rise to a vanishing coefficient of viscosity. Now, when $f(\theta)$ as given above approaches to a value $\frac{1}{3}$, the scale factor R(t) changes. In fact, for $f(\theta) \simeq \frac{1}{3}$, we have from (530)

$$\frac{\dot{n}}{n} = \frac{3}{2}\frac{\dot{\theta}}{\theta} = -\frac{3}{2t} \tag{675}$$

in the case of a FRW universe other than the trivial case $\dot{\theta} = 0$ which corresponds to a usual exponential inflation. This equation gives rise to the time behaviour of the number density as $n \propto t^{-\frac{3}{2}}$. Again, as $\rho \propto t^{-2}$ for such FRW universes, any such universe can no longer be a matter-dominated one with the energy density$\rho = mn$ because the time behaviour of mass is given by $m = \overline{m}(3 + 2\alpha\theta)$ with $\theta \propto t^{-1}$. Thus, the universe must be a radiation-dominated one with

$$\rho \simeq \rho_\gamma \simeq E_\gamma n \ , \quad n \simeq n_\gamma$$

Consequently, γ must be large and hence $\frac{\dot{\theta}}{\theta^2} = -\frac{2}{3}$. This gives

$$R(t) = t^{\frac{1}{2}} \tag{676}$$

Also, we have $\rho \propto E_\gamma^4 \propto \frac{1}{R^4(T)}$ as $E_\gamma \propto \frac{1}{R(t)} \propto t^{-\frac{1}{2}}$. $n \propto E_\gamma^3 \propto \frac{1}{R^3(t)}$. Hence $\frac{\dot{n}}{n} = -\theta$ or $\frac{\dot{N}}{N} = 0$ and consequently

$$P_c = 0 \tag{677}$$

That is, the particle creation stops. As it occurs when$f(\theta) = \frac{1}{3}$, we find from (674) and (667) that the corresponding epoch is given by $t = \alpha$. Thus, we see that the universe transits to the radiation-dominated FRW era with the usual cosmology (that is, with no particle creation) at $t = \alpha$. Now, for a viscous fluid model of the universe the governing equations (530) for the evolution of the early era can be modified to have the following equations:

$$\frac{\dot{\rho}}{\rho} = 2\frac{\dot{\theta}}{\theta} = 3k\zeta(t) + (1+f(\theta))\frac{\dot{n}}{n} = \frac{1+f(\theta)}{f(\theta)}\frac{E_\gamma}{E_\gamma+\gamma(\gamma-1)m}\left[\frac{\dot{m}}{m} + \frac{\dot{E}_\gamma}{E_\gamma}\frac{\gamma(\gamma-1)m}{E_\gamma} + \frac{\dot{\gamma}}{\gamma}\left(2 - \frac{\gamma_m}{E_\gamma}\right)\right] - \frac{3k\zeta(t)}{f(\theta)} \tag{678}$$

together with equations (520), (672), and (673). From these equations it also follows that the equation of state with $f(\theta)$ given by (674) gives rise to the zero viscous pressure (i.e., $\zeta(t) = 0$) for the 'creation era' matter-dominated evolution of the very early universe. On the other hand, for large γ, we have from (672), (673),(527) and (678) the following relations:

$$\frac{\dot{\rho}}{\rho} = 2\frac{\dot{\theta}}{\theta} = -\frac{4\theta}{3} = \frac{1+f(\theta)}{f(\theta)}\frac{\dot{E}_\gamma}{E_\gamma} - \frac{3k\zeta(t)}{f(\theta))} = 3k\zeta(t) + (1+f(\theta))\frac{\dot{n}}{n} \tag{679}$$

$$\rho_m << \rho_\gamma \ , \quad \rho \simeq \rho_\gamma \ , \quad n_\gamma >> n_m \ , \quad n \simeq n_\gamma \quad and\ consequently\ \rho \simeq E_\gamma n \tag{680}$$

The last relation implies that the universe is radiation-dominated and from (679) it follows that

$$R(t) \propto t^{\frac{1}{2}} \tag{681}$$

In the perfect fluid model it was pointed out above that when $f(\theta) = \frac{1}{3}$ (at the epoch time α.) the particle creation stops ($p_c = 0$), that is, the radiation-dominated FRW era begins at this epoch with the usual cosmology. Here, in the viscous fluid case if the coefficient of viscosity $\zeta(t)$ is given by

$$\zeta(t) = \frac{\theta}{3k}\left(f(\theta) - \frac{1}{3}\right) \tag{682}$$

Then, we have because of (681),

$$\widehat{p} = \frac{\rho}{3} = p + p_c - \zeta\theta = \rho f(\theta) + p_c - \zeta\theta$$

or,

$$p_c = -\rho\left(f(\theta) - \frac{1}{3}\right) + \zeta\theta = -\frac{\theta^2}{3k}\left(f(\theta) - \frac{1}{3}\right) + \zeta\theta = 0$$

[using (661)]
Also, it follows from (679) and (682) that

$$3k\zeta(t) = (1 + f(\theta))\frac{\dot{E}_\gamma}{E_\gamma} + \frac{4\theta}{3}f(\theta) - \theta\left(f(\theta) - \frac{1}{3}\right)$$

or, we have

$$\frac{\dot{E}_\gamma}{E_\gamma} = \theta \ or \ E_\gamma \propto \frac{1}{R} \propto t^{-\frac{1}{2}}$$

and hence

$$\rho \propto E_\gamma^4 \ and \ n \propto E_\gamma^3$$

Thus, in this imperfect fluid model of the Universe $p_c = 0$ before the epoch time $t = \alpha$, when $f(\theta) > \frac{1}{3}$. With a positive coefficient of viscosity $\zeta(t)$, the Universe becomes a radiation dominated FRW Universe with no creation of massive particle just before the epoch α. At the epoch α, $f(\theta) = \frac{1}{3}$ and consequently $\zeta(t) = 0$. The era with the nonzero $\zeta(t)$ just before the epoch α is the 'transition era' when matter-dominated 'creation era' transits into the radiation-dominated FRW era with no creation of massive particles. But, it follows from (670) that the specific entropy per particle is created only in this transition era because of the development of a nonzero viscous pressure in this period. We get from (670) and (682) the following equation for the creation of specific entropy per particle:

$$\dot{\sigma} = \frac{\theta^3}{3knT}\left(f(\theta) - \frac{1}{3}\right) \tag{683}$$

We can now use the following cosmological invariant as the universe evolved with the standard cosmology after the transition era:

$$RT = R_0T_0 = C = 1.18 \times 10^{29}u \qquad (1 < u < 1.8) \tag{684}$$

Where R_0 and T_0 are the present values of the scale factor and the temperature of the universe respectively. Then, since $n = \frac{N}{R^3}$ with constant N (as there is no particle creation in the transition era as well as in the subsequent evolution of the universe) we have from (683) and (684),

$$\dot{\sigma} = \frac{\theta^3 C^3}{3kNT^4}\left(f(\theta) - \frac{1}{3}\right) \tag{685}$$

Using the expression for $f(\theta)$ from (674) and remembering that the transition era is a radiation-dominated FRW universe with $\theta = \frac{3}{2t}$, we find

$$\dot{\sigma} = \frac{9C^3}{8kNT^4t^3}\left(\frac{\alpha}{2t+\alpha} - \frac{1}{3}\right)$$

Also, it is easy to see from (684) that

$$\frac{T_\alpha}{T} = \left(\frac{t}{\alpha}\right)^{\frac{1}{2}} \qquad (\ since\ \ R(t)\ \propto\ t^{\frac{1}{2}}\ ) \tag{686}$$

where T_α is the universe temperature at $t = \alpha$. Using this we finally get

$$\dot{\sigma} = \frac{9C^3}{8kN\alpha^2T_\alpha^4}\frac{1}{t}\left(\frac{\alpha}{2t+\alpha} - \frac{1}{3}\right) \tag{687}$$

Now, if the transition era which is also the era of creation of specific entropy per particle begins at the epoch $\alpha' \leq \alpha$, then the specific entropy per particle, σ, created in this era (α', α) is given by

$$\sigma = \frac{9C^3}{8kN\alpha^2T_\alpha^4}\int_{\alpha'}^{\alpha}\left[\frac{1}{t}\left(\frac{\alpha}{2t+\alpha} - \frac{1}{3}\right)\right]dt = \frac{9C^3}{8kN\alpha^2T_\alpha^4}\left[\frac{2}{3}\ln\frac{\alpha}{\alpha'} - \ln\frac{3\alpha}{2\alpha'+\alpha}\right] \tag{688}$$

We can here use the value of the total particle number N from (644) of the previous chapter where we have considered quantum creation of particles. There was also found in (646) the universe temperature T_α at the epoch α. The value of the constant C is given in (684). Using these values and noting that $k = 8\pi G = \frac{1}{m_{pl}^2} = t_{pl}^2$ in the natural unit where m_{pl}^2 and t_{pl}^2 are the Planck mass and time respectively, we find from (688)

$$\sigma = 9.4 \times 10^{10}\left[\frac{2}{3}\ln\frac{\alpha}{\alpha'} - \ln\frac{3\alpha}{2\alpha'+\alpha}\right] \tag{689}$$

Clearly, the specific entropy per particle thus produced is depended on the duration of the transition era. We can, of course, find an upper limit of it by noting that before an epoch, say $\alpha' = 0.01\alpha$ the universe must have been a matter-dominated one because $f(\theta) \simeq 1$ for $t \leq \alpha' = 0.01\alpha$. Thus, with the beginning of the transition era at $\alpha' = 0.01\alpha$, we find from (689) $\sigma = 1.87 \times 10^{11}$. This is an upper limit because we get lesser values if the transition era starts after the epoch $\alpha' = 0.01\alpha$. For example, if we take $\alpha' = 0.5\alpha$ we get a good estimate of σ as $\sigma = 5.33 \times 10^9$.

The produced specific entropy per particle in the transition period due to the bulk viscous pressure remained constant subsequently till the present state of the universe. In fact, after the epoch time α there remains no bulk viscosity and the universe evolves with the standard cosmology. Therefore this value of σ corresponds to the present value of it.

6.2: MODIFIED GENERAL RELATIVITY

The theoretical foundations of covariant energy momentum conservation in curved space-time were put into doubt by Rastall(1972). Consequently, he modified it with an assumption that the divergence of the energy-momentum tensor might be dependent on the curvature. Actually, it was thought to be proportional to the gradient of the scalar curvature. This scalar curvature, of course, vanishes for flat space-time and thus, in special relativity the conservation of energy-momentum is restored. The modified general relativity(MGR) thus obtained is found to be equivalent to the gravitational field equations derived recently by Al-Rawaf and Taha (1996a,b) with the use of conventional heuristic methods(Wienberg, 1972) but not requiring energy-momentum conservation. The field equations of the MGR of Rastall, Al-Rawaf and Taha can not be derivable from a variational principle. However, a prototype of this MGR can be derived from a variational principle, but it contains a variable gravitational 'constant' (Smalley, 1984). Interestingly, the formulation of Al-Rawaf and Taha contains two independent fundamental constants, one of which is the usual Newton's constant and the other is an adjustable parameter η satisfying $0 < \eta \leq 1$. This constant η may be regarded as the characteristic of non-Newtonian regime. The standard general relativity (GR) can be achieved with the value of this parameter $\eta = 1$. In the formulation of MGR by Al-Rawaf and Taha the modified field equations were derived with the use of conventional approach and were of the following form:

$$G_{\mu\nu} = -8\pi G T_{\mu\nu} \tag{690}$$

where $G_{\mu\nu}$ is a gravitational tensor on which the usual condition that $G_{\mu\nu}$ contains only terms that are linear in the second, or quadratic in the first, derivatives of the metric tensor is imposed. This led to the following form of $G_{\mu\nu}$:

$$G_{\mu\nu} = \alpha R_{\mu\nu} + \beta R g_{\mu\nu} \tag{691}$$

where α and β are constants. $R = g^{\mu\nu} R_{\mu\nu} = g^{\mu\nu} g^{\lambda\rho} R_{\lambda\mu\nu\rho}$ is the curvature invariant. Another requirement is that in the limit of weak stationary field the non-relativistic Newtonian equation

$$\nabla^2 g_{00} = -8\pi G T_{00} \tag{692}$$

should be satisfied. For this, one should have

$$\beta = \frac{\alpha(\alpha - 2)}{2(3 - 2\alpha)} \quad , \quad \alpha \neq 0 \, , \frac{3}{2} \tag{693}$$

With the substitution of (693) in (691) and then in (690) the modified field equations result. They are given by

$$R_{\mu\nu} - \frac{1}{2}\gamma R g_{\mu\nu} = -\frac{8\pi G}{\alpha} T_{\mu\nu} \tag{694}$$

where

$$\gamma = \frac{(2 - \alpha)}{(3 - 2\alpha)} \tag{695}$$

The standard GR is regained when $\alpha = 1$. It is to be noted that the conservation condition

$$T^{\mu}_{\nu\,;\,\mu} = 0 \tag{696}$$

has not been explicitly imposed and one finds from (694)

$$T^{\mu}\nu\,;\,\mu = \frac{\gamma - 1}{2} \frac{\alpha}{8\pi G} R_{,\,\nu} \tag{697}$$

Also, by contracting (694) we have

$$R = \frac{8\pi G}{\alpha} \frac{1}{2\gamma - 1} T \qquad where \;\; T = g^{\mu\nu} T_{\mu\nu} \tag{698}$$

Then, from (697) it follows that (by using (698))

$$T^{\mu\nu}_{;\mu} = -\frac{1}{2}(1 - \alpha) T_{,\,\nu} \tag{699}$$

It has been pointed out by Al-Rawaf and Taha from the above relation that a viable covariantly conserved system would necessarily possess a traceless energy-momentum tensor. Conversely, it follows that MGR coincides with standard GR for all systems having a traceless energy momentum tensor because for such systems R = 0 and hence

$$R_{\mu\nu} = -\frac{8\pi G}{\alpha} T_{\mu\nu} \tag{700}$$

Of course, here G has been replaced by $\frac{G}{\alpha}$ and any observable consequence for this change can give rise to a possible determination of the parameter α. Again, the field equations (694) can be written in the form:

$$R_{\mu\nu} - \frac{1}{2} R g_{\mu\nu} = -\frac{8\pi G}{\alpha} \theta_{\mu\nu} \tag{701}$$

with a 'modified' energy-momentum tensor $\theta_{\mu\nu}$ given by

$$\theta_{\mu\nu} = T_{\mu\nu} + \frac{1}{2}(1-\alpha) T g_{\mu\nu} \tag{702}$$

Wwhich shows that, within standard GR, the energy-momentum tensor is modified by the addition of an extra term. This extra term indicates a matter-gravitation interaction involving a new universal constant α. Obviously, the total energy-momentum tensor, $\theta_{\mu\nu}$, is covariantly conserved. Similar views regarding non conservation of energy-momentum tensor have been adopted in the recently proposed theories. (Özer and Taha, 1986, Freese et el, 1987, Chen and Wu, 1990, Silveria and Waga, 1994, Abdel-Rahaman,1995) of decaying vacuum cosmologies in which a time-dependent cosmological 'constant' has been introduced. These theories are also not deducible from variational principle. There, in stead of matter energy-momentum conservation, a sum of tensors corresponding to this usual energy-momentum and a 'vacuum energy-momentum' was considered as to be conserved. One can regard such conservation in MGR also(as we have seen above) and, in fact, Al-Rawaf and Taha (1996 a,b) have shown that MGR can be moulded into a model with a variable cosmological 'constant'. Now, with a change of the parameter given by

$$\eta = \frac{2\alpha}{(3-\alpha)} \tag{703}$$

the modified field equations (694) become

$$R_{\mu\nu} - \frac{4-\eta}{6(2-\eta)} R g_{\mu\nu} = -\frac{8\pi G}{3}\left(1 + \frac{2}{\eta}\right) T_{\mu\nu} \tag{704}$$

As $\alpha \neq 0, \frac{3}{2}, \eta \neq 0, 2$. In fact, the parameter η is now supposed to satisfy

$$0 < \eta \leq 1 \tag{705}$$

Clearly, $\alpha = 1$ or $\eta = 1$ corresponds to GR.

An interesting observational consequence derived from MGR applied to a matter-dominated Robertson-Waker (RW) universe is that it can resolve the conflict between the ages of the oldest stars in our galaxy and that of the universe itself, as derived from the measurement of Hubble constant from recent observations(Pierce et el, 1994). In fact, as shown by Al-Rawaf and Taha (1996 a, b), there remains no such problem with the values of the parameter $\eta \leq 0.6$ and the present value of the matter density parameter Ω lying on a range $0.1 \leq \Omega \leq 0.25$. Recently, Abdel-Rahaman (1997) has applied MGR in radiation-dominated era of the universe and discussed the implications of the nucleosynthesis constraints for the age of the universe. He has also shown the consistency of matter-dominated model considered in the framework of MGR with the neoclassical cosmological tests. There, it is also shown that the baryon asymmetry in the early universe was significantly smaller than that at present. In the following section the recent consideration (De,1999) of the very early universe in the framework of MGR, which is taken as a thermodynamically open system in the sense of Prigogine(1989) will be presented.

6.3: EARLY UNIVERSE WITH MGR

Here, we shall consider spatially flat($k=0$) RW universe with the following metric

$$ds^2 = dt^2 - a^2(t)(dr^2 + r^2 d\theta^2 + r^2 sin^2\theta d\phi^2) \tag{706}$$

(using the natural unit $c = \hbar = 1$)

Also, the energy-momentum tensor for the universe as an adiabatic perfect fluid is given by

$$T_{\mu\nu} = -pg_{\mu\nu} + (\rho + p)u_\mu u_\nu \tag{707}$$

where ρ and p are the density and pressure respectively and $u_\mu = (1,0,0,0)$ is a unit vector in the time direction. With these, we get from the field equations (704), the following equations for the cosmology in MGR

$$\frac{\dot{a}^2}{a^2} = \frac{k}{3\eta}[\rho - (1-\eta)p] \tag{708}$$

$$\frac{\ddot{a}}{a} = -\frac{k}{6\eta}[\eta\rho + (4-\eta)p] \tag{709}$$

where $k = 8\pi G$ and dot(.) represents differentiation with respect to time. The following Bianchi identity follows from (708) and (709) :

$$\frac{d}{da}[a^3(\rho + (\eta - 1)p)] + a^2[(\eta - 1)\rho + (5 - 2\eta)p] = 0 \tag{710}$$

or, in the differential form,

$$d[a^3(\rho + (\eta - 1)p)] + \frac{1}{3}[(\eta - 1)\rho + (5 - 2\eta)p]d(a^3) = 0 \tag{711}$$

For $\eta = 1$, we get the usual conservation law for GR, that is,

$$d(\rho a^3) + pd(a^3) = 0 \tag{712}$$

In Prigogine(1989) this conservation law has been modified for the universe considered as a thermodynamically open system.Specifically, the equation (712) is modified to the following thermodynamic energy conservation law for a homogeneous and isotropic universe:

$$d(\rho a^3) + pd(a^3) - \frac{h}{n}d(na^3) = dQ \tag{713}$$

where $h = \rho + p$ is the enthalpy per unit volume and $n = \frac{N}{V}$ where N is the number of particles in a given volume V, that is, the comoving volume given by $V = a^3$. For adiabatic transformation we have $dQ = 0$ and we get

$$d(\rho a^3) + pd(a^3) - \frac{h}{n}d(na^3) = 0 \tag{714}$$

From this conservation law it follows that

$$\dot{\rho} = \frac{\dot{n}}{n}(\rho + p) \tag{715}$$

which replaces the usual Einstein equation (Bianchi identity for homogeneous and isotropic universe), that is,

$$\dot{\rho} = -3H(\rho + p) \tag{716}$$

where H is the Hubble function given by $H = \frac{\dot{a}}{a}$. Of course, the other Einstein equation corresponding to the equation (708) with $\eta = 1$ remains to be valid for this case of thermodynamically open universe. This equation is

$$k\rho = 3H^2 \tag{717}$$

One can make an alternative interpretation of the conservation law (714) or (715) by retaining the usual form of the conservation law (Bianchi identity) with a phenomenological pressure $\hat{p}$. That is,

$$d(\rho a^3) = -\hat{p}d(a^3) \tag{718}$$

where the two pressures $\hat{p}$ and p are related by

$$\hat{p} = p + p_c \tag{719}$$

Here, p_c represents a pressure, negative or zero, and corresponds to the creation of particles. In fact, when $p_c = 0$ the creation of particles stops and in this case $\hat{p} = p$. Consequently, the conventional law of conservation holds or, in other words, the usual Einstein equations of GR hold. The pressure p_c is given by

$$p_c = -\frac{\rho + p}{3H}\frac{\dot{S}}{S} \tag{720}$$

where S is the entropy. Also, it is shown that (Prigogine, 1989)

$$\frac{\dot{S}}{S} = \frac{\dot{N}}{N} = \frac{\dot{n}}{n} + \theta \tag{721}$$

where $\theta = 3H$ is the expansion scalar.

Now, the Bianchi identity (711) of MGR for the early universe considered as a thermodynamically open system to account for the creation of particles is to be modified. This modification is made in the same manner as above,that is, as for GR. For adiabatic transformation ($dQ = 0$) the conservation law (711) is, now, changed into the following form:

$$d[a^3(\rho + (\eta - 1)p)] + \frac{1}{3}[(\eta - 1)\rho + (5 - 2\eta)p]d(a^3) - \frac{h}{n}d(na^3) = 0 \tag{722}$$

The equation (708) remains valid for this case of thermodynamically open early universe and becomes the usual Einstein equation (717) for GR of $\eta = 1$. From (722), it follows that

$$\dot{\rho} + (\eta - 1)\dot{p} - (\rho + p)\left[\frac{\dot{n}}{n} - (\eta - 1)\frac{\theta}{3}\right] = 0 \tag{723}$$

Now, in the particle production era of the very early universe the energy density $\rho = \rho_m + \rho_\gamma$ should be dominated by the matter density ρ_m, that is, $\rho_m \gg \rho_\gamma$, the radiation energy density. Consequently, we can take

$$\rho = mn \tag{724}$$

where m is the particle mass. Here, for simplicity, the masses of all types of particles created are assumed to be the same. As before, the mass of the particle has been taken to be epoch dependent. It is given by

$$m = \bar{m}(1 + 2\alpha H(t)) \tag{725}$$

where $\alpha = 0.26 \times 10^{-23}$ sec. and $\bar{m}$ represents the 'inherent' mass of the particle. As pointed out earlier, this inherent mass is equal to the present mass of the particle with a very high degree of accuracy.For massless particles(that is, for particles with no inherent mass) the corresponding relation is

$$m = 2\alpha\hat{m}H(t) \tag{726}$$

where $\hat{m}$ is the mass of the particle at the epoch time $t = \alpha$.

Now, the equations (708), (723), (724) and (725) together with an equation of state given as

$$p = F(\rho) \equiv \rho f(\theta) \tag{727}$$

are the governing equations that describe the early evolution of the thermodynamically open universe with MGR.

Using (724) and the equation of state (727), we have from (723)

$$(\eta - 2)f(\theta)\frac{\dot{\rho}}{\rho} + (\eta - 1)f'(\theta)\dot{\theta} + (1 + f(\theta))\left[\frac{\dot{m}}{m} + (\eta - 1)\frac{\theta}{3}\right] = 0 \tag{728}$$

Again, from (725) it follows that

$$\frac{\dot{m}}{m} = \frac{2\alpha\dot{\theta}}{3 + 2\alpha\theta} \tag{729}$$

and from (708), by using (727) one can get

$$\rho = \frac{\eta\theta^2}{3k[1 + (\eta - 1)f(\theta)]} \tag{730}$$

Consequently, we have

$$\frac{\dot{\rho}}{\rho} = \frac{2\dot{\theta}}{\theta} - \frac{(\eta - 1)f'(\theta)\dot{\theta}}{1 + (\eta - 1)f(\theta)} \tag{731}$$

With the use of equations (729) and (731) one can arrive from (728) at the following equation:

$$(\eta - 2)f(\theta)\left[\frac{2\dot{\theta}}{\theta} - \frac{(\eta - 1)f'(\theta)\dot{\theta}}{1 + (\eta - 1)f(\theta)}\right] + (\eta - 1)f'(\theta)\dot{\theta} + (1 + f(\theta))\left[(\eta - 1)\frac{\theta}{3} + \frac{2\alpha\dot{\theta}}{3 + 2\alpha\theta}\right] = 0 \tag{732}$$

Again, as it is pointed out earlier that the creation era of the very early universe is matter-dominated, the relation (724) holds for the energy density and consequently it follows that

$$\rho a^3 = Nm \tag{733}$$

By using (725) and (730), we get the number of particles in a comoving volume $V = a^3$ in terms of the expansion scalar θ as follows:

$$N = \frac{\eta\theta^2 a^3}{k\bar{m}(3+2\alpha\theta)[1+(\eta-1)f(\theta)]} \tag{734}$$

The equations (732) and (734) now describe the 'creation-era' early universe in MGR formulation.

6.4: MILD INFLATION IN THE EARLY UNIVERSE

Let us first recall the case of GR in the thermodynamically open early universe. For the creation-era matter-dominated early universe the following equations can be deduced(De, 1993a) from the governing equations (715), (717), (724) and (725) if the equation of state (727) is incorporated:

$$\frac{\dot{\rho}}{\rho} = \frac{2\dot{\theta}}{\theta} = \frac{\dot{n}}{n}(1+f(\theta)) = \frac{\dot{m}}{m}\frac{1+f(\theta)}{f(\theta)}$$

$$with\ \frac{\dot{m}}{m} = \frac{2\alpha\dot{\theta}}{3+2\alpha\theta} \tag{735}$$

These equations have a trivial solution

$$\dot{\theta} = \dot{\rho} = \dot{n} = \dot{m} = 0$$

which represents the usual inflation.Apart from this trivial solution, these equations can not determine the expansion scalar θ. On the other hand, if $\dot{\theta} \neq 0$ then one can determine the equation of state, that is, the function $f(\theta)$ can be specified. It is given by

$$f(\theta) = \frac{\alpha\theta}{3+\alpha\theta} \tag{736}$$

The expansion scalar θ, however, may be specified with the contention that $\hat{p} = 0$ in equation (718) as this era of the universe is matter-dominated. This gives

$$a(t) \propto t^{2/3} \tag{737}$$

and

$$p_c = -p = -f(\theta)\rho \tag{738}$$

This negative pressure p_c is responsible for the particle creation in the era $t \ll \alpha$. On the other hand, as t increased $f(\theta)$ decreased and in fact, $f(\theta) \to 0$ as t become large. It was argued in De(1993a) that the universe around the epoch, became radiation-dominated RW universe, the usual universe according to the standard cosmology and with no particle creation(see also section 6.1).In fact, the particles at this transition epoch became relativistic and contributed to the radiation-energy density, that is, $\rho = \rho_\gamma$,$f(\theta) = 1/3$,$p_c = 0$ and consequently $a(t) \propto t^{1/2}$.

Now, for MGR we rewrite the equation (732) in the following form:

$$2\dot{\theta}\left[\frac{\alpha(1+f(\theta))}{3+2\alpha\theta} - \frac{f(\theta)}{\theta}\right] + (\eta-1)\left[f(\theta)\dot{\theta}\left(\frac{2}{\theta} - \frac{(\eta-2)f'(\theta)}{1+(\eta-1)f(\theta)}\right) + f'(\theta)\dot{\theta} + \frac{\theta}{3}(1+f(\theta))\right] = 0 \tag{739}$$

Obviously, for $\eta = 1$, that is, for GR, we arrive at the same results as stated above, that is either $\dot{\theta} = 0$ or the specified equation of state $p = f(\theta)\rho$, with $f(\theta)$ given by (736). The usual inflationary solution is represented by the trivial case $\dot{\theta} = 0$. Of course, the actual physical conditions that might determine the turn-on and turn-off phases of this inflationary paradigm are still remaining obscure. Although, it is now believed that it is similar to a symmetry breaking phase transition. The dominance of the vacuuum energy field may be responsible for triggering off a symmetry breaking at the GUT energy scale.On the other hand, the switch-off phenomenon of the inflationary phase is assumed to be occurred due to a second phase transition in which the symmetry of the weak nuclear force and electrostatic force was broken.

From the equation (738) it is clear that one can determine, in principle, the expansion scalar θ and consequently the scale factor a(t) if one knows $f(\theta)$, that is, the equation of state. Conversely, if one knows the character of expansion of the universe(either a(t) or $\theta(t)$),the equation of state can be obtained from this equation. It is also clear that if $\eta = 1$ or, if $\eta \to 1$, the equation of state can be specified and is given by (736). Thus, one may suppose the same functional dependence $f(\theta)$ on the expansion scalar for all values of η in (0,1]. In fact, if one supposes that $f(\theta)$ is independent of the parameter η then it can be shown that$f(\theta)$ must be given by (736). In this connection it is also noted that if $f(\theta) = -1$ then the usual inflationary solution $\dot{\theta} = 0$ follows from (738). The equation of state $p = f(\theta)\rho$ with $f(\theta)$ as in (736) provides us the following equation for θ which decides the nature of the early evolution of the universe:

$$f(\theta)\dot{\theta}\left(\frac{2}{\theta} - \frac{(\eta-2)f'(\theta)}{1+(\eta-1)f(\theta)}\right) + f'(\theta)\dot{\theta} + \frac{\theta}{3}(1+f(\theta)) = 0 \tag{740}$$

This equation is valid for all values of η which satisfies $0 < \eta < 1$. Since it holds for $\eta \to 1$ one can take it to be valid for the case $\eta = 1$, that is, for GR. For the present case, the assumption of zero phenomenological pressure, that is, $\hat{p} = 0$ is not necessary in finding out the expansion scalar θ. The equation (739) gives the expansion scalar for all values of η in (0,1]. Again, by using the expression for $f(\theta)$ from (736) in (739) we arrive at the following equation:

$$\dot{\theta}\left[\frac{3}{\theta} - \frac{4\alpha}{3+2\alpha\theta} - \frac{\alpha}{(1-\eta)(3+\alpha\theta)} + \frac{\eta^2\alpha}{(1-\eta)(3+\eta\alpha\theta)} + \frac{1}{\alpha}\right] \quad \mathit{for}\ \eta \neq 1 \tag{741}$$

and

$$\frac{\dot{\theta}}{\theta}\left[\frac{2\alpha}{3+2\alpha\theta} + \frac{3\alpha}{(3+\alpha\theta)^2}\right] + \frac{1}{3} = 0 \quad \mathit{for}\ \eta = 1 \tag{742}$$

One can find the solution of the equation (740) by adjusting the constant of integration suitably. In fact, an integrating constant has, here, been taken to ensure that $\alpha\theta$ be very large for the epoch times $t \ll \alpha$. Thus, for $t < \alpha$ we can have

$$\frac{4(\alpha\theta)^3(3+\eta\alpha\theta)^{\frac{\eta}{1-\eta}}}{(3+2\alpha\theta)^2(3+\alpha\theta)^{\frac{1}{1-\eta}}} = \left(\eta^{\frac{\eta}{1-\eta}}\right)e^{-\frac{t}{\alpha}} \tag{743}$$

For $t \ll \alpha$, for which $\alpha\theta$ is very large, we have from (742)

$$\left(1 - \frac{3}{\alpha\theta} - \frac{3}{(1-\eta)\alpha\theta}\right)^{1+\frac{3}{(1-\eta)\alpha\theta}} = 1 - \frac{t}{\alpha}$$

or,

$$1-\frac{3}{\alpha\theta}-\frac{3}{(1-\eta)\alpha\theta}+\frac{3}{(1-\eta)\alpha\theta}=1-\frac{t}{\alpha}$$

Therefore,

$$\theta=\frac{3}{t}\ or,\ H=\frac{1}{t} \tag{744}$$

This gives the scale factor

$$a(t)\propto t \tag{745}$$

This scale factor indicates a 'mild inflation' in the very early stage of evolution of the universe. For $\eta=1$, we get from (741) the solution for θ. By adjusting the constant of integration in the same manner as above, we can get the following relation which determine θ for $t<\alpha$:

$$\frac{4(\alpha\theta)^3}{(3+2\alpha\theta)^2(3+\alpha\theta)}exp\left(\frac{3}{3+\alpha\theta}\right)=exp(-\frac{t}{\alpha}) \tag{746}$$

It is, here, to be noted that we can also arrive at the above solution from(742) by making $\eta\to 1$. From (745), we can find θ for $t\ll\alpha$. In fact, for this case of $\eta=1$, that is, for GR we get the same expansion scalar θ and scale factor a(t) as in (743) and (744) respectively. Thus, for all values of the parameter η in the range $0<\eta\leq 1$ we have the 'mild inflation' phase before the time scale α.

Now, from the conservation law (722) in MGR for the early universe as a thermodynamically open system with adiabatic transformation, it is obvious that the pressure p_c is given by

$$p_c=-\frac{(\rho+p)\frac{d}{dt}(na^3)}{n\frac{d}{dt}(a^3)} \tag{747}$$

(cf. equations (714), (720) and (721))

Using (724),(729), and (731) we have from (716),

$$p_c=-\frac{\rho(1+f(\theta))}{\theta}\left[\dot{\theta}\left(\frac{2}{\theta}-\frac{(\eta-1)f'(\theta)}{1+(\eta-1)f(\theta)}-\frac{2\alpha}{3+2\alpha\theta}\right)+\theta\right] \tag{748}$$

Again, from the equation (739) we find

$$\frac{(1+f(\theta))f'(\theta)\dot{\theta}}{1+(\eta-1)f(\theta)}=-\left[\frac{2f(\theta)\dot{\theta}}{\theta}+\frac{\theta}{3}(1+f(\theta))\right] \tag{749}$$

Using this, we have from (747), the following expression for the pressure p_c:

$$p_c=-\frac{\rho}{\theta}\left[(1+f(\theta))\left(\frac{2\dot{\theta}}{\theta}\frac{3+\alpha\theta}{3+2\alpha\theta}+\theta+(\eta-1)\frac{\theta}{3}\right)+(\eta-1)\frac{2f(\theta)\dot{\theta}}{\theta}\right] \tag{750}$$

Finally, by using the expression for $f(\theta)$ from (736), we obtain

$$p_c=-\frac{\rho}{3+\alpha\theta}\left[\frac{2\dot{\theta}}{\theta^2}(3+\eta\alpha\theta)+\frac{\eta+2}{3}(3+2\alpha\theta)\right] \tag{751}$$

It is evident from this equation that for $\theta = \frac{3}{t}$, that is, for the era $t \ll \alpha$, the pressure $p_c = -\frac{4\rho}{3}$. Due to this negative pressure the particle creation had been continued in the time scale $t \ll \alpha$. As the time t became closer to the epoch time α, the function $f(\theta)$ changed from its value almost equal to one to a value less than unity. In fact, it is evident from (742) and (745) that $\alpha\theta$ became closed to zero as the time t became larger than α. To find out the behaviour of the pressure p_c in respect to the changes in $\alpha\theta$ we make use of the equation(740) into (750) to get

$$p_c = -\rho\left[\frac{1}{3}(\eta+2)\left(\frac{3+2\alpha\theta}{3+\alpha\theta}\right) - \frac{2}{\alpha\theta}\left(\frac{3+\eta\alpha\theta}{3+\alpha\theta}\right)\left(3 - \frac{4\alpha\theta}{3+2\alpha\theta} - \frac{\alpha\theta}{(1-\eta)(3+\alpha\theta)} + \frac{\eta^2\alpha\theta}{(1-\eta)(3+\eta\alpha\theta)}\right)^{-1}\right] \quad (752)$$

For very small $\alpha\theta$,p_c becomes positive and consequently there should be no creation of particles due to this pressure. Even for$\alpha\theta = O(1)$, for example if $\alpha\theta = 0.9$, the pressure $p_c \geq 0$ for values of η either near zero or near unity. For $\alpha\theta = 0.3$, $p_c \geq 0$ for all values of η in its range $0 < \eta \leq 1$. Again, it is apparent from (740) and (745) that when $\alpha\theta$ becomes of the order of unity, the time t is around the time scale α. Thus, for all values of η in (0,1], $p_c \geq 0$ around the epoch time α and hence the creation of particles brought about by the pressure p_c must have been stopped around that epoch time. Consequently, the usual cosmology either with MGR or GR (the case $\eta = 1$) must have set in after the epoch time α. In fact, the universe then became a radiation-dominated RW universe since it was no longer remained thermodynamically open and matter-dominated with particle creation. Of course, one has a different situation for $\eta \neq 1$ or $\eta = 1$ (the case of GR), although, as pointed out earlier, the universe described in MGR framework has no observational discrepancy from the present state of the universe (Abdel-Rahman, 1997).

An important aspect of this 'mild inflation' is that in this case there are no problems with its turn-on and turn-off phases. As we have seen above that this inflation is automatically turned-off in the time scale α. On the other hand, it was turned-on in a planck-order time scale as discussed in earlier chapter and also in the previous paper (De, 1993b). It is shown there that very massive particles (of masses more than fifty times of planck mass m_{pl}) might have been created quantum mechanically in an 'anisotropically perturbed' Minkowski space-time for a duration of the planck-order time. These created very massive particles can make the Minkowski space-time unstable (Nardone, 1989) and thrust it into an expansion phase, the beginning phase of the expanding universe at this planck-order time. In this very early expanding phase of the universe quantum creation of particles in its curved space-time was also been considered in that article and in the previous chapter. This consideration can, indeed, give rise to the radiation energy density, universe-temperature, baryon number etc. at the transition epoch α, in agreement with the observational facts. The 'particle creation era' expanding phase of the universe which had been turned-on from the Minkowski space-time must be a thermodynamically open system which is being considered here. The nature of expansion has been found here to be a mild inflation. Thus, the universe has been switched on to a mild inflationary stage from the planck-order time upto an epoch time α , when it automatically transited into the radiation-dominated RW universe. It is an interesting fact that the highly massive particles created in this very early period (at the planck-order time scale) may be the primordial black holes or even, the known elementary particles such as, muons, electrons and massive neutrinos whose masses were, in fact, to the order of more than fifty times of planck mass owing to the epoch dependence of particle masses as mentioned above.

In the present consideration of mild inflation, the number of particles created during this phase may be estimated. If $N(\hat{t})$ is the number of particles in the comoving volume a^3 at the time $\hat{t}$ (a planck-order time) of turning-on the inflation and $N(\alpha')$ is that at the epoch $\alpha' < \alpha$, the turning off time, then we can have the following relation from the equation (734):

$$\frac{N(\alpha')}{N(\hat{t})} = \frac{\theta^2(\alpha')a^3(\alpha')(3+2\alpha\theta(\hat{t}))[1+(\eta-1)f(\theta(\hat{t}))]}{\theta^2(\hat{t})a^3(\hat{t})(3+2\alpha\theta(\alpha'))[1+(\eta-1)f(\theta(\alpha'))]} \quad (753)$$

Here, α' is the epoch time up to which the universe was matter-dominated, that is, the relation (724) for the energy density was valid. The time interval from α' to α can be regarded as the transition era from this matter dominated thermodynamically open universe to the radiation-dominated RW universe. From (752) we have, by using the expression for the expansion scalar from (743),

$$\frac{N(\alpha')}{N(\hat{t})} \simeq \left(\frac{\alpha'}{\hat{t}}\right)^2 \tag{754}$$

The time $\hat{t}$ represents the epoch time at which the masses of the particles were 54 times the Planck mass. It was found that $\hat{t} = 0.05t_{pl}$. If we take $\alpha' = 0.1\alpha$, we find an estimate of the number of particles at the epoch α' in comparison with that at the epoch $\hat{t}$. That is given by

$$N(\alpha') \simeq 10^{40} N(\hat{t}) \tag{755}$$

Thus, during the phase of mild inflation the number of particles has increased by a factor of 10^{40}. After this the nature of the expansion scalar changed and the particle creation stopped. In fact, the phenomenological pressure $\hat{p}$, in the MGR framework can be shown to be given by

$$\hat{p} = -\rho\left[(1 + \frac{2\dot{\theta}}{\theta^2}) + (\eta - 1)\left(\frac{1}{3} + \frac{2f(\theta)}{1 + f(\theta)}\frac{\dot{\theta}}{\theta^2}\right)\right] \tag{756}$$

With this pressure the conservation law (Bianchi identity) takes the usual form (718). When the creation of particles due to pressure p_c stopped around $t = \alpha$ on the onset of the radiation-dominated RW era with $\theta = \frac{3}{2t}$, the pressure $\hat{p}$ changed and, indeed, $\hat{p} = \frac{\rho}{3}$ at $t = \alpha$ together with $f(\theta) = \frac{1}{3}$.

As in the case of GR considered in the previous chapter (section 2 of chapter V) one can calculate the ratio of photon to particle number or the specific entropy per particle at the transition epoch α. The result for this ratio is the same as that is given in (540) or (547). In the GR formulation this ratio remains constant after the transition from the creation to radiation era because the particle creation has ceased afterwards. However, in the MGR framework the specific entropy per particle (baryon) does not remain constant after the epoch α to account for the present value of it.Although the particle creation due to the pressure p_c stopped at $t = \alpha$ but the creation phenomenon continued in the subsequent era governed by standard cosmology in MGR. In fact, the particle creation continued due to the nonconservation of matter under MGR. It is shown by Abdel-Rahman (1997) that the specific entropy per baryon remains proportional to $T^{1-\eta}$ in the standard cosmology, that is, the baryon asymetry was much smaller in the early universe than at present. Thus,

$$\frac{\sigma_0}{\sigma_\alpha} = \left(\frac{T_0}{T_\alpha}\right)^{1-\eta} \tag{757}$$

where σ_α and T_α are the corresponding quantities at the epoch time α. However, for a value of η in the range $0.9 < \eta \leq 1$, the present value of the specific entropy σ_0 per baryon is $\geq 10^8$, which is still within the observational limits. In the following section the production of specific entropy per baryon as an effect of bulk viscosity in the early universe will be discussed in the framework of MGR.

6.5: BULK VISCOSITY IN THE EARLY UNIVERSE UNDER MGR

In section 1, the effect of bulk viscous pressure in the early universe in the framework of GR was considered. Presently, we shall investigate its effect in the MGR formulation of the early universe as a thermodynamically open system. As before we shall consider bulk viscosity under equilibrium thermodynamics. In contrast to the extended irreversible thermodynamics(Cases-Vazquez et al(1984)) the equilibrium thermodynamics has shortcoming. But justification of dealing the bulk viscosity with the equilibrium thermodynamics can be made for the present case with MGR (as well as with GR that has been considered earlier) for the following reason: [see also De (2002)]

The equation for bulk viscous pressure Π in the extended irreversible thermodynamics is given by

$$\Pi = -\zeta\theta - \tau\dot{\Pi} \tag{758}$$

where τ is the relaxation time. This relaxation time can be taken as

$$\tau = \frac{\zeta(t)}{\rho} \tag{759}$$

(Pavon et al (1991); see also Sudharsan and Johri(1994)). Since $\rho = \frac{\theta^2}{3K}$, we have from (758) the following equation for Π:

$$\Pi = -\zeta\theta - \frac{3K\zeta(t)}{\theta^2}\dot{\Pi} \tag{760}$$

For the GR case the coefficient of bulk viscosity is given by (682). The bulk viscous pressure is nonzero just before the epoch time α , when the universe goes through radiation-dominated RW phase. The coefficient of bulk viscosity is proportional to the difference $f(\theta)-\frac{1}{3}$ in that era and this difference is very small in the transition era just before $t=\alpha$. In fact, at $t=\alpha$, this difference becomes equal to zero. We shall see in the following discussion that the same is also true for the present case of MGR. Thus, if we take the average value of the factor $f(\theta)-\frac{1}{3}$ over this transition era to be equal to l which is obviously very small then, for GR case we have from (682)

$$\zeta(t) = \frac{\theta l}{3K} \tag{761}$$

Since for radiation-dominated RW era we have

$$\frac{\dot{\theta}}{\theta^2} = -\frac{2}{3} \tag{762}$$

we have from (760) and (761)

$$\frac{d\Pi}{d\theta} - \frac{1}{2k\zeta(t)}\Pi = \frac{\theta}{2k} \tag{763}$$

or

$$\frac{d\Pi}{d\theta} - \frac{3}{2\theta l}\Pi = \frac{\theta}{2k} \tag{764}$$

The solution of this equation is given by

$$\Pi = -\frac{\theta^2 l}{3k(1-\frac{4}{3}l)} = -\zeta(t)\frac{\theta}{1-\frac{4}{3}l} \tag{765}$$

Since l is very small we have

$$\Pi \simeq -\zeta(t)\theta \tag{766}$$

This corresponds to the equilibrium solution for the viscous pressure. The same conclusion is also valid for MGR case as we shall see that the coefficient of viscosity for this case differs from that of GR case by only a numerical factor.

Now, in the case of bulk viscosity under equilibrium thermodynamics the stress-energy tensor is given by

$$T_{\mu\nu} = (\rho + p + p_c - \zeta\theta)u_\mu u_\nu - (\rho + p_c - \zeta\theta)g_{\mu\nu} \tag{767}$$

Consequently, the Einstein equation (708) and the conservation law (722) for the thermodynamically open universe are modified into the following forms:

$$\theta^2 = \frac{3K}{\eta}\left[\rho - (1-\eta)(p - \zeta\theta)\right] \tag{768}$$

$$d(a^3[\rho - (1-\eta)(p-\zeta\theta)]) + \frac{1}{3}[(\eta-1)\rho + (5-2\eta)(p-\zeta\theta)]d(a^3) - \frac{h}{n}d(na^3) = 0 \tag{769}$$

From equations (766) and (767), we arrive at

$$\frac{\eta}{3k}(2\dot{\theta} + \theta^2)\theta + \frac{\theta}{3}[(\eta-1)\rho + (5-2\eta)(p-\zeta\theta)] - (p+\rho)(\frac{\dot{n}}{n} + \theta) = 0 \tag{770}$$

When the particle creation in the open universe stops, $\frac{\dot{N}}{N} = 0$, that is, $\frac{\dot{n}}{n} + \theta = 0$ and the universe goes through the usual radiation-dominated era (of course, in the framework of MGR and with bulk viscosity) for which $\theta = \frac{3}{2t}$, and consequently $\dot{\theta} + \frac{2\theta^2}{3} = 0$ and we have from (768),

$$\frac{\eta\theta^2}{3k} = (\eta-1)\rho + (5-2\eta)(p-\zeta\theta) \tag{771}$$

Now, using equation (766) and the equation of state (727) we find, from (769)

$$\zeta(t) = \frac{1}{\theta}[f(\theta) - \frac{1}{3}]\rho \tag{772}$$

Again from (766), we get

$$\rho = \frac{\frac{\eta\theta^2}{3k} - (1-\eta)\zeta\theta}{1-(1-\eta)f(\theta)} \tag{773}$$

From equations (770) and (771), by eliminating ρ , we finally arrive at the following expression for the coefficient of bulk viscosity:

$$\zeta(t) = \frac{\eta\theta}{(2+\eta)k}\left[f(\theta) - \frac{1}{3}\right] \tag{774}$$

Now, the entropy equation in this case is given by

$$T\frac{\dot{S}}{V} = \zeta\theta^2 + \frac{TS}{V}\frac{\dot{N}}{N} \tag{775}$$

where $V = a^3(t)$ is the comoving volume. This equation is the same as in (665) for the GR case and consequently the same equation for the specific entropy production per particle , that is, equation (670) follows. From this equation it follows that the specific entropy per particle,σ , remains constant when $\varsigma = 0$. Note that when particle creation stops, $\dot{N} = 0$, the relation (670) remains valid.When $f(\theta) = \frac{1}{3}$ at the epoch α at the onset of the radiation-dominated RW era, $\varsigma(t)$ becomes zero and σ remains constant afterward. But in the transition era before that epoch time, $f(\theta) > \frac{1}{3}$ and production of specific entropy per particle is possible. In fact, in the GR formulation of the early universe as a thermodynamically open universe with bulk viscosity considered earlier in section 6.1 we have seen that for the equation of state (727) where $f(\theta)$ is given by (736) , the coefficient of bulk viscosity must vanish during the 'particle production' (matter dominated) RW era until the transition epoch. In this era , the creation of particles and entropy production continue with constant specific entropy per particle σ. In the present MGR formulation we consider the case of nonzero $\varsigma(t)$ in the transition epoch just before the epoch time α. The equation for σ follows from (772) and (670):

$$\dot{\sigma} = \frac{\eta\theta^3}{(2+\eta)kTn}\left[f(\theta) - \frac{1}{3}\right] \tag{776}$$

Now, we can use the standard cosmological invariant (684) (which is valid from this transition period to the present era) and note that as particle creation stops , $N = nR^3$ remains constant afterward. Then, from (774) we have

$$\dot{\sigma} = \frac{\eta\theta^3 C^3}{(2+\eta)kNT^4}\left[f(\theta) - \frac{1}{3}\right] \tag{777}$$

An estimate of specific entropy produced per particle during the short transition period specified as ($\alpha - \Delta\alpha,\ \alpha$) can be obtained. This estimate for the period $\Delta\alpha$ just before α is given by

$$\Delta\sigma = \frac{\eta\theta_\tau^3 cC^3}{(2+\eta)kNT_\tau^4}\left[f(\theta_\tau) - \frac{1}{3}\right]\Delta\alpha \tag{778}$$

where the subscript τ represents the values for the corresponding quantities at some intermediate epoch τ of the transition period given by $\tau = \alpha - r\Delta\alpha, 0 < r < 1.$. Now, from the cosmological invariant given in (684) it also follows that

$$\frac{T_\alpha}{T_\tau} = \left(\frac{\tau}{\alpha}\right)^{1/2} \tag{779}$$

In natural units , $T_\alpha \simeq 10^{22} cm^{-1}$. Also, $k = 8\pi G = \frac{1}{m_{pl}^2} = t_{pl}^2$ in these units, m_{pl} and t_{pl} being , respectively , the Planck mass and time. Putting $\frac{\Delta\alpha}{\alpha} = q$, we have from (776), by using (736),(684) and (777)

$$\begin{aligned}\Delta\sigma &= \left(\frac{\alpha}{2\tau+\alpha} - \frac{1}{3}\right)\frac{27\eta C^3\Delta\alpha}{8(\eta+2)\tau t_{pl}^2 NT_\alpha^4\alpha^2} = \left(\frac{1}{3-2rq} - \frac{1}{3}\right)\frac{27\eta C^3 q}{8(\eta+2)Nt_{pl}^2 T_\alpha^4\alpha^2(1-rq)} \\ &= 2.09\times 10^{12}\left[\frac{nq}{(\eta+2)(1-rq)}\left(\frac{1}{3-2rq} - \frac{1}{3}\right)\right]\end{aligned} \tag{780}$$

where we have used the total particle (baryon) number of the present universe , given by

$$N_b = 2.45\times 10^{78} u^3 \quad (1 < u < 1.8) \tag{781}$$

as calculated in De (1993b). There the matter creation was considered quantum mechanically for the initial anisotropic fluctuation of Minkowski space-time and subsequently in the matter dominated RW era up to the epoch time α (see also the previous chapter). With the specification of r and q (which specify the period of transition) as $r = 0.5$ and $q = \frac{1}{3}$ we find an estimate of the specific entropy produced per particle as

$$\Delta\sigma \simeq 3.48 \times 10^{10}\eta/(\eta + 2) \tag{782}$$

As pointed out above, the effect of viscous pressure vanishes after the epoch α when $f(\theta) = \frac{1}{3}$. In the GR case for which $\eta = 1$, the specific entropy per particle is , from (780)

$$\Delta\sigma \simeq 1.16 \times 10^{10} \tag{783}$$

which remains constant afterward until the present state of the universe. This value is also very close to that in (547). If the transition period is a bit shorter, say $q = \frac{1}{4}$,the specific entropy per particle is (in the GR case) 6×10^9 . The interesting point is that this observable quantity of the present universe was produced in the transition period just before the epoch time α. For the MGR case, however, as pointed out earlier, this quantity decreases.Only for a value of η in the range $0.9 < \eta \leq 1$ its present value is within the observational limits. For a smaller η, the transition era might have been much longer to account for a production of specific entropy per particle. There can be calculated from (775) using (777) and (779), and for the transition period from the epoch time τ to the epoch α. We have

$$\begin{aligned}\sigma &= \tfrac{\eta C^3}{(2+\eta)kN} \int_\tau^\alpha \tfrac{\theta^3}{T^4}\left[f(\theta) - \tfrac{1}{3}\right] dt = 2.09 \times 10^{12}\tfrac{\eta}{\eta+2}\int_\tau^\alpha \tfrac{1}{t}\left[\tfrac{\alpha}{2t+\alpha} - \tfrac{1}{3}\right] dt \\ &= 2.09 \times 10^{12}\tfrac{\eta}{\eta+2}\left[\tfrac{2}{3}\ln\tfrac{\alpha}{\tau} - \ln\tfrac{3\alpha}{2\tau+\alpha}\right]\end{aligned} \tag{784}$$

The present value of the specific entropy per particle is given by

$$\sigma_0 = 2.09 \times 10^{12}\frac{\eta}{2+\eta}\left(\frac{2}{3}ln\frac{\alpha}{\tau} - ln\frac{3\alpha}{2\tau+\alpha}\right)\left(\frac{T_0}{T_\alpha}\right)^{1-\eta} \tag{785}$$

However, a much smaller value of η from the range $0.9 < \eta \leq 1$ cannot account for the present value σ_0 within its observational limits. For example, for a transition period from $\tau = 10^{-26}$ sec to α , η should be equal to 0.8 to have $\sigma_0 \simeq 10^8$, the observational lower limit. Even a greater transition period cannot give rise to a value of $\eta \leq 0.75$ if $\sigma_0 \geq 10^8$. Therefore, from this observational fact, it is evident that the parameter η should lie in the range $0.75 \leq \eta \leq 1$.

It was pointed out in section 6.1 that in the GR formulation the usual inflation for which $\dot{\theta} = 0$ cannot produce specific entropy per particle. There the usual inflation was found as a trivial solution of the evaluation equation for the early universe. In the MGR formulation $\dot{\theta} = 0$ is not a trivial solution of the evolution equation for the early universe. But an usual inflation solution $\dot{\theta} = 0$ of the equation (738) or (739) can arise if one takes a fixed value of $f(\theta)$, given by

$$f(\theta) = -1$$

and hence

$$p = f(\theta)\rho = -\rho$$

In this case we have from (766) and (768)

$$\frac{\eta\theta^2}{3K} = (2-\eta)\rho + (1-\eta)\zeta\theta \tag{786}$$

and

$$\frac{\eta\theta^2}{3K} = (2-\eta)\rho + \frac{5-2\eta}{3}\zeta\theta \tag{787}$$

where we have used the relation (784). From these two equations it follows that

$$(1-\eta)\varsigma\theta - \frac{5-2\eta}{3}\varsigma\theta = 0. \tag{788}$$

or,

$$(2+\eta)\varsigma\theta = 0. \tag{789}$$

Since $\eta \neq -2$, we must have $\varsigma(t) = 0$. Consequently, from (670), it follows that $\dot{\sigma} = 0$. Thus, for the usual inflation in the MGR formulation σ remains constant.

In the MGR framework with the introduction of bulk viscous pressure the production of specific entropy per particle is possible. The fact that bulk viscosity can account for the production of σ was long ago conceived by Zeldovich(1970) and subsequently considered by others(see Desikan (1997) for references). The present consideration of specific entropy per particle in the early universe with MGR gives rise to a value within the observational limits for η in the range $0.75 \leq \eta \leq 1$. For $\eta = 1$, that is, for GR, the value of σ obtained from bulk viscosity is found to be almost the same as that obtained from the former consideration(see chapter V) of matter and radiation energy densities at the epoch α when the particles become relativistic and contribute to the radiation energy density. Thus, the present consideration limits η to the range $0.75 \leq \eta \leq 1$.

It is interesting to note that for the lowest value of η in the above mentioned range the present value of the matter density parameter Ω_0(which is the ratio of the present matter density to the critical density) given by $\Omega_0 = \eta$ (Abdel-Rahaman,1997) is reduced to 75% of its value from a standard spatially flat RW universe with GR ($\eta = 1$). However, this lowest value can only increase the age of the universe marginally; specifically, by a factor $\frac{3}{(2+\eta)} = 1.09$.

In the previous chapter (and also in De, 1993b), the quantum production of very large massive particles(masses $\geq 53.3m_{pl}$) was considered. These created particles at the Plank order time give rise to the energy density at the epoch from which the expansion stage of the universe begins. From this energy density one can calculate the energy density at the time the mild inflationary phase is turned off, that is, at the transition epoch α. It is found to be (from (594))

$$\rho(\alpha) = 6.14 \times 10^{90} cm^{-4}$$

From the ratio of the energy densities at α, as in equation(541), the matter-energy density ρ_m at α can be obtained. With the value of γ given by (538) we have

$$\rho_m(\alpha) = 6.14 \times 10^{72} cm^{-4}$$

From this, the number density of the particles (baryons) can be found and it is given by $n_m \simeq 10^{59} cm^{-3}$. We can find the total particle number at the transition epoch with the use of the volume of the universe at that epoch obtained from the standard cosmological invariant (684). In fact, from this invariant one can find

$$R(\alpha) \simeq 5 \times 10^6 \; ucm., \quad (1 < u < 1.8)$$

Consequently, the total particle number is $\geq 10^{79}$ and therefore, the total entropy $S \geq 10^{88}$. In the GR formalism, these quantities remain constant after the epoch α when the standard cosmology sets in. Thus, these values of the total particle number and entropy correspond to their present values also. Now, as discussed by Blau and Guth (1987), this large value of entropy per comoving volume might be regarded as an alternative statement of the flatness problem. In the standard cosmological models it is a matter of setting the large value of S as an initial condition, but in the present case, it achieves such a large value because of the creation phenomenon. The large value of the particle number (baryon number) of the universe is also the outcome of this model. Thus, there remains no flatness problem in the present consideration of the early universe.

6.6: CAUSAL HOMOGENIZATION CONDITION

Challane- Tlemsani and Le Denmat (1995) investigated the early universe as the viscous cosmic fluid in the framework of the extended irreversible thermodynamics. With the supposition of the coefficient of bulk viscosity $\varsigma(t) \propto \rho^{\frac{1}{2}}$, they obtained the scale factor $R(t) \propto t$ as the solution.This solution corresponds to a singular origin of the universe. Of course, in there it is shown that the causal homogenization condition of the hypersurface $t = \bar{t}$ (a given time)

$$\int_{\bar{t}}^{t_0} \frac{dt}{R(t)} \geq \int_{t}^{\bar{t}} \frac{dt}{R(t)} \tag{790}$$

where t_0 is the present time, is satisfied for the time $\bar{t} = t^*$ when the phase $R(t) \propto t$ ends. when the phase $R(t) \propto t$ ends. In (790) the functions $R(t)$ represent the scale factors for the corresponding periods within the limits of integrations. Thus, whatever the scale factor may be for the interval (t^*, t_0), the condition (790) is satisfied as the left hand side diverges. Although, the problem of switching off for the phase $R(t) \propto t$ still remains likewise for the case of usual exponential inflation where, as pointed out earlier, extra physical process is necessary for it. In fact, when the viscosity is negligible the universe transmits to the radiation-dominated RW phase.

The mild inflation ($R(t) \propto t$) considered in the former sections needs no such additional physical processes for switching on and off. Also, the the conservation law (Bianchi identity) takes the usual form (718), that is,

$$d(\rho V) = -\hat{p} dV \tag{791}$$

with the phenomenological pressure $\hat{p}$ given in (6.4.21). For $t < \alpha$, we have $f(\theta) \simeq 1$ and $\frac{\dot{\theta}}{\theta^2} = -\frac{1}{3}$ we have from (755),

$$\hat{p} = -\frac{1}{3}\rho \tag{792}$$

Since $\rho V = mnV = mN = M =$ total mass in the comoving volume $V = R(t)^3$, we have from (791) and (792),

$$\frac{dM}{M} = \frac{1}{3}\frac{dV}{V} = \frac{dR}{R}$$

Consequently,

$$M \propto R(t) \propto t$$

or,

$$M = \frac{\hat{M}}{\hat{t}} t \tag{793}$$

where $\hat{M}$ is the total mass in V at the time $\hat{t}$ (the Planck order time when the mild inflation is switched on). It is interesting to note that if one supposes that there is no 'switching on' time of this inflationary phase then it is obvious from (793) that at $t = 0$, $M = 0$. That is, the universe originates from a 'no mass' (or 'no particle') state and, of course, from a singularity. But, the fact is that before the time $\hat{t}$, the universe (the space-time) was in a 'quantum state' , a Minkowskian space-time perturbed anisotropically. This state of the space-time creates matter with very high massive particles (masses $\geq 53.3\ m_{pl}$). The time-scale of the damped anisotropic perturbation is of the order of $\hat{t}$. Due to the instability of the Minkowski space-time caused by these created very massive particles the universe begins its expansion phase, the mild inflation at this time scale, that is, at $\hat{t}$. Subsequently the anisotropy is damped out because of the back reaction and the universe returns instantaneously to isotropy. (Hu and Parker, 1978; Hartle and Hu, 1979 and references therein). Now, if we adopt the view point as in Prigogine (1989) that "time precedes creation" then the space-time before the "duration" of the anisotropic perturbation was Minkowskian. The condition of causal homogenization (790) for the present case becomes

$$\int_{-\infty}^{\bar{t}} \frac{dt}{R(t)} = \int_{-\infty}^{\hat{t}} \frac{dt}{R(t)} + \int_{\hat{t}}^{\bar{t}} \frac{dt}{R(t)} \geq \int_{\bar{t}}^{t_0} \frac{dt}{R(t)} \tag{794}$$

The scale factor R(t) in the range $-\infty < t \le \hat{t}$ is, in fact, corresponds to the anisotropically perturbed Minkowski space-time. This is given as

$$R(t) = R(\hat{t}) \; + \; damped \;\; anisotropic \;\; perturbation \tag{795}$$

where R($\hat{t}$) is a constant and is equal to the value of scale factor R(t) of the mild inflation at its beginning, that is, at $t = \hat{t}$. Obviously, the integral $\int_{-\infty}^{\hat{t}} \frac{dt}{R(t)}$ diverges and thus the condition (793) is satisfied for $\bar{t} = \alpha$, the switching off time of the mild inflation. The hypersurface $t = \alpha$ is, thus, entirely homogenized causally. Interestingly, if the universe originates, as pointed out above, from a "no matter" or "no particle" state but a space-time singularity then also the condition of homogenization is satisfied because for mild inflation the integral $\int_0^{\alpha} \frac{dt}{R(t)}$ diverges. However, for the present case it is due to the fact that the universe was Minskowskian before its expansion, there appears no horizon problem that is generally encountered by the standard model.

6.7: COSMOLOGICAL CONSTANT PROBLEM

Wienberg (1989) described the cosmological constant problem as a veritable crisis in Physics. The problem is connected with the fact that the theoretical estimations from the theories of elementary particles for the cosmological constant are exceeding observational limits by as many as 120 order of magnitude. In order to resolve this problem of very 'near to zero' value of cosmological constant numerous attempts have already been made (for review, see Wienberg, 1989). These attempts are based on considerations such as supersymmetry, supergravity and superstring theories, anthropic principle, adjustment mechanism, changing gravity approaches and quantum cosmology. Presently, we shall discuss this problem of 'zero' cosmological constant in the context of our previous consideration of the early universe (see also De, 1995). To this end, it is the changing gravity approach (Buchmuller and Dragon, 1988a, b; Van der Bij et al. 1982; Wienberg, 1983; Wilezek and Zee 1983; see also Wienberg, 1989) which is most suitable and promising in this situation. In this approach the cosmological constant appears as a constant of integration, unrelated to any parameter in the action for gravity and matter. On the contrary, the problem of cosmological constant arises because the effective vacuum energy contributes to the effective cosmological constant and the theoretical estimates for it from the vacuum energy exceed observational limits by about 118 - 120 orders of magnitude. By changing the rules of classical general relativity one can delink it from the vacuum energy. The action for gravity and matter is taken as the usual one which is

$$I[\psi, g] = \frac{1}{16\pi G} \int d^4x \sqrt{-g} R + I_M[\psi, g]. \tag{796}$$

ψ being a set of matter fields that appear in the matter action I_M. This matter action I_M may include a possible cosmological constant term $\frac{\lambda}{8\pi G} \int \sqrt{-g} d^4x$. As in conventional general relativity, general covariance is also maintained here. But this changed theory of gravity differs from the conventional theory because of the fact that here not all the components of the metric $g_{\mu\nu}$ are the dynamical variables. On the other hand, it is supposed here that the determinant g is not dynamical and consequently the action is to be stationary only with respect to variations in the metric that maintain the determinant fixed, that is, for which

$$g^{\mu\nu} \delta g_{\mu\nu} = 0. \tag{797}$$

This leads to the following field equations

$$R^{\mu\nu} - \frac{1}{4} g^{\mu\nu} R = -8\pi G \left(\widetilde{T}^{\mu\nu} - \frac{1}{4} g^{\mu\nu} \widetilde{T} \right) \tag{798}$$

where $\widetilde{T} = \widetilde{T}_{\lambda}^{\lambda}$, $\widetilde{T}^{\mu\nu}$ being the variational derivative of I_M with respect to $g_{\mu\nu}$. The field equations (798) are just the traceless part of the Einstein field equations. For the present case of the early universe taken as a thermodynamically open system, $\widetilde{T}^{\mu\nu}$ represents the phenomenological energy-momentum tensor. It corresponds to the creation of particles and becomes the usual energy-momentum tensor when the creation of particles stops. For the creation era the phenomenological energy-momentum tensor in a perfect fluid model is given by

$$\widetilde{T}^{\mu\nu} = T^{\mu\nu} + \hat{T}^{\mu\nu} \tag{799}$$

where

$$T^0_0 = \rho, T^i_j = -p\delta^i_j. \tag{800}$$

and

$$\hat{T}^0_0 = 0, \hat{T}^i_j = -p_c\delta^i_j. \tag{801}$$

Here, ρ and p are the energy density and the true thermodynamic pressure respectively. As pointed out earlier, the creation of matter is due to a supplementary pressure p_c, which is a part of the phenomenological pressure $\widetilde{p}$.

In fact

$$\hat{p} = p + p_c \tag{802}$$

where p_c is given in (720). By using the equation of state (720), the mass relation (725) and the evolution equations (6.4.1) it follows that

$$p_c = -\rho\left(1 + f(\theta) + \frac{2\dot{\theta}}{\theta^2}\right) = -\rho\left(1 + f(\theta) + \frac{4\alpha\overline{m}\dot{m}}{3(m-\overline{m})^2}\right) \tag{803}$$

where $\bar{m}$ is the inherent mass of the particle and is equal to its present mass to a very high degree of accuracy. Also, it was discussed earlier that the pressure p_c becomes zero at the epoch $t = \alpha$ when the equation of state is given by $p = f(\theta)\rho = \frac{\varrho}{3}$ at the onset of the radiation era. Actually, p_c is negative or zero according as the presence on absence of particle production. Thus, in the absence of particle production $\hat{T}^{\mu\nu}$ becomes zero and consequently we have $\tilde{T}^{\mu\nu} = T^{\mu\nu}$.

Now, as this formulation maintains general covariance, the energy-momentum tensor satisfies the usual conservation property:

$$\tilde{T}^{\mu\nu}_{;\mu} = 0. \tag{804}$$

Again, when the Bianchi identities

$$(R^{\mu\nu} - \frac{1}{2}g^{\mu\nu}R)_{;\mu} = 0. \tag{805}$$

are taken into consideration a nontrivial consistency condition is to be satisfied. This consistency condition is not necessary for usual Einstein field equations because these are consistent with the equations (803) and (804). The consistency condition for the present case takes the following form:

$$\frac{1}{4}\partial_\mu R = 8\pi G\left(\frac{1}{4}\partial_\mu \tilde{T}\right) \tag{806}$$

from which it follows that

$$R - 8\pi G\tilde{T} = 4\Lambda. \tag{807}$$

where 4Λ appears as the constant of integration. This constant Λ can now be identified with the cosmological constant because one can recover the Einstein field equations

$$R^{\mu\nu} - \frac{1}{2}g^{\mu\nu}R + \Lambda g^{\mu\nu} = -8\pi G\tilde{T}^{\mu\nu}. \tag{808}$$

from the present field equations (798) with the use of equation (806). Clearly, this cosmological constant has nothing to do with any terms in the action or vacuum fluctuations. Actually, the equations (798) do not involve any cosmological constant and the contribution of vacuum fluctuations automatically cancels on the right hand side of these equations. It is interesting to note that one can derive the equations (798) from (807). In fact, by contraction, from (807) we can arrive at the equation (806) and then by using this equation into (807) we can find the equation (798). In this case, of course, Λ is not a constant of integration but it is the cosmological constant.

Now, the important consequence of this changing rule of gravity is that the field equations have flat-space solutions in the absence of matter and radiation and with no creation of particles. Thus, in order to resolve the cosmological constant problem the only question now remains, to wit; why should we choose the flat-space solution (in the absence of matter and radiation)? The answer to it is not an intricate one in the context of the consideration of the early universe as discussed in the previous chapter.

Since, in this case, the universe has originated from the flat space, free of matter and radiation, that is, Minkowski space-time through anisotropic perturbation therein, then the situation appears as trivial in resolving the problem. It is, indeed, that a flat space solution is a 'reality' just before the 'creation' of the universe. As discussed earlier, the produced heavy massive particles in the Minkowski space-time anisotropically perturbed for a very short period (to the order of Planck time) thrust the space-time into its expanding stage. The back-reaction effect essentially makes the universe instantaneously to isotropy (Hu and Parker 1978; Hartle and Hu, 1979; Lukash et al. 1976). With the creation of particles in the mild inflationary stage for a brief period up to the epoch time $t = \alpha$, the universe enters into the radiation era with the standard cosmology. Thus, this present consideration of the early universe with the changing gravity can incorporate the 'zero' cosmological constant adequately.

REFERENCES

[1] Abdel-Rahaman,A.M.M.(1995) General Relativity and Gravitation, **27**, 573.

[2] Abdel-Rahaman,A.M.M.(1997) General Relativity and Gravitation, **29**, 1329.

[3] Abramo, L.R.W.and Lima, J.A.S. (1996) Classical and Quantum Gravity, **13**, 2953.

[4] Al-Rawaf,A.S. and Taha,M.O. (1996a) Physics LettersB,**366**, 69.

[5] Al-Rawaf,A.S. and Taha,M.O. (1996b)General Relativity and Gravitation, **28**, 935.

[6] Blau, S.K. and Guth,A.H.(1987) In Three Hundred Years of Gravitation, S.W. Hawking and W. Israel, eds, Cambridge University Press, Cambridge, p-524.

[7] Buchmüller, W. and Dragon, N. (1988a) University of Hannover Preprint no. ITP-UH 1/88.

[8] Buchmüller, W. and Dragon, N. (1988b)Physics Letters B, **207**, 292.

[9] Calvao, M.O., Lima, J.A.S. and Waga, I. (1992), Physics Letters A, **162**, 223.

[10] Cases-Vazquez, J., Jou, D., and Lebon, G., eds(1984) Recent Development of Nonequilibrium Thermodynamics (Lecture notes in Physics,199), Springer-Verlag, Berlin.

[11] Challane-Tlemsani, S. and Le Denmat, G. (1995), International Journal of Modern Physics D, **4**, 549.

[12] Chen, W. and Wu, Y.S (1990), Physical Review D, **41**, 695

[13] De, S.S. (1993a) International Journal of Theoretical Physics, **32**, 1603.

[14] De, S.S. (1993b) Communications in Theoretical Physics, **2**, 249.

[15] De, S.S. (1995) Communications in Theoretical Physics,**4**, 115.

[16] De, S.S. (1999) International Journal of Theoretical Physics, **38**, 2419.

[17] De, S.S. (2002) International Journal of Theoretical Physics, **341**, 137.

[18] Desikan, K. (1997) General Relativity and Gravitation, **29**, 435.

[19] Freese, K., Adams, F.C., Frieman, J.A. and Mottola,E. (1987), Nuclear Physics B, **287**, 797.

[20] Gunzig, E. , Maartens, R. and Nesteruk, A.V. (1998), Classical and Quantum Gravity, **15**, 928.

[21] Hu, B.L. and Parker, L. (1978) Physical Review D, **17**, 933.

[22] Hartle, J.B. and Hu, B.L. (1979), Physical Review D, **20**, 1772.

[23] Johri, V.B. and Sudharsan, R. (1988), Physics Letters A, **132**, 316.

[24] Luhash,V.N. , Novikov,I.D. , Starobinsky,A.A. and Zel'dovich,Ya,B. (1976), Nuovo Cimento B, **35**, 293

[25] Nardone, P. (1989) International Journal of Theoretical Physics, **28**, 935.

[26] Özer, M. and Taha, M.O. (1986), Physics Letters B, **171**, 363.

[27] Padmanabhan, T. and Chitre, S.M. (1987), Physics Letters A, **120**, 433.

[28] Pavon, D. , Bafaluy, J. and Jou, D. (1991), Classical and Quantum Gravity, **8**, 347.

[29] Pierce, M.J. , et al. (1994) , Nature, **371**, 385.

[30] Prigogine, I. (1989), International Journal of Theoretical Physics, **28**, 927.

[31] Rastall, P. (1972), Physical Review D, **6**, 3357.

[32] Silveira, V. and Waga, I. (1994), Physical Review D, **50**, 4890.

[33] Smalley L.L. (1984), Nuovo Cimento B, **80**, 42.

[34] Sudharsan, R. and Johri, V.B.(1994), General Relativity and Gravitation, **26**, 41.

[35] Van der Bij, J.J. , Van Dam, H. and Ng, Y.J. (1982), Phisica, **116A** , 307.

[36] Waga, I. , Falcao, R.C. and Chandra, R. (1986), Physical Review D, **33**, 1839.

[37] Weinberg, S. (1971), Astrophysical Journal, **168**, 175.

[38] Weinberg, S. (1972), Gravitation and Cosmology, Wiley, New York.

[39] Weinberg, S. (1983), Unpublished remarks at the Workshop on "Problems in Unification and Supergravity" La Jolla Institute, 1983.

[40] Weinberg, S. (1989), Review of Modern Physics, **61**, 1.

[41] Wilezek, F. and Zee, A. (1983), Unpublished work quoted by Zee (1985) in High Energy Physics : Proceedings of the Annual Orbis Scientiae, edited by S.L. Mintz and A. Perlmutter, (Plenum, New York).

[42] Zel'dovich, Ya. B. (1962), Soviet Physics - JETP, **16**, 1163.

[43] Zel'dovich, Ya. B. (1970), JETP Letters, **12**, 307.

CHAPTER SEVEN:

A MODIFIED RIEMANNIAN GEOMETRY: LYRA GEOMETRY

7.1: INTRODUCTION

Einstein's General Theory of Relativity is geometrized theory of gravitation based on Riemannian geometry. Shortly, after the discovery of Einstein's theory, several modification of Riemannian geometry (or generalization of Riemannian geometry) have developed to solve the problems such as unification of gravitation with electromagnetism, problems arising when the gravitational field is coupled to matter fields, singularities of standard cosmology etc. In 1918, Weyl suggested a generalization of Riemannian geometry by introducing a new scalar field that accompanies the metric field and changes the scale of length measurements. This new modification of Riemannian geometry was not attracted to the readers because, it was based on non integrability of length transfer. Also this modification of Riemannian geometry criticized by Einstein as this implies that spectral frequencies of atoms depend on their past histories which has never been observed. In 1951, Lyra suggested another modification of Riemannian geometry (may be also considered as the modification of Weyl geometry) by introducing a gauge function in the structureless manifold which removes the non integrability condition of a vector under parallel transport. Also, this modification of Riemannian geometry has taken care of the notion of Einstein's principle of geometrization of gravitation i.e. here both the scalar and vector fields have more or less intrinsic geometrical significance.

7.2: FOUNDATION OF LYRA GEOMETRY

The differential of geometrical structure of a manifold is determined is as follows:
(i) Let PP' be two neighbouring points P(x^i) and $P'(x^i + dx^i)$. An affine connection characterized by its components $\Gamma^{\mu}_{\alpha\beta}$, which are defined by the change due to infinitesimal parallel transfer of a vector ξ^{μ} from P to P'

$$d\xi^{\mu} = -\Gamma^{\mu}_{\alpha\beta}\xi^{\alpha}dx^{\beta} \tag{809}$$

(ii) a metrical connection characterized by the metric fundamental tensor $g_{\alpha\beta}$ which is defined by the measure of length ξ of ξ^{μ} ,

$$\xi = g_{\alpha\beta}\xi^{\alpha}\xi^{\beta} \tag{810}$$

Riemannian Geometry is characterized by the following assumptions:

(i) symmetry of $\Gamma^{\mu}_{\alpha\beta}$ in the lower two indices i.e.,

$$\Gamma^{\mu}_{\alpha\beta} = \Gamma^{\mu}_{\beta\alpha} \tag{811}$$

(ii) length of a vector remains unaltered under the parallel transfer i.e.,

$$\delta\xi = \delta(g_{\alpha\beta}\xi^{\alpha}\xi^{\beta}) = 0 \tag{812}$$

Using the above two conditions, the Riemannian connection is uniquely determined by the metric tensor

$$\Gamma^{\mu}_{\alpha\beta} = \left\{ \begin{matrix} \mu \\ \alpha \quad \beta \end{matrix} \right\}$$

Here the second quantities are the Christoffel symbols of the second kind.
Weyl generalized the notion of Riemannian metrical connection by introducing the concept of non integrability of length transfer i.e. the length ξ of any vector ξ^μ is assumed to change due to infinitesimal parallel transfer from P(x^i) to $P'(x^i + dx^i)$ as

$$\delta\xi = -\xi\phi_\mu dx^\mu \tag{813}$$

where ϕ_μ is a vector function characterized the manifold. Thus a Weyl manifold is characterized by two independent quantities $g_{\alpha\beta}$ and ϕ_μ.
Note: If ϕ_μ is the gradient of a scalar function, then the change in the length ξ of any vector ξ^μ in going from one point to another point is independent of the path followed. At every point of Weyl manifold, there is a possibility of change of scalar (i.e. a recalibration) or a gauge transformation. The concept of gauge transformation is coming from the non integrability of length transfer. Now, making a gauge transformation i.e. recalibrating all lengths,

$$\xi \longrightarrow \overline{\xi} = \lambda(x^\mu)\xi, \tag{814}$$

$g_{\alpha\beta}$ and ϕ_μ transform as

$$g_{\alpha\beta} \longrightarrow \overline{g}_{\alpha\beta} = \lambda g_{\alpha\beta} \quad ; \quad \phi_\mu \longrightarrow \overline{\phi}_\mu = \phi_\mu - \frac{\lambda,_\mu}{\lambda} \tag{815}$$

where, $\lambda,_\mu = \frac{\partial\lambda}{\partial x^\mu}$.

[$\xi \longrightarrow \overline{\xi}$ implies, $ds \longrightarrow \overline{ds}$. Now, $\overline{ds}^{\,2} = \overline{g}_{\mu\nu}dx^\mu dx^\nu$ and $ds^2 = g_{\mu\nu}dx^\mu dx^\nu$.
Since, $\overline{\xi} = \lambda(x^\mu)\xi$, it follows that $\overline{ds}^{\,2} = \lambda(x^\mu)ds^2$. In other words, $\overline{g}_{\mu\nu} = \lambda(x^\mu)g_{\mu\nu}$.
Again, let after the gauge transformation, we have, $d\overline{\xi} = -\overline{\xi}\,\overline{\phi}_\mu dx^\mu$.
Now, equation (814) implies $d\overline{\xi} = \xi\lambda,_\mu dx^\mu + d\xi\lambda$.
Using (813), one gets, $-\xi\lambda\,\overline{\phi}_\mu dx^\mu = \xi\lambda,_\mu dx^\mu - \lambda\xi\phi_\mu dx^\mu$. This yields $\overline{\phi}_\mu = \phi_\mu - \frac{\lambda,_\mu}{\lambda}$]

Using equations (813) and (809), one can get the components of affine connection $\Gamma^\mu_{\alpha\beta}$ of a Weyl manifold as:

$$\Gamma^\mu_{\alpha\beta} = \left\{ \begin{matrix} \mu \\ \alpha \quad \beta \end{matrix} \right\} + \frac{1}{2}\left(\delta^\mu_\alpha\phi_\beta + \delta^\mu_\beta\phi_\alpha - g_{\alpha\beta}\phi^\mu\right) \tag{816}$$

where $\phi^\mu = g^{\mu\lambda}\phi_\lambda$.
For $n = 4$, the Weyl curvature scalar R^W(Ricci scalar) can be calculated from (8) as

$$R^W = R + 3\phi^\alpha{}_{;\,\alpha} + \frac{3}{2}\phi_\alpha\phi^\alpha \tag{817}$$

where, R is the Riemannian curvature scalar and $\phi^\alpha_{;\alpha}$, the covariant derivative of ϕ^α with respect to the Christoffel symbols $\left\{ \begin{matrix} \mu \\ \alpha \quad \beta \end{matrix} \right\}$.

$$\left[R^W = g^{\lambda\nu}R_{\lambda\nu} = g^{\lambda\nu}\left(\Gamma^\alpha_{\lambda\beta}\Gamma^\beta_{\mu\nu} - \Gamma^\mu_{\lambda\nu}\Gamma^\beta_{\mu\beta} + \Gamma^\mu_{\lambda\mu,\,\nu} - \Gamma^\mu_{\lambda\nu,\,\mu}\right)\right]$$

The most serious objection (or drawbacks) of Weyl's hypothesis is non integrability of length transfer (i.e., connection is not metric preserving).
Three decades after Weyl, Lyra (1951) has suggested a modification of Riemannian geometry (also may be considered as modification of Weyl geometry) which removes the non integrability of length transfer of a vector under parallel transport (i.e. connection is metric preserving) by introducing a gauge function

in the structureless manifold. According to Lyra, the displacement vector PP' between two neighbouring points P(x^i) and $P'(x^i + dx^i)$ has the components

$$\xi^\mu = x^0(x^\mu)dx^\mu \tag{818}$$

where $x^0(x^\mu)$ is a non zero gauge function. The coordinate system x^μ and the gauge function x^0 form a reference system (x^0, x^μ). The transformation to a new reference system $(x^{\overline{0}}, x^{\overline{\mu}})$ is given by

$$x^\mu \longrightarrow x^{\overline{\mu}} = x^{\overline{\mu}}(x^\lambda) \qquad ; \qquad x^0 \longrightarrow x^{\overline{0}} = x^{\overline{0}}(x^0, x^\mu) \tag{819}$$

with

$$A^{\overline{\mu}}_{\mu} \equiv \frac{\partial x^{\overline{\mu}}}{\partial x^\mu} \quad , \quad det\ A^{\overline{\mu}}_{\mu} \neq 0 \ ; \ A^{\mu}_{\overline{\mu}} \equiv \frac{\partial x^{\mu}}{\partial x^{\overline{\mu}}} \quad , \quad det\ A^{\mu}_{\overline{\mu}} \neq 0 \ \ and \ \ \frac{\partial x^{\overline{0}}}{\partial x^0} \neq 0.$$

In view of above transformation (819), a multicomponent tensor $\xi^{\rho_1\rho_2 \ldots\ldots \rho_s}_{\sigma_1\sigma_2\ldots\ldots\sigma_r}$ transforms as follows:

$$\xi^{\overline{\rho}_1\overline{\rho}_2 \ldots\ldots \overline{\rho}_s}_{\overline{\sigma}_1\overline{\sigma}_2\ldots\ldots\overline{\sigma}_r} = \lambda^{s-r} A^{\overline{\rho}_1}_{\rho_1} \ldots\ldots A^{\overline{\rho}_s}_{\rho_s} A^{\sigma_1}_{\overline{\sigma}_1} \ldots\ldots A^{\sigma_r}_{\overline{\sigma}_r} \xi^{\rho_1\rho_2 \ldots\ldots \rho_s}_{\sigma_1\sigma_2\ldots\ldots\sigma_r} \tag{820}$$

Here the factor λ^{s-r} where $\lambda = \frac{x^{\overline{0}}}{x^0}$, arises as a consequence of the introduction of the gauge function. In any general reference system, the coefficients of the affine connection $\Gamma^\mu_{\alpha\beta}$ (which have been arisen as consequence of general coordinate transformation) are determined in the following manner: Let us consider that, in a coordinate system (x^μ), a vector ξ^μ is constant i.e. $\frac{\partial \xi^\mu}{\partial x^\lambda} = 0$. Then, in another local coordinate system $\xi^{\overline{\mu}}$, we have in usual case ($\xi^\mu = A^\mu_{\overline{\mu}}\ \xi^{\overline{\mu}}$),

$$\frac{\partial \xi^{\overline{\mu}}}{\partial x^{\overline{\lambda}}} + \Gamma^{\overline{\mu}}_{\overline{\nu}\ \overline{\lambda}}\ \xi^{\overline{\nu}} = 0 \tag{821}$$

where

$$\Gamma^{\overline{\mu}}_{\overline{\nu}\ \overline{\lambda}} = -A^{\mu}_{\overline{\nu}}\ A^{\overline{\mu}}_{\mu\ ,\ \overline{\lambda}} \quad ; \quad A^{\overline{\mu}}_{\mu\ ,\ \overline{\lambda}} = \frac{\partial A^{\overline{\mu}}_{\mu}}{\partial x^{\overline{\lambda}}} \tag{822}$$

Alternatively, we say the consequence of the fact $\xi^\mu = constant$ that the equation (821) is valid in all coordinate systems, but $\Gamma^\mu_{\nu\lambda} = 0$ in a particular system x^μ.
It is shown by Lyra (1951) and Sen (1957, 1958) that in any general reference system the coefficients of the generalized affine connection characterized not only by $\Gamma^\mu_{\nu\lambda}$ but also the function ϕ_α, which arises due to the introduction of the gauge function x^0 into the structureless manifold.
A vector ξ^μ in Lyra geometry transforms as

$$\xi^{\overline{\mu}} = \lambda\ A^{\overline{\mu}}_{\mu}\ \xi^\mu \tag{823}$$

If $\xi^\mu{}_{,\ \lambda} = \frac{\partial \xi^\mu}{\partial x^\lambda} = 0$ in the reference system (x^0, x^μ) , then in the reference system $(x^{\overline{0}}, x^{\overline{\mu}})$, we have

$$\left(\frac{1}{x^{\overline{0}}}\right)\ \xi^{\overline{\mu}}{}_{,\ \overline{\lambda}} + \Gamma^{\overline{\mu}}_{\overline{\nu}\ \overline{\lambda}}\ \xi^{\overline{\nu}} - \frac{1}{2}\ \phi_{\overline{\lambda}}\ \xi^{\overline{\mu}} = 0 \tag{824}$$

where

$$\Gamma^{\overline{\mu}}_{\overline{\nu}\ \overline{\lambda}}\ \xi^{\overline{\nu}} = -\left(\frac{1}{x^{\overline{0}}}\right)\ A^{\mu}_{\overline{\nu}}\ A^{\overline{\mu}}_{\mu\ ,\ \overline{\lambda}} \ ; \ \phi_{\overline{\lambda}} = \left(\frac{1}{x^0}\right)\ \frac{\partial \ln \lambda^2}{\partial x^{\overline{\lambda}}} \tag{825}$$

Note that $A^{\overline{\mu}}_{\mu}\ A^{\mu}_{\overline{\nu}} = \delta^{\overline{\mu}}_{\overline{\nu}}$ and hence by partial differentiating with respect to $x^{\overline{\nu}}$, one gets, $A^{\overline{\mu}}_{\mu,\ \overline{\nu}}\ A^{\mu}_{\overline{\nu}} = -A^{\overline{\mu}}_{\mu}\ A^{\mu}_{\overline{\nu},\ \overline{\nu}}$. Accordingly, $\Gamma^{\overline{\mu}}_{\overline{\nu}\ \overline{\lambda}}$ is symmetrical in $\overline{\nu}$ and $\overline{\lambda}$. Similar to Riemannian geometry, the parallel transfer of a vector ξ^μ in Lyra geometry is given by

$$\delta\xi^\mu = -\widetilde{\Gamma}^\mu_{\alpha\beta}\ \xi^\alpha\ x^0\ dx^\beta \tag{826}$$

where

$$\widetilde{\Gamma}^{\mu}_{\alpha\beta} = \Gamma^{\mu}_{\alpha\beta} - \frac{1}{2}\,\delta^{\mu}_{\alpha}\,\phi_{\beta} \tag{827}$$

Note that $\widetilde{\Gamma}^{\mu}_{\alpha\beta}$ are not symmetric although $\Gamma^{\mu}_{\alpha\beta} = \Gamma^{\mu}_{\beta\alpha}$. Thus the components of the generalized affine connection in Lyra geometry are characterized by $\Gamma^{\mu}_{\alpha\beta}$ and ϕ_{β}.

The metric or the measure of length of the displacement vector $\xi^{\mu} = x^0\ dx^{\mu}$ between two points P(x^i) and $P'(x^i + dx^i)$ is invariant under both coordinate and gauge transformations and is given by

$$ds^2 = g_{\mu\nu}\ x^0\ dx^{\mu}\ x^0\ dx^{\nu} \tag{828}$$

where $g_{\mu\nu}$ is a second rank symmetric tensor. The parallel transfer of a vector in Lyra geometry is integrable i.e. the length of a vector is conserved under parallel transport as in Riemannian geometry but in contrast to Weyl's geometry ; i.e.,

$$\delta(g_{\mu\nu}\ \xi^{\mu}\ \xi^{\nu}) = 0 \tag{829}$$

From equations (826) and (829), it follows that

$$\Gamma^{\mu}_{\alpha\beta} = \left(\frac{1}{x^0}\right)\left\{ \begin{matrix} \mu \\ \alpha\quad\beta \end{matrix} \right\} + \frac{1}{2}\left(\delta^{\mu}_{\alpha}\phi_{\beta} + \delta^{\mu}_{\beta}\phi_{\alpha} - g_{\alpha\beta}\phi^{\mu}\right) \tag{830}$$

where $\phi^{\mu} = g^{\mu\lambda}\phi_{\lambda}$.
One can note that the components of affine connection in Weyl's geometry is identical with $\Gamma^{\mu}_{\alpha\beta}$ in (830) except the factor $(\frac{1}{x^0})$, not, however, $\widetilde{\Gamma}^{\mu}_{\alpha\beta}$. Hence, we have a vector field with intrinsic geometrical significance in Lyra geometry, apart from the inconvenience of non integrability of length transfer of Weyl's geometry. Thus Lyra geometry is characterized by the two fundamental entities $g_{\alpha\beta}$ and ϕ_{λ}.

7.3: LYRA MANIFOLD AS A MODIFIED RIEMANNIAN MANIFOLD

Let us consider a connected second countable Housdorff space M. We are concerning with Lyra formulation and its essential notion is a reference system.
The local reference system on M is a triplet (U_i, ψ_i, f_i) where
(1) U_i is an open subset of M
(2)ψ_i is a homomorphism of U_i onto an open subset of R^n
(3) $f_i : U_i \longrightarrow R - \{0\}$ is non zero gauge function on U_i

We call M is an n dimensional C^k Lyra manifold if $\exists$ a collection of local reference system $\{(U_i, \psi_i, f_i)\ ;\ i \in I,$ *an index set*$\}$ on M such that
(1) U_i cover M i.e. the union of U_i is equal to M, $\cup_{i\in I} U_i = M$
(2) Whenever $U_i \cap U_j \neq \varphi$, the maps $\psi_i\ o\ \psi_j^{-1}$ and $\psi_j\ o\ \psi_i^{-1}$ are C^k on their domains of definition.

(3) For each $i \in I$, the map $f_i\ o\ \psi_i^{-1}$ is a C^k function on $\psi_i(U_i)$.

Note: If the functions $\psi_i\ o\ \psi_j^{-1}$, $\psi_j\ o\ \psi_i^{-1}$ and $f_i\ o\ \psi_i^{-1}$ are continuous and can be differentiable as many times as you like i.e. the functions are C^{∞}, then we say M is a C^{∞} Lyra manifold or Lyra smooth manifold. Here, $\{U_i, \psi_i\}$ are atlas of M.
A C^k or (C^{∞}) curve on a Lyra manifold M is a mapping $\sigma : [a, b] \subset R \longrightarrow R$ such that $\psi\ o\ \sigma : R \longrightarrow R^n$ is C^k (or C^{∞}) (cf. figure 9)

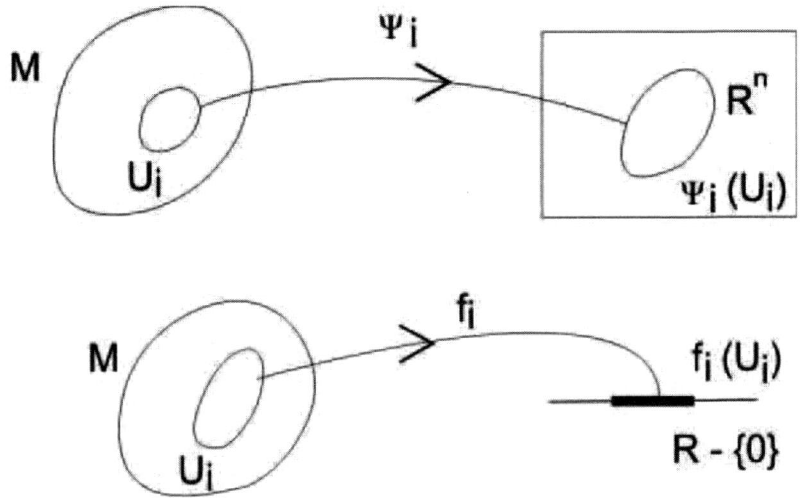

Figure 6: ψ_i is a homomorphism of U_i onto an open subset of R^n. $f_i : U_i \longrightarrow R - \{0\}$ is non zero gauge function on U_i.

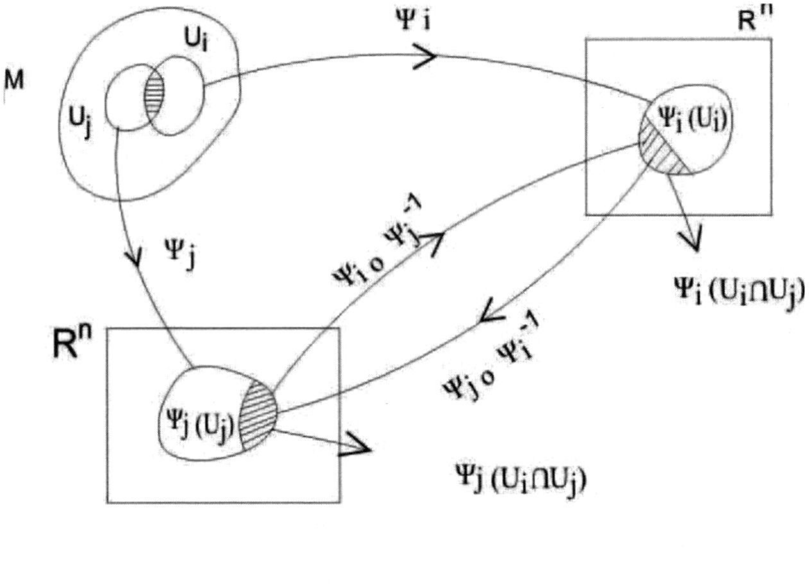

Figure 7: The maps $\psi_i\ o\ \psi_j^{-1}$ and $\psi_j\ o\ \psi_i^{-1}$ defined on shaded regions are must be C^k

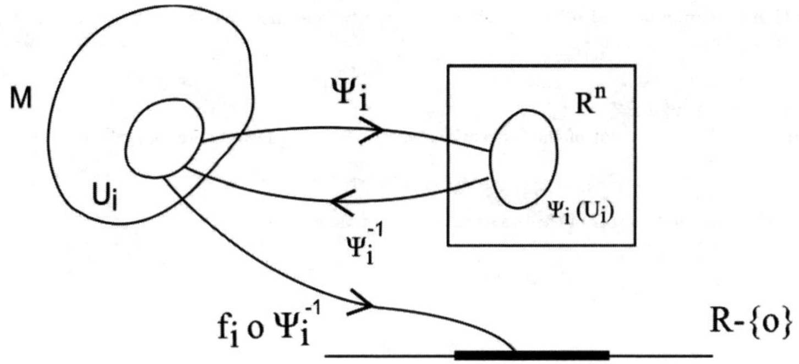

Figure 8: the map $f_i \ o \ \psi_i^{-1}$ is a C^k function on U_i.

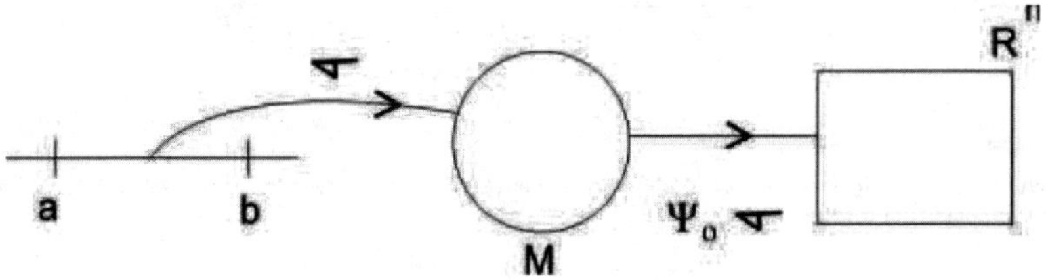

Figure 9:

Now, one can introduce tangent space $T_p(M)$ at a point p on a Lyra Manifold M in the usual manner. Consider a reference system (U_i, ψ_i, f_i) on M. For $p \in U_i$, let $\psi_i(p) = \{x^1(p), ..., x^n(p)\} = x^\mu$ be the local coordinate of p. Now, we denote gauge function $f_i \; o \; \psi_i^{-1} \; : \; U_i \longrightarrow R - \{0\}$ as $x^0 \; : \; x^\mu \in M \longrightarrow x^0(x^\mu)$ in terms of local coordinate. Thus n coordinate x^μ together with the gauge function $x^0 = f_i \; o \; \psi_i^{-1}$ characterize the local reference system (= coordinate system + gauge) in local coordinates. Let us consider another local reference system (U_i', ψ_i', f_i') such that $U_i \cap U_j \neq \varphi$.

Under transformation of local reference (x^0, x^μ) to the new reference system $(x^{\overline{0}}, x^{\overline{\mu}})$, the basis vectors transform as

$$e_{\overline{\mu}} = \lambda^{-1} \; A^\mu_{\overline{\mu}} \; e_\mu \tag{831}$$

where $\lambda = \frac{x^{\overline{0}}}{x^0}$ and $A^\mu_{\overline{\mu}} = \frac{\partial x^\mu}{\partial x^{\overline{\mu}}}$.

Consequently, the component of the tangent vector $\xi = \xi^\mu \; e_\mu$ in local reference system transforms as follows:

$$\xi^{\overline{\mu}} = \lambda \; A^{\overline{\mu}}_\mu \; \xi^\mu \tag{832}$$

We note that the Lie brackets of the basis vectors do not vanish:

$$[\; e_\mu \; , \; e_\nu \;] = \frac{1}{2} \left(\delta^\alpha_\mu \; \tilde{\phi}_\nu - \delta^\alpha_\nu \; \tilde{\phi}_\mu \right) \; e_\alpha \tag{833}$$

where

$$\tilde{\phi}_\alpha = (x^0)^{-1} \; [\; \ln(x^0)^2 \;]_{\; , \; \alpha} \tag{834}$$

Elements of the cotangent space $T^*_P(M)$ are linear functions from $T_p(M) \longrightarrow R$. The natural basis in $T^*_P(M)$ [dual to $e_\alpha(p)$] is

$$e^\alpha(p) = x^0 \; dx^\alpha \tag{835}$$

Here the basis 1-forms are given by the condition

$$e^\mu(e_\nu) = \gamma^\mu_\nu \tag{836}$$

Let T_p and T^*_p are tangent and cotangent space of an n dimensional smooth Lyra manifold M at p, then, the inner product

$$< . \; , \; . > \; : \; T^*_p \times T^p \longrightarrow R$$

is defined by

$$< \; w \mid v \; >= w(v) = w_\mu v^\mu \tag{837}$$

A tensor of type (r,s) is a multilinear mapping of r copies of T^*_p and s copies of T^p into R:

$$T : \; \overbrace{T^*_p \; \times \; T^*_p \; \times ..... \times \; T^*_p} \; \times \overbrace{T_p \; \times \; T_p \; \times .... \times \; T_p} \longrightarrow R$$

Then an element t in an (r,s) type tensor space T^r_s can be expressed as

$$t = t^{\nu_1 ..... \nu_s}_{\mu_1 ..... \mu_r} \; e^{\mu_1} \otimes ...... \otimes e^{\mu_r} \; \otimes e_{\nu_1} \otimes .... \otimes e_{\nu_s} \tag{838}$$

Then under a transformation from a reference system (U_i, ψ_i, x_0) to a new reference system $(\overline{U}_i, \overline{\psi}_i, \overline{x}_0)$, the components of a tensor t of type (r,s) transform as

$$\begin{aligned} t^{\overline{\mu}_1 \; ..... \; \overline{\mu}_r}_{\overline{\nu}_1 ..... \overline{\nu}_s} &= \lambda^{r-s} \frac{\partial x^{\overline{\mu}_1}}{\partial x^{\nu_1}} \; ..... \; \frac{\partial x^{\overline{\mu}_r}}{\partial x^{\nu_r}} \cdot \frac{\partial x^{\delta_1}}{\partial x^{\overline{\nu}_1}} \; ...... \; \frac{\partial x^{\delta_s}}{\partial x^{\overline{\nu}_s}} \\ &= \lambda^{r-s} A^{\overline{\mu}_1}_{\nu_1} \; ...... \; A^{\overline{\mu}_r}_{\nu_r} \; . \; A^{\delta_1}_{\overline{\nu}_1} \; .... \; A^{\delta_s}_{\overline{\nu}_s} \; t^{\nu_1 \; ..... \; \nu_r}_{\delta_1 \; ..... \; \delta_s} \end{aligned} \tag{839}$$

where $A^{\overline{\mu}}_\nu \equiv \frac{\partial x^{\overline{\mu}}}{\partial x^\nu}$ and $A^\mu_{\overline{\nu}} \equiv \frac{\partial x^\mu}{\partial x^{\overline{\nu}}}$.

The components of the metric tensor g is a symmetric tensor field of type (0,2) on Lyra Manifold M and is denoted in a local reference system (i.e. in the natural basis) by

$$g_{\mu\nu} = g(e_\mu, e_\nu) \tag{840}$$

and the classical metric has the following form

$$ds^2 = g_{\mu\nu} e_\mu \otimes e_\nu = (x^0)^2 \; g_{\mu\nu} \; dx^\mu dx^\nu \tag{841}$$

In an usual way, one can define affine connection (∇) on C^k Lyra manifold.
Let us consider a bilinear mapping

$$\nabla \; : \; T(M) \times T(M) \longrightarrow T(M)$$

which satisfies the following conditions where T(M) denotes the linear space of all C^k vector fields in Lyra manifold M (or more specifically, tangent space in M):
For any $f \in C^k(M)$ and arbitrary $X, Y, Z \in T(M)$

$$\begin{aligned} &(i) \qquad\qquad \nabla_{fX} Y = f \nabla_X Y \\ &(ii) \; \nabla_X (fY) = f \nabla_X Y + (Xf) Y \\ &(iii) \; \nabla_X (Y + Z) = \nabla_X Y + \nabla_X Z \end{aligned} \tag{842}$$

$[\nabla(X,Y) \equiv \nabla_X Y]$
The bilinear mapping satisfying the above conditions is known as Linear Lyra connection on M.
The torsion of ∇ is the mapping

$$Tor_\nabla(X,Y) \equiv B(X,Y) = \nabla_X Y - \nabla_Y X - [X,Y] = \frac{1}{2}[\phi(Y)X - \phi(X)Y] \tag{843}$$

where ϕ is a given 1-form and [ ] is the Lie bracket of X, Y. With the help of this connection ∇, we can define the covariant derivative of any tensor field with respect to a vector field Z which preserves the tensor type.
The covariant derivative of any second order symmetric tensor field of type (0,2) is given by

$$(\nabla_Z \; g)(X,Y) \equiv A(X,Y,Z) = Z(g(X,Y)) - g(\nabla_Z X, Y) - g(\nabla_Z Y, X) \tag{844}$$

A connection ∇ is said to be metric preserving or satisfying the condition of metricity or to have integrable length transfer if

$$\nabla_Z \; g(X,Y) = 0 \;\; , \qquad for \;\; all \;\; Z \in T(M) \tag{845}$$

A connection ∇ is said to be torsion free if

$$Tor_\nabla(X,Y) = 0 \tag{846}$$

In Riemannian Geometry, the connection ∇ is both Torsion free and length transfer is integrable. In Weyl geometry the connection ∇ is torsion free but not metric preserving. In Lyra Geometry, the connection ∇ is metric preserving but not torsion free.
In a local reference system (x^μ, x^0), the Lyra connection ∇ is given by

$$\nabla_{e_\beta} e_\alpha = \widetilde{\Gamma}^\mu_{\alpha\beta} \; e_\mu \tag{847}$$

where

$$e_\mu(p) = \left((x^0)^{-1} \frac{\partial}{\partial x^\mu} \right) \tag{848}$$

in $T_p(M)$ are the natural basis.
Result: In a local reference system the Lyra connection components are given by

$$\widetilde{\Gamma}^\mu_{\alpha\beta} = \left(\frac{1}{x^0} \right) \left\{ \begin{array}{cc} \mu & \\ \alpha & \beta \end{array} \right\} + \frac{1}{2} \left(\delta^\mu_\beta \; \phi_\alpha - g_{\alpha\beta} \; \phi^\mu \right) \tag{849}$$

Lyra set

$$\widetilde{\Gamma}^{\mu}_{\alpha\beta} = \Gamma^{\mu}_{\alpha\beta} - \frac{1}{2}\delta^{\mu}_{\alpha}\, \phi_{\beta} \tag{850}$$

so that the above equation takes the form

$$\Gamma^{\mu}_{\alpha\beta} = \left(\frac{1}{x^0}\right)\left\{ \begin{matrix} \mu \\ \alpha \quad \beta \end{matrix} \right\} + \frac{1}{2}\left(\delta^{\mu}_{\alpha}\, \phi_{\beta} + \delta^{\mu}_{\beta}\, \phi_{\alpha} - g_{\alpha\beta}\, \phi^{\mu}\right) \tag{851}$$

where

$$\phi_{\alpha} = \phi(e_{\alpha}) + \tilde{\phi}_{\alpha} \ \ and \ \ \phi^{\mu} = g^{\mu\alpha}\phi_{\alpha} \tag{852}$$

with

$$\tilde{\phi}_{\alpha} = (x^0)^{-1}\ [\ \ln(x^0)^2\ ]\ ,\ _{\alpha} \equiv -2\partial_{\alpha}\left(\frac{1}{x^0}\right) \tag{853}$$

Proof:

Suppose that $\nabla_Z(X,Y) = A(X,Y,Z)$ and $Tor_{\nabla}(X,Y) \equiv B(X,Y) = \frac{1}{2}[\phi(Y)X - \phi(X)Y]$ for all $X, Y, Z \in T(M)$. Then one can get the following expression expression for Lyra connection as

$2g(\nabla_X Y, Z) = X(g(Y,Z)) + Y(g(Z,X)) - Z(g(X,Y)) - g([X,Z],Y) - g([Y,X],Z)$

$+\, g([Z,Y],X) - \phi(Z)g(X,Y) + \phi(Y)g(X,Z)$

(Using $A(X,Y,Z) = 0$, since the length transfer is integrable in Lyra Geometry)

Now, putting $X = e_{\alpha}$, $Y = e_{\beta}$, $Z = e_{\nu}$ in the above expression, we get

$$2\widetilde{\Gamma}^{\mu}_{\beta\alpha}\, g_{\mu\nu} = \left(\frac{1}{x^0}\right)(\partial_{\alpha}g_{\beta\nu} + \partial_{\beta}g_{\nu\alpha} - \partial_{\nu}g_{\alpha\beta}) + (\phi(e_{\beta}) + \tilde{\phi}_{\beta})g_{\alpha\nu} - (\phi(e_{\nu}) + \tilde{\phi}_{\nu})g_{\alpha\beta}$$

Here, we have used the following expressions:

$\nabla_X Y = \nabla_{e_\alpha} e_{\beta} = \widetilde{\Gamma}^{\mu}_{\beta\alpha}\, e_{\mu}$, $g(e_{\alpha}, e_{\beta}) = g_{\alpha\beta}$, $[X,Y] = [e_{\alpha}, e_{\beta}] = \frac{1}{2}\left(\delta^{\mu}_{\alpha}\, \tilde{\phi}_{\beta} - \delta^{\mu}_{\beta}\, \tilde{\phi}_{\alpha}\right)\, e_{\mu}$,
$\phi_{\alpha} = \phi(e_{\alpha}) + \tilde{\phi}_{\alpha}$, $\tilde{\phi}_{\alpha} = (x^0)^{-1}\ [\ \ln(x^0)^2\ ]\ ,\ _{\alpha} \equiv -2\partial_{\alpha}\left(\frac{1}{x^0}\right)$, $g(\nabla_X Y, Z) = g\left(\Gamma^{\mu}_{\beta\alpha}\, e_{\mu}, e_{\nu}\right) = \widetilde{\Gamma}^{\mu}_{\beta\alpha}\, e_{\mu}\, g_{\mu\nu}$

According to Lyra, if we set $\widetilde{\Gamma}^{\mu}_{\alpha\beta} = \Gamma^{\mu}_{\alpha\beta} - g_{\alpha\beta}\, \phi^{\mu}$, one can get finally,

$$\Gamma^{\mu}_{\alpha\beta} = \left(\frac{1}{x^0}\right)\left\{ \begin{matrix} \mu \\ \alpha \quad \beta \end{matrix} \right\} + \frac{1}{2}\left(\delta^{\mu}_{\alpha}\, \phi_{\beta} + \delta^{\mu}_{\beta}\, \phi_{\alpha} - g_{\alpha\beta}\, \phi^{\mu}\right)$$

Hence the proof.

Now, we derive the transformation formula for $\widetilde{\Gamma}^{\mu}_{\alpha\beta}$. In an other reference system $(x^{\overline{0}}, x^{\overline{\mu}})$, we have

$$\nabla_{e_{\overline{\beta}}} e_{\overline{\alpha}} = \nabla_{\lambda^{-1}\, A^{\beta}_{\overline{\beta}}\, e_{\beta}}(\lambda^{-1}\, A^{\alpha}_{\overline{\alpha}}\, e_{\alpha}) = \widetilde{\Gamma}^{\overline{\mu}}_{\overline{\alpha}\overline{\beta}}\, e_{\overline{\mu}} \tag{854}$$

Applying conditions (842), one gets,

$$\widetilde{\Gamma}^{\overline{\mu}}_{\overline{\alpha}\overline{\beta}} = \lambda^{-1}\, A^{\overline{\mu}}_{\mu}\, A^{\alpha}_{\overline{\alpha}}\, A^{\beta}_{\overline{\beta}}\, \widetilde{\Gamma}^{\mu}_{\alpha\beta} + \left(x^0\right)^{-1} A^{\overline{\mu}}_{\nu}\, A^{\nu}_{\overline{\alpha}\ ,\ \overline{\beta}} - \frac{1}{2}\, \left(x^{\overline{0}}\right)^{-1}\, \delta^{\overline{\mu}}_{\overline{\alpha}}\, A^{\nu}_{\overline{\beta}}\, (\ln\lambda^2), \nu \tag{855}$$

Using $\widetilde{\Gamma}^{\mu}_{\alpha\beta} = \Gamma^{\mu}_{\alpha\beta} - \delta^{\mu}_{\alpha}\, \phi_{\beta}$, one gets the following transformation laws as

$$\widetilde{\Gamma}^{\overline{\mu}}_{\overline{\alpha}\overline{\beta}} = \lambda^{-1} \left[A^{\overline{\mu}}_{\mu} A^{\alpha}_{\overline{\alpha}} A^{\beta}_{\overline{\beta}} \widetilde{\Gamma}^{\mu}_{\alpha\beta} + (x^0)^{-1} A^{\overline{\mu}}_{\nu} A^{\nu}_{\overline{\alpha}\,,\,\overline{\beta}} \right] \tag{856}$$

It is to be noted that the ϕ_α do not transform as (831). Rather under gauge transformation ($x^0 \longrightarrow x^{\overline{0}}$), it would be

$$\phi_\alpha \longrightarrow \phi_{\overline{\alpha}} = \lambda^{-1} A^{\alpha}_{\overline{\alpha}} \left[\phi_\alpha + \left(\frac{1}{x^0}\right) \partial_\alpha (\ln \lambda^2)\right] \tag{857}$$

7.4: GEODESICS IN LYRA MANIFOLD

Let $x^\tau = x^\tau(s)$ be a curve whose tangential vector $\xi^\tau = \frac{x^0 dx^\tau}{ds}$ is transferred parallel to itself. In Lyra geometry, the autoparallels are the curves $x^\tau(s)$ whose tangent vectors $\xi^\tau = \frac{x^0 dx^\tau}{ds}$ satisfy the differential equations

$$\frac{d\xi^\tau}{ds} + \widetilde{\Gamma}^\tau_{\nu\mu} \xi^\nu \xi^\mu = 0 \tag{858}$$

where, $\widetilde{\Gamma}^\tau_{\nu\mu}$ are the Lyra connection components. Then the auto parallel curves of the Lyra connection ∇ in a local reference system is given by

$$\frac{x^0 d^2 x^\tau}{ds^2} + \left[\left(\frac{1}{x^0}\right)\left\{ \begin{matrix} \tau \\ \nu \quad \mu \end{matrix} \right\} + \frac{1}{2}\left(\delta^\tau_\nu\, \phi_\mu + \delta^\tau_\mu\, \phi_\nu - g_{\nu\mu}\, \phi^\tau\right)\right] \frac{dx^\nu}{ds}\frac{dx^\mu}{ds} (x^0)^2 - \frac{1}{2}(\phi_\nu - \tilde{\phi}_\nu) \frac{dx^\nu}{ds}\frac{dx^\tau}{ds} (x^0)^2 = 0 \tag{859}$$

where

$$\tilde{\phi}_\alpha = (x^0)^{-1} \left[\, \ln (x^0)^2 \,\right]_{,\,\alpha} \equiv -2\partial_\alpha \left(\frac{1}{x^0}\right)$$

A geodesic equation of the metrical Lyra connection is an extreme curve $x^\mu = x^\mu(t)$ given by

$$\delta \int ds = \delta \int \left[(x^0)^2\, g_{\mu\nu}\, \frac{dx^\mu}{dt}\frac{dx^\nu}{dt}\right]^{\frac{1}{2}} dt = 0$$

i.e. by

$$\delta\left[\int L\; dt\right] = 0$$

where

$$L = \left[(x^0)^2\, g_{\mu\nu}\, \frac{dx^\mu}{dt}\frac{dx^\nu}{dt}\right]^{\frac{1}{2}}$$

The Euler Lagrange equation for the geodesics are

$$\frac{d}{dt}\left[\frac{\partial L}{\partial \dot{x}^\nu}\right] - \frac{\partial L}{\partial x^\nu} = 0 \qquad where\ \dot{x}^\nu = \frac{dx^\nu}{dt}$$

. Here,

$$\frac{\partial L}{\partial \dot{x}^\nu} = \frac{1}{2}\left[(x^0)^2\, g_{\mu\nu}\, \frac{dx^\mu}{dt}\frac{dx^\nu}{dt}\right]^{-\frac{1}{2}} (x^0)^2\, g_{\mu\nu}\, \dot{x}^\mu = \frac{1}{2}\left(\frac{ds}{dt}\right)^{-1} (x^0)^2\, g_{\mu\nu}\, \dot{x}^\mu$$

$$\frac{\partial L}{\partial x^\nu} = \frac{1}{2}\left[(x^0)^2\, g_{\mu\nu}\, \frac{dx^\mu}{dt}\frac{dx^\nu}{dt}\right]^{-\frac{1}{2}} \dot{x}^\mu\, \dot{x}^\nu\, \{(x^0)^2\, g_{\mu\nu}\}_{,\,\nu} = \frac{1}{2}\left(\frac{ds}{dt}\right)^{-1} \dot{x}^\mu\, \dot{x}^\nu\, \{(x^0)^2\, g_{\mu\nu}\}_{,\,\nu}$$

Substituting these in Euler Lagrange equation and putting $t = s$, we get the geodesics equation of the metrical Lyra connection as

$$\frac{d^2 x^\tau}{ds^2} + \left\{ \begin{matrix} \tau \\ \nu \quad \mu \end{matrix} \right\} \frac{dx^\nu}{ds}\frac{dx^\mu}{ds} + \frac{x^0}{2}\left(\delta^\tau_\mu \tilde{\phi}_\nu + \delta^\tau_\nu \tilde{\phi}_\mu - \tilde{\phi}^\tau g_{\nu\mu}\right) \frac{dx^\mu}{ds}\frac{dx^\nu}{ds} = 0 \tag{860}$$

where

$$\tilde{\phi}^{\alpha} = g^{\alpha\nu}\tilde{\phi}_{\nu} \quad and \quad \tilde{\phi}_{\alpha} = (x^0)^{-1}\ [\ \ln(x^0)^2\ ]\ ,\ _{\alpha} \equiv -2\partial_{\alpha}\left(\frac{1}{x^0}\right)$$

One can note that the equation (858) of auto parallel of the Lyra affine connection is not be a curve of extremal length given by $\delta\int ds = 0$.
The nature is sharp contrast to the Riemannian Geometry where auto parallels associated with the affine connection coincide with the geodesics which arise from the metric. It is seen that these types of curves will be the same if

$$\tilde{\phi}_{\alpha} = \phi_{\alpha} \tag{861}$$

The above condition (860) is invariant under gauge transformation because $\tilde{\phi}_{\alpha}$ transforms exactly as ϕ_{α} when $x^0 \longrightarrow x^{\bar{0}}$.

7.5: CURVATURE TENSOR

The curvature tensor $K^{\mu}_{\lambda\alpha\beta}$ of lyra Geometry is defined in the same manner as the curvature tenor $R^{\mu}_{\lambda\alpha\beta}$ of Riemannian Geometry. Thus a curvature tensor of a Lyra connection ∇ can be regarded as a mapping

$$K : T(M) \times T(M) \times T(M) \longrightarrow T(M)$$

given by

$$(X, Y, Z) \longrightarrow K(X, Y)Z = \nabla_X\nabla_Y Z - \nabla_Y\nabla_X Z - \nabla_{[X,Y]}Z \tag{862}$$

In a local reference system with natural basis vectors, if we set

$$K(e_{\alpha}, e_{\beta})e_{\lambda} = K^{\beta\mu}_{\lambda\alpha\beta}e_{\mu}\ , \tag{863}$$

then the components of the curvature tensor are given by

$$K^{\mu}_{\lambda\alpha\beta} = (x^0)^{-2}\left[\frac{\partial\left(x^0\,\widetilde{\Gamma}^{\mu}_{\lambda\beta}\right)}{\partial x^{\alpha}} - \frac{\partial\left(x^0\,\widetilde{\Gamma}^{\mu}_{\lambda\alpha}\right)}{\partial x^{\beta}}\right] + \widetilde{\Gamma}^{\mu}_{\rho\alpha}\,\widetilde{\Gamma}^{\rho}_{\lambda\beta} - \widetilde{\Gamma}^{\mu}_{\rho\beta}\,\widetilde{\Gamma}^{\rho}_{\lambda\alpha} \tag{864}$$

The contracted curvature tensor (similar to Ricci tensor) is obtained by setting $\mu = \beta$ in (863) as

$$K_{\lambda\alpha} = K^{\mu}_{\lambda\alpha\mu} = (x^0)^{-2}\left[\frac{\partial\left(x^0\,\widetilde{\Gamma}^{\mu}_{\lambda\mu}\right)}{\partial x^{\alpha}} - \frac{\partial\left(x^0\,\widetilde{\Gamma}^{\mu}_{\lambda\alpha}\right)}{\partial x^{\mu}}\right] + \widetilde{\Gamma}^{\mu}_{\rho\alpha}\,\widetilde{\Gamma}^{\rho}_{\lambda\mu} - \widetilde{\Gamma}^{\mu}_{\rho\mu}\,\widetilde{\Gamma}^{\rho}_{\lambda\alpha} \tag{865}$$

The Lyra contracted curvature scalar is given by

$$K = g^{\lambda\alpha}K_{\lambda\alpha} = \frac{R}{(x^0)^2} + \frac{1}{4}(n-2)(n-1)\phi^{\alpha}(\phi_{\alpha} + \tilde{\phi}_{\alpha}) + \frac{(n-1)}{x^0}\phi^{\alpha}_{\ ;\ \alpha} \tag{866}$$

where R is the Riemannian curvature scalar. Here the covariant derivative of a tensor field components in a local reference system is same as the corresponding formula in Riemannian Geometry except that $\Gamma^{\mu}_{\alpha\beta}$ is replaced by $\widetilde{\Gamma}^{\mu}_{\alpha\beta}$ and $\frac{\partial}{\partial x^{\alpha}}$ by $(x^0)^{-1}\frac{\partial}{\partial x^{\alpha}}$ i.e.

$$F^{\lambda.....}_{\sigma.....\ ;\ \mu} = \frac{1}{x^0}\,\frac{\partial}{\partial x^{\mu}}\,F^{\lambda.....}_{\sigma.....} \ + \ \widetilde{\Gamma}^{\lambda}_{\mu\alpha}\,F^{\alpha.....}_{\sigma.....} \ - \ \widetilde{\Gamma}^{\alpha}_{\mu\sigma}\,F^{\lambda.....}_{\alpha.....} \ - \ ..... \tag{867}$$

In particular, the covariant derivative for a scalar field (ϕ), vector fields (A^{ν} *or* A_{ν}) and Fundamental tensor (g_{ij})in Lyra Geometry can be defined as

$$\phi_{\ ;\ \mu} = \frac{1}{x^0}\,\frac{\partial\phi}{\partial x^{\mu}} \tag{868}$$

$$A^{\nu}_{;\,\mu} = \frac{1}{x^0}\,\frac{\partial A^{\nu}}{\partial x^{\mu}} \;+ \widetilde{\Gamma}^{\nu}_{\mu\alpha}\,A^{\alpha} \tag{869}$$

$$A_{\nu\,;\,\mu} = \frac{1}{x^0}\,\frac{\partial A_{\nu}}{\partial x^{\mu}} \;-\; \widetilde{\Gamma}^{\nu}_{\mu\alpha}\,A_{\alpha} \tag{870}$$

$$g_{ij\,;\,k} = \frac{1}{x^0}\,\frac{\partial g_{ij}}{\partial x^k} \;- g_{rj}\widetilde{\Gamma}^{r}_{ik} - g_{ir}\widetilde{\Gamma}^{r}_{jk} \tag{871}$$

Since the parallel transfer of a vector in Lyra geometry is integrable i.e. the connection is metric preserving, the last expression implies

$$g_{ij\,;\,k} = 0 \; i.e. \; \frac{1}{x^0}\,\frac{\partial}{\partial x^k}\,g_{ij} \;\; - g_{rj}\widetilde{\Gamma}^{r}_{ik} - g_{ir}\widetilde{\Gamma}^{r}_{jk} = 0$$

If we choose normal gauge $x^0 = 1$, then for $n = 4$, equation (866) becomes

$$K = R + \frac{3}{2}\,\phi^{\alpha}\,\phi_{\alpha} + 3\phi^{\alpha}_{\;;\,\alpha} \tag{872}$$

which is identical with the corresponding curvature scalar of Weyl's Geometry (see equation (817)).

In a local reference system, the components of the Lyra torsion tensor are found to be (using equation (843))

$$\tau^{\rho}_{\mu\nu} = \widetilde{\Gamma}^{\rho}_{\mu\nu} - \widetilde{\Gamma}^{\rho}_{\nu\mu} - \frac{1}{x^0}\,\left[\delta^{\rho}_{\mu}\,\partial_{\nu} - \delta^{\rho}_{\nu}\,\partial_{\mu}\right]\ln(x^0) \tag{873}$$

The trace is given by

$$\tau_{\mu} = \tau^{\rho}_{\mu\rho} \tag{874}$$

.

7.6: DKP FIELD IN LYRA GEOMETRY

It is known that the Duffin-Kemmer-Petiau (DPK) equation is a first order relativistic equation describing the fields of spin 0 and 1. We will consider Massive as well as Massless DKP field in Lyra Geometry.

Case - I . Massive DKP field in Lyra Geometry:

The Lagrangian density for the massive DKP field in Minkowski space (M^4) is given by

$$L = \frac{i}{2}\,\overline{\psi}\,\beta^a\,\partial_a\psi - \frac{i}{2}\,\partial_a\overline{\psi}\,\beta^a\,\psi - m\,\overline{\psi}\,\psi \tag{875}$$

Here $\overline{\psi} = \psi^{\dagger}\,\eta^0, \quad \eta^0 = 2(\beta^0)^2 - 1$ and β^a are matrices satisfying the usual DKP algebra, given by

$$\beta^a\,\beta^b\,\beta^c + \beta^c\,\beta^b\,\beta^a = \beta^a\,\eta^{bc} + \beta^c\,\eta^{ba} \tag{876}$$

where, a = 1, 2, 3 are spatiotemporal Minkowski indexes and η^{ab} is the metric tensor of Minkowski spacetime with signature $(+,-,-,-)$.
The matrices β^a have no inverse as these are the singular matrices which have only three irreducible representations of dimensions 1, 5 and 10. The last two have physical significance as these are correspond

to fields of spin 0 and 1 respectively where as the first one is trivial. From the above Lagrangian , one gets the massive DKP wave equation in Minkowski space as

$$i\beta^a \, \partial_a \psi - m \, \psi = 0 \tag{877}$$

Now, we will construct the DKP equation in Lyra spacetime (L^4) and for this purpose we will use the standard formalism of tetrads (or vierbeins):

A set of four vector fields $e^a_\mu(x)$ is known as a tetrad if for each x in L^4, it satisfy the following relations:

$$\eta^{ab} = e^a_\mu(x) \; e^b_\nu(x) \; g^{\mu\nu}(x) \tag{878}$$

$$g_{\mu\nu}(x) = e^a_\mu(x) \; e^b_\nu(x) \; \eta_{ab} \tag{879}$$

and

$$\eta_{ab} = e^\mu_a(x) \; e^\nu_b(x) \; g_{\mu\nu}(x) \tag{880}$$

$$g^{\mu\nu}(x) = e^\mu_a(x) \; e^\nu_b(x) \; \eta^{ab} \tag{881}$$

where

$$e^\mu_a(x) = g^{\mu\nu}(x) \; \eta_{ab} \; e^b_\nu(x) \tag{882}$$

$$e^a_\nu \; e^b_\nu = \delta^a_b \quad ; \quad e \; = \; det \; (e^\mu_a) \; = \; \sqrt{-g} \; , \; g \; = \; det(g_{\mu\nu}) \tag{883}$$

Here the Latin indexes being raised and lowered by the Minkowski metric η^{ab} and the Greek ones by the metric $g^{\mu\nu}$ of the manifold L^4.

The components B^{ab} in M^4 of a tensor $B^{\mu\nu}$ defined on L^4 are given by

$$B^{ab} = e^a_\mu \; e^b_\nu \; B^{\mu\nu} \quad , \quad B_{ab} = e^\mu_a \; e^\nu_b \; B_{\mu\nu} \tag{884}$$

or inversely

$$B^{\mu\nu} = e^\mu_a \; e^\nu_b \; B^{ab} \quad , \quad B_{\mu\nu} = e^a_\mu \; e^b_\nu \; B_{ab} \tag{885}$$

One can note that

$$A_\mu \; B^\mu = A_a \; B^a \tag{886}$$

Here each point coordinates x^μ in Lyra spacetime L^4 with metric $g_{\mu\nu}$, one can attach a Minkowski spacetime M^4 with metric η_{ab} whose point coordinates are labeled by x^a. The projection of all tensor quantities in M^4 onto L^4 are done via the tetrad fields e^μ_a. One can also define the matrices β^μ through the contraction with the tetrad fields e^μ_a as

$$\beta^\mu = e^\mu_a \; \beta^a \tag{887}$$

These β^μ satisfy the the following generalized DKP algebra

$$\beta^\mu \; \beta^\nu \; \beta^\alpha + \beta^\alpha \; \beta^\nu \; \beta^\mu = \beta^\mu \; g^{\nu\alpha} + \beta^\alpha g^{\nu\mu} \tag{888}$$

where $g^{\mu\nu}$ is Lyra metric tensor.
The Lagrangian density for the DKP field minimally coupled to the Lyra manifold is given by

$$L = \sqrt{-g}\left[\frac{i}{2} \; \overline{\psi} \; \beta^\mu \; \nabla_\mu \psi - \frac{i}{2} \; \nabla_\mu \overline{\psi} \; \beta^\mu \; \psi - m \; \overline{\psi} \; \psi\right] \tag{889}$$

where ∇_μ is the Lyra covariant derivative associated to the affine connection $\Gamma^\mu_{\alpha\beta}$.
Now, we try to establish the covariant derivative of the DKP field ψ in Lyra manifold and for this purpose

the normal procedure of analyzing the bahaviour of the field under Lorentz Transformation is followed as:

$$\psi(x) \longrightarrow \psi'(x) = U(x)\ \psi(x) \tag{890}$$

where U is a spin representation of the Lorentz group characterizing the DKP field. Now it is desired to describe a spin connection S_μ in such a way that the object

$$\nabla_\mu \psi = \frac{1}{x^0}\frac{\partial \psi}{\partial x^\mu} + S_\mu \psi \tag{891}$$

transforms like a DKP field in (891)

$$\nabla_\mu \psi \longrightarrow (\nabla_\mu \psi)' = U(x)\ \nabla_\mu \psi \tag{892}$$

This implies

$$\left(\frac{1}{x^0}\ \partial_\mu \psi + S_\mu \psi\right)' = U\left(\frac{1}{x^0}\ \partial_\mu \psi + S_\mu \psi\right)$$

or,

$$\left(\frac{1}{x^0}\ \partial_\mu \psi' + S'_\mu\ \psi'\right) = U\left(\frac{1}{x^0}\ \partial_\mu \psi + S_\mu \psi\right)$$

or,

$$\frac{1}{x^0}\ \partial_\mu [U(x)\ \psi(x)] + S'_\mu\ U(x)\ \psi(x) = U(x)\left(\frac{1}{x^0}\ \partial_\mu \psi(x) + S_\mu \psi(x)\right)$$

or,

$$S'_\mu\ U(x)\ \psi(x) = U(x)\ S_\mu \psi(x) - \frac{1}{x^0}\ \partial_\mu U(x) \psi(x)$$

Thus, finally we get S_μ transforms as

$$S'_\mu = U(x)\ S_\mu\ U^{-1}(x) - \frac{1}{x^0}\ \partial_\mu U(x)\ U^{-1}(x) \tag{893}$$

It is known that $\psi\overline{\psi}$ is a scalar and therefore covariant derivative of $\psi\overline{\psi}$ in Lyra manifold is

$$\nabla_\mu(\psi\overline{\psi}) = \frac{1}{x^0}\ \partial_\mu(\psi\overline{\psi})$$

Using equation(891), one gets,

$$\nabla_\mu \overline{\psi} = \frac{1}{x^0}\frac{\partial \overline{\psi}}{\partial x^\mu} - \overline{\psi} S_\mu \tag{894}$$

The DKP current is defined by $\overline{\psi}\ \beta^\nu\ \psi$ and taking Lyra covariant derivative of it, one obtains

$$\begin{aligned}\nabla_\mu(\overline{\psi}\ \beta^\nu\ \psi) &= \frac{1}{x^0}\ \frac{\partial(\overline{\psi}\ \beta^\nu\ \psi)}{\partial x^\mu} + \widetilde{\Gamma}^\nu_{\mu\lambda}(\overline{\psi}\ \beta^\lambda\ \psi) \\ &= (\nabla_\mu \overline{\psi})\ \beta^\nu\ \psi + \overline{\psi}(\nabla_\mu \beta^\nu)\ \psi + \overline{\psi}\ \beta^\nu\ (\nabla_\mu \psi)\end{aligned} \tag{895}$$

Using covariant derivatives of DKP fields given in equations (891) and (894), one obtain the following expression for the covariant derivative of β^ν as

$$\nabla_\mu \beta^\nu = \frac{1}{x^0}\frac{\partial \beta^\nu}{\partial x^\mu} + \widetilde{\Gamma}^\nu_{\mu\lambda}\ \beta^\lambda + S_\mu\ \beta^\nu - \beta^\nu\ S_\mu \tag{896}$$

One of the immediate consequences of the metricity condition, $\nabla_\mu g_{\alpha\beta} = 0$ in Lyra Manifold is the absolute parallelism of the tetrad i.e.

$$\nabla_\mu e^\nu_a = \frac{1}{x^0}\frac{\partial e^\nu_a}{\partial x^\mu} + \widetilde{\Gamma}^\nu_{\mu\lambda}\ e^\lambda_a + \omega^b_{\mu a}\ e^\nu_b = 0 \tag{897}$$

where $\omega_{\mu ab}$ is the spin connection coefficients. Here the tetrad fields e^{μ}_{a} and its inverse e^{a}_{μ} related to the spacetime metric given in equation (882). Using equation (897) and writing β^{ν} in terms of tetrad fields $\beta^{\nu} = e^{\nu}_{a}\beta^{a}$, we get from equation (896) as

$$\omega^{b}_{\mu a}\ e^{\nu}_{b}\ \beta^{a} = S_{\mu}\ \beta^{\nu} - \beta^{\nu}\ S_{\mu} = [S_{\mu}\ ,\ \beta^{\nu}] \tag{898}$$

A particular solution to this equation is given by

$$S_{\mu} = \frac{1}{2}\omega_{\mu ab}S^{ab} \tag{899}$$

where

$$S^{ab} = [\beta^{a}\ ,\ \beta^{b}] \tag{900}$$

Thus covariant derivative of the DKP field is valid in Lyra Manifold.

From the above Lagrangian (889) , one can define the action as

$$S = \int d^{4}x\ (x^{0})^{4}\ \sqrt{-g}\left[\frac{i}{2}\ \overline{\psi}\ \beta^{\mu}\ \nabla_{\mu}\psi - \frac{i}{2}\ \nabla_{\mu}\overline{\psi}\ \beta^{\mu}\ \psi - m\ \overline{\psi}\ \psi\right] \tag{901}$$

or,

$$S = \int d^{4}x\ (x^{0})^{4}\ \sqrt{-g}\left[\frac{i}{2}\ \overline{\psi}\ \beta^{a}e^{\mu}_{a}\left(\frac{1}{x^{0}}\ \frac{\partial\psi}{\partial x^{\mu}} + S_{\mu}\psi\right) - \frac{i}{2}\ \left(\frac{1}{x^{0}}\ \frac{\partial\overline{\psi}}{\partial x^{\mu}} - \overline{\psi}S_{\mu}\right)\ e^{\mu}_{a}\beta^{a}\ \psi - m\ \overline{\psi}\ \psi\right] \tag{902}$$

To derive the equations of motion of the massive DKP field within the framework of Lyra Geometry, we take the total variation of the action integral (902) as follows:

$$\begin{aligned}
\delta S &= \int_{\Omega} d^{4}x 4(x^{0})^{3}\ e\ L\ \delta x^{0} - \int_{\Omega} d^{4}x(x^{0})^{4}\ e\left[-\tfrac{i}{2(x^{0})^{2}}\ \overline{\psi}\ \beta^{\mu}\ \partial_{\mu}\psi + \tfrac{i}{2(x^{0})^{2}}\ \overline{\psi}\ \beta^{\mu}\ \partial_{\mu}\overline{\psi}\right]\delta x^{0}\\
&+ \int_{\Omega} d^{4}x(x^{0})^{4}\ (\delta e)L + \int_{\Omega} d^{4}x(x^{0})^{4}e\left[\tfrac{i}{2}\ \overline{\psi}\ \delta\ \beta^{\mu}\ \nabla_{\mu}\psi - \tfrac{i}{2}\ \nabla_{\mu}\overline{\psi}\ \delta\ \beta^{\mu}\ \psi\right]\\
&+ \int_{\Omega} d^{4}x(x^{0})^{4}e\left[\tfrac{i}{2}\overline{\psi}\beta^{\mu}\delta S_{\mu}\psi + \tfrac{i}{2}\overline{\psi}\delta S_{\mu}\beta^{\mu}\psi\right] + \int_{\Omega} d^{4}x(x^{0})^{4}e\delta\overline{\psi}\left[\tfrac{i}{2}\beta^{\mu}\nabla_{\mu}\psi + \tfrac{i}{2}S_{\mu}\beta^{\mu}\psi - m\psi\right]\\
&- \int_{\Omega} d^{4}x(x^{0})^{4}e\left[\tfrac{i}{2}\nabla_{\mu}\overline{\psi}\beta^{\mu} - \tfrac{i}{2}\overline{\psi}\beta^{\mu}S_{\mu} - m\overline{\psi}\right]\delta\psi + \int_{\Omega} d^{4}x(x^{0})^{4}e\left[\tfrac{i}{2x^{0}}\overline{\psi}\beta^{\mu}\delta\partial_{\mu}\psi - \tfrac{i}{2x^{0}}\delta\partial_{\mu}\overline{\psi}\beta^{\mu}\ \psi\right]
\end{aligned} \tag{903}$$

If one assumes the functional variation only in the massive DKP field, then, $\delta x^{0} = \delta e^{\mu}_{b} = \delta\omega_{\mu ab} = 0$. Note that $\delta e = -ee^{b}_{\mu}\delta e^{\mu}_{b}$. Considering $[\delta, \partial_{\mu}] = 0$, the equation (903) takes the form as

$$\begin{aligned}
\delta S &= \int_{\Omega} d^{4}x(x^{0})^{4}e\delta\overline{\psi}\left[\tfrac{i}{2}\beta^{\mu}\nabla_{\mu}\psi - m\psi + \tfrac{i}{2}S_{\mu}\beta^{\mu}\psi\right]\\
&- \int_{\Omega} d^{4}x(x^{0})^{4}e\left[\tfrac{i}{2}\nabla_{\mu}\overline{\psi}\beta^{\mu} + m\overline{\psi} - \tfrac{i}{2}\overline{\psi}\beta^{\mu}S_{\mu}\right]\delta\psi\\
&+ \int_{\Omega} d^{4}x\partial_{\mu}\left[\tfrac{i}{2}e(x^{0})^{3}\left(\overline{\psi}\beta^{\mu}\delta\psi - \delta\overline{\psi}\beta^{\mu}\ \psi\right)\right]
\end{aligned} \tag{904}$$

Now following the action principle, we get the the generator of the variation of the massive DKP field,

$$G_{\partial\psi} = \int_{\Omega} d\sigma_{\mu}\left[\frac{i}{2}e(x^{0})^{3}\left(\overline{\psi}\gamma\beta^{\mu}\delta\psi - \delta\overline{\psi}\beta^{\mu}\ \gamma\ \psi\right)\right] \tag{905}$$

and the Euler Lagrange equations for the massive DKP fields in Lyra Manifold are

$$\frac{i}{2}\beta^{\mu}\left[\nabla_{\mu}+S_{\mu}\gamma\right]\psi-m\psi=0 \tag{906}$$

$$-\frac{i}{2}\nabla_{\mu}\overline{\psi}\beta^{\mu}+\frac{i}{2}\overline{\psi}\beta^{\mu}S_{\mu}-m\overline{\psi}=0 \tag{907}$$

Here $S_{\mu}=\frac{1}{2}\omega_{\mu ab}S^{ab}$ is the trace of the Lyra torsion tensor. Note that the metricity condition (897) gives a relationship between spin connections and affine connections as

$$\omega_{\mu}^{ab}=e^{\nu b}\,\widetilde{\Gamma}^{\rho}_{\mu\nu}\,e^{a}_{\rho}-\frac{1}{x^{0}}\,e^{\nu b}\,\frac{\partial e^{a}_{\nu}}{\partial x^{\mu}} \tag{908}$$

Case - II . Massless DKP field in Lyra Geometry:

It is known that the massive DKP theory does not go over the massless DKP theory in the zero mass limit. To develop massless DKP theory in Lyra Manifold, at first, we consider the Harish Chandra Lagrangian density in Minkowski spacetime M^4 given by

$$L=i\,\overline{\psi}\,\gamma\,\beta^{a}\,\partial_{a}\psi-i\,\partial_{a}\overline{\psi}\,\beta^{a}\,\gamma\,\psi-\overline{\psi}\,\gamma\,\psi \tag{909}$$

As before β^a matrices obey the algebraic relations (i.e. usual DKP algebra)

$$\beta^{a}\,\beta^{b}\,\beta^{c}+\beta^{c}\,\beta^{b}\,\beta^{a}=\beta^{a}\,\eta^{bc}+\beta^{c}\,\eta^{ba} \tag{910}$$

and γ is a singular matrix satisfying

$$\beta^{a}\,\gamma+\gamma\beta^{a}=\beta^{a}\quad,\quad\gamma^{2}=\gamma \tag{911}$$

From the above Lagrangian, one gets the massless DKP wave equation as

$$i\beta^{a}\,\partial_{a}\psi-\gamma\,\psi=0 \tag{912}$$

It is noted that the above equation describe a set of four free assless gauge fields including the massless Klein-Gordon-Fock and the Maxwell electromagnetic fields. The above equations can be generalized to Lyra spacetimes through tetrads formalism as before. Let us consider a Lyra spacetime L^4 with metric $g_{\mu\nu}$ whose point coordinates are denoted by x^{μ}. To each point in L^4, we consider a tangent in Minkowski spacetime M^4 with metric η_{ab} whose point coordinates are denoted by x^a.

All the tensor quantities defined on M^4 can be projected into L^4 via tetrad fields $e^{\mu}_{a}(x)$. As before, the tetrad field and its inverse related to the spacetime metric in the following equations:

$$g_{\mu\nu}(x)=e^{a}_{\mu}(x)\;e^{b}_{\nu}(x)\;\eta_{ab} \tag{913}$$

$$g^{\mu\nu}(x)=e^{\mu}_{a}(x)\;e^{\nu}_{b}(x)\;\eta^{ab} \tag{914}$$

where

$$e^{a}_{\mu}(x)=g_{\mu\nu}(x)\;\eta^{ab}\;e^{\nu}_{b}(x) \tag{915}$$

$$e\;=\;det\;(e^{a}_{\mu})\;=\;\sqrt{-g}\;\;,\;\;g\;=\;det(g_{\mu\nu}) \tag{916}$$

Here the Latin indexes being raised and lowered by the Minkowski metric η^{ab} and the Greek ones by the metric $g^{\mu\nu}$ of the manifold L^4.
The Lagrangian density (906) of the massless DKP field minimally coupled to the Lyra Manifold is

$$L=\sqrt{-g}\left[i\,\overline{\psi}\,\gamma\,\beta^{\mu}\,\nabla_{\mu}\psi-i\,\nabla_{\mu}\overline{\psi}\,\beta^{\mu}\,\gamma\,\psi-\overline{\psi}\,\gamma\,\psi\right] \tag{917}$$

where ∇_μ is the Lyra covariant derivative associated to the affine connection $\Gamma^\mu_{\alpha\beta}$.
In the previous section, it is shown that the covariant derivative of the DKP field in Lyra Manifold is well defined. By the introduction of the tetrad fields, the resulting action takes the form as

$$S = \int d^4x (x^0)^4 \sqrt{-g} \left[i\, \overline{\psi}\, \gamma\, e^\mu_a\, \beta^a\, \nabla_\mu \psi - i\, \nabla_\mu \overline{\psi}\, \beta^a\, e^\mu_a\, \gamma\, \psi - \overline{\psi}\, \gamma\, \psi \right] \tag{918}$$

Now to find the equation of motion, one can use the classical version of the Schwinger Action Principle which is the most general version of the usual variational principles. From the classical point of view the Schwinger Action Principle can be expressed as

$$\delta S = \delta \int_\Omega d^4x (x^0)^4 L = \int_{\partial\Omega} d\sigma_\mu\, G^\mu \tag{919}$$

where S is the classical action and G^μ are the generators of the cannonical transformations.
Using the expression of the Lyra covariant derivative of the DKP field, the above action (915) takes the following form as

$$S = \int d^4x (x^0)^4 \sqrt{-g} \left[i\overline{\psi}\gamma e^\mu_a \beta^a \left(\frac{1}{x^0} \frac{\partial \psi}{\partial x^\mu} + S_\mu \psi \right) - i \left(\frac{1}{x^0} \frac{\partial \overline{\psi}}{\partial x^\mu} - \overline{\psi} S_\mu \right) \beta^a e^\mu_a\, \gamma\, \psi - \overline{\psi}\gamma\psi \right] \tag{920}$$

To derive the equations of motion of the massless DKP field within the framework of Lyra Geometry, we take the total variation of the action integral (917) as follows:

$$\begin{aligned}
\delta S &= \textstyle\int_\Omega d^4x 4 (x^0)^3\, e\, L\, \delta x^0 - \int_\Omega d^4x (x^0)^4\, e \left[-\frac{i}{(x^0)^2}\, \overline{\psi}\, \gamma\, \beta^\mu\, \partial_\mu \psi + \frac{i}{(x^0)^2}\, \overline{\psi}\, \gamma\, \beta^\mu\, \partial_\mu \overline{\psi} \right] \delta x^0 \\
&+ \textstyle\int_\Omega d^4x (x^0)^4\, (\delta e) L + \int_\Omega d^4x (x^0)^4 e \left[i\, \overline{\psi}\, \gamma\, \delta\, \beta^\mu\, \nabla_\mu \psi - i\, \nabla_\mu \overline{\psi}\, \delta\, \beta^\mu\, \gamma\, \psi \right] \\
&+ \textstyle\int_\Omega d^4x (x^0)^4 e \left[i\gamma\, \overline{\psi} \beta^\mu \delta S_\mu \psi + i\overline{\psi} \delta S_\mu \beta^\mu \gamma \psi \right] + \int_\Omega d^4x (x^0)^4 e \delta\overline{\psi} \left[i\gamma\beta^\mu \nabla_\mu \psi - \gamma\psi + i S_\mu \beta^\mu \gamma \psi \right] \\
&- \textstyle\int_\Omega d^4x (x^0)^4 e \left[i\nabla_\mu \overline{\psi} \beta^\mu \gamma + \overline{\psi}\gamma - i\overline{\psi}\gamma\beta^\mu S_\mu \right] \delta\psi + \int_\Omega d^4x (x^0)^4 e \left[\frac{i}{x^0} \overline{\psi}\gamma\beta^\mu \delta\partial_\mu \psi - \frac{i}{x^0} \delta\partial_\mu \overline{\psi} \beta^\mu\, \gamma\, \psi \right]
\end{aligned} \tag{921}$$

If one assumes the functional variation only in the massless DKP field, then, $\delta x^0 = \delta e^\mu_b = \delta\omega_{\mu ab} = 0$. Note that $\delta e = -e e^b_\mu \delta e^\mu_b$. Considering $[\delta, \partial_\mu] = 0$, and after an integration the equation (918) takes the form as

$$\begin{aligned}
\delta S = &\textstyle\int_\Omega d^4x (x^0)^4 e \delta\overline{\psi} \left[i\gamma\beta^\mu \nabla_\mu \psi - \gamma\psi + i S_\mu \beta^\mu \gamma\psi \right] \\
&- \textstyle\int_\Omega d^4x (x^0)^4 e \left[i\nabla_\mu \overline{\psi} \beta^\mu \gamma + \overline{\psi}\gamma - i\overline{\psi}\gamma\beta^\mu S_\mu \right] \delta\psi \\
&+ \textstyle\int_\Omega d^4x \partial_\mu \left[ie (x^0)^3 \left(\overline{\psi}\gamma\beta^\mu \delta\psi - \delta\overline{\psi}\beta^\mu\, \gamma\, \psi \right) \right]
\end{aligned} \tag{922}$$

Now following the Schwinger action principle, we get the the generator of the variation of the massless DKP field,

$$G_{\partial\psi} = \int_\Omega d\sigma_\mu \left[ie (x^0)^3 \left(\overline{\psi}\gamma\beta^\mu \delta\psi - \delta\overline{\psi}\beta^\mu\, \gamma\, \psi \right) \right] \tag{923}$$

and the Euler Lagrange equations for the massless DKP fields in Lyra Manifold are

$$i\beta^\mu \nabla_\mu \psi + i\beta^\mu \gamma \left[-\nabla_\mu + S_\mu \right] \psi - \gamma\psi = 0 \tag{924}$$

$$i\nabla_\mu \overline{\psi} \beta^\mu + i \left[-\nabla_\mu - S_\mu \right] \overline{\psi}\gamma\beta^\mu + \overline{\psi}\gamma = 0 \tag{925}$$

[we have used the result $\beta^\mu \gamma + \gamma \beta^\mu = \beta^\mu$]

It is also noted that if we vary only the background Lyra manifold, then one can get energy momentum density tensor as well as spin density tensor in Lyra background.

If we consider the variation of tetrad field only , then the above variation of action (918) takes the form as

$$\delta S = \int_\Omega d^4x (x^0)^4 e \left[i\, \overline{\psi}\, \gamma\, \beta^a\, \nabla_\mu \psi - i\, \nabla_\mu \overline{\psi} \beta^a\, \gamma\, \psi - e^a_\mu L \right] \delta e^\mu_a \tag{926}$$

[we have used $\delta \beta^\mu = \delta(\beta^a e^\mu_a) = \beta^a \delta e^\mu_a$]

If one defines the energy momentum density tensor as $T^a_\mu \equiv \frac{1}{(x^0)^4} \frac{\delta S}{\delta e^\mu_a}$, then we get,

$$T^a_\mu = i\, \overline{\psi}\, \gamma\, \beta^a\, \nabla_\mu \psi - i\, \nabla_\mu \overline{\psi} \beta^a\, \gamma\, \psi - e^a_\mu L \tag{927}$$

From this, one can write, $T^\nu_\mu = e^\nu_a\, T^a_\mu$.

Again, if we consider the variation of the components of the spin connection only , then the above variation of action (918) takes the form as

$$\delta S = \int_\Omega d^4x (x^0)^4 e \left[i\gamma\, \overline{\psi} \beta^\mu \delta S_\mu \psi + i \overline{\psi} \delta S_\mu \beta^\mu \gamma \psi \right] \tag{928}$$

Using $S_\mu = \frac{1}{2}\omega_{\mu ab} S^{ab}$, the above equation takes the form as

$$\delta S = \int_\Omega d^4x (x^0)^4\; e\; \frac{1}{2} \delta\omega_{\mu ab} \left[i\gamma\, \overline{\psi} \beta^\mu S^{ab} \psi + i \overline{\psi} S^{ab} \beta^\mu \gamma \psi \right] \tag{929}$$

As above, if one defines the spin density tensor as, $S^{\mu ab} \equiv \frac{2}{(x^0)^4} \frac{\delta S}{\delta \omega_{\mu ab}}$, then we get,

$$S^{\mu ab} = i\gamma\, \overline{\psi} \beta^\mu S^{ab} \psi + i \overline{\psi} S^{ab} \beta^\mu \gamma \psi \tag{930}$$

Finally, if we consider the variation of the scale function x^0 only, then by using the definition of energy momentum and the spin density tensors, the above action (918) yields

$$\delta S = \int_\Omega d^4x (x^0)^4 e \left[T^a_\mu e^\mu_a - \frac{1}{2} S^{\mu ab} \omega_{\mu ab} \right] \frac{\delta x^0}{x^0} \tag{931}$$

Thus one gets the relationship between the traces of energy momentum and the spin density tensors.

$$T^a_\mu e^\mu_a - \frac{1}{2} S^{\mu ab} \omega_{\mu ab} = 0 \tag{932}$$

This is known as trace relation. This identity is used to constraint the form of the spin connection $\omega_{\mu ab}$ in a given content of matter.

7.7: SPINORIAL FIELD IN LYRA GEOMETRY

It is known that the spinorial (Dirac) field (in particular, say, spin $\frac{1}{2}$ fermion) ψ possesses the Lagrangian density in Minkowski spacetime M^4 is given by

$$L = \frac{i}{2}\, \overline{\psi}\, \gamma^a\, \partial_a \psi - \frac{i}{2}\, \partial_a \overline{\psi}\, \gamma^a\, \psi - m\, \overline{\psi}\, \psi \tag{933}$$

where $\overline{\psi}$ is the Dirac adjoint of ψ (i.e. $\overline{\psi} = \psi^{\dagger}\gamma^{0}$) and γ^{a} are Dirac matrices that satisfy the anti commutation relations:

$$\{\gamma^{a}, \gamma^{b}\} = \gamma^{a}\gamma^{b} + \gamma^{b}\gamma^{a} = 2\eta^{ab} \tag{934}$$

where a = 0, 1, 2, 3 are the spatiotemporal Minkowski indexes and η^{ab} is the metric tensor of Minkowski spacetime with signature $(+,-,-,-)$. Using the variation of $\overline{\psi}$ in the action $S = \int L d^{4}x$, one can get Dirac equation for a particle of mass m as

$$i\,\gamma^{a}\,\partial_{a}\psi - m\,\psi = 0 \tag{935}$$

It is noted that the above equation describing the spinor field in Minkowski spacetime M^{4}. Now, we try to generalize the above equation in Lyra spacetime L^{4} through the tetrad formalism as before. To derive the equation of motion of the spinor field in Lyra Manifold, we use Schwinger Action Principle. The Lagrangian density of the spinor field minimally coupled to the Lyra Manifold is given by

$$L = \frac{i}{2}\,\overline{\psi}\,\gamma^{\mu}\,\nabla_{\mu}\psi - \frac{i}{2}\,\nabla_{\mu}\overline{\psi}\,\gamma^{\mu}\,\psi - m\,\overline{\psi}\,\psi \tag{936}$$

From the above Lagrangian (933), one can define the cation as

$$S = \int_{\Omega} dx^{4}(x^{0})^{4}\sqrt{-g}\left[\frac{i}{2}\,\overline{\psi}\,\gamma^{\mu}\,\nabla_{\mu}\psi - \frac{i}{2}\,\nabla_{\mu}\overline{\psi}\,\gamma^{\mu}\,\psi - m\,\overline{\psi}\,\psi\right] \tag{937}$$

The matrices γ^{μ} is defined through the contraction with the tetrad fields e^{μ}_{a} as $\gamma^{\mu} = e^{\mu}_{a}\gamma^{a}$. These γ^{μ} satisfy the generalized Dirac matrix algebra

$$\{\gamma^{\mu}, \gamma^{\nu}\} = 2g^{\mu\nu} \tag{938}$$

Here ∇_{μ} is the Lyra covariant derivative associated to the affine connection $\Gamma^{\mu}_{\alpha\beta}$.
Now, we try to establish the covariant derivative of the spinor field ψ in Lyra manifold and for this purpose the normal procedure of analyzing the behaviour of the field under Lorentz Transformation is followed as:
Assume a Lorentz Transformation L(x). Suppose a spinor field ψ is described in first frame and ψ' is that the transformed frame. There exists is a local relationship between ψ and ψ' so that the observer in the second frame may reconstruct ψ' when ψ is given. We assume the relation is linear i.e.

$$\psi(x) \longrightarrow \psi'(x) = U[L(x)]\;\psi(x) \tag{939}$$

where U is an one half spin representation of the Lorentz group, in particular, U is a non singular 4×4 matrix, characterizing the spinor field. Now it is desired to describe a spin connection S_{μ} in such a way that the object

$$\nabla_{\mu}\psi = \frac{1}{x^{0}}\,\frac{\partial\psi}{\partial x^{\mu}} + S_{\mu}\psi \tag{940}$$

transforms like a spinor field in (936)

$$\nabla_{\mu}\psi \longrightarrow (\nabla_{\mu}\psi)' = U(x)\;\nabla_{\mu}\psi \tag{941}$$

This implies

$$\left(\frac{1}{x^{0}}\;\partial_{\mu}\psi + S_{\mu}\psi\right)' = U\left(\frac{1}{x^{0}}\;\partial_{\mu}\psi + S_{\mu}\psi\right)$$

After some algebraic manipulation, one obtains, the transformation of S_{μ} as

$$S'_{\mu} = U(x)\;S_{\mu}\;U^{-1}(x) - \frac{1}{x^{0}}\;\partial_{\mu}U(x)\;U^{-1}(x) \tag{942}$$

It is known that $\psi\overline{\psi}$ is a scalar and therefore covariant derivative of $\psi\overline{\psi}$ in Lyra manifold is

$$\nabla_{\mu}(\psi\overline{\psi}) = \frac{1}{x^{0}}\;\partial_{\mu}(\psi\overline{\psi})$$

Using equation(891), one gets,

$$\nabla_\mu \overline{\psi} = \frac{1}{x^0}\frac{\partial \overline{\psi}}{\partial x^\mu} - \overline{\psi} S_\mu \tag{943}$$

The Fermi current is defined by $\overline{\psi}\ \beta^\nu\ \psi$ and taking Lyra covariant derivative of it, one obtains

$$\begin{aligned}\nabla_\mu(\overline{\psi}\ \gamma^\nu\ \psi) &= \frac{1}{x^0}\frac{\partial(\overline{\psi}\ \gamma^\nu\ \psi)}{\partial x^\mu} + \widetilde{\Gamma}^\nu_{\mu\lambda}(\overline{\psi}\ \gamma^\lambda\ \psi)\\ &= (\nabla_\mu \overline{\psi})\ \gamma^\nu\ \psi + \overline{\psi}(\nabla_\mu \gamma^\nu)\ \psi + \overline{\psi}\ \gamma^\nu\ (\nabla_\mu \psi)\end{aligned} \tag{944}$$

Using covariant derivatives of spinor fields given in equations (891) and (894), one obtain the following expression for the covariant derivative of β^ν as

$$\nabla_\mu \gamma^\nu = \frac{1}{x^0}\frac{\partial \beta^\nu}{\partial x^\mu} + \widetilde{\Gamma}^\nu_{\mu\lambda}\ \gamma^\lambda + S_\mu\ \gamma^\nu - \gamma^\nu\ S_\mu \tag{945}$$

One of the immediate consequences of the metricity condition, $\nabla_\mu g_{\alpha\beta} = 0$ in Lyra Manifold is the absolute parallelism of the tetrad i.e.

$$\nabla_\mu e^\nu_a = \frac{1}{x^0}\frac{\partial e^\nu_a}{\partial x^\mu} + \widetilde{\Gamma}^\nu_{\mu\lambda}\ e^\lambda_a + \omega^b_{\mu a}\ e^\nu_b = 0 \tag{946}$$

where $\omega_{\mu ab}$ is the spin connection coefficients. Here the tetrad fields e^μ_a and its inverse e^a_μ related to the spacetime metric as

$$e^a_\mu = g_{\mu\nu}(x)\eta^{ab} e^\nu_b(x) \tag{947}$$

Using equation (942) and writing γ^ν in terms of tetrad fields $\gamma^\nu = e^\nu_a \gamma^a$, we get from equation (943) as

$$\omega^b_{\mu a}\ e^\nu_b\ \gamma^a = S_\mu\ \gamma^\nu - \gamma^\nu\ S_\mu = [S_\mu\ ,\ \gamma^\nu] \tag{948}$$

A particular solution to this equation is given by

$$S_\mu = \frac{1}{2}\omega_{\mu ab}\Sigma^{ab} \tag{949}$$

where

$$\Sigma^{ab} = \frac{1}{4}[\gamma^a\ ,\ \gamma^b] \tag{950}$$

Thus covariant derivative of the spinor field is valid in Lyra Manifold.

Using the expression of the Lyra covariant derivative of the spinor field, the above action (924) takes the following form as

$$S = \int d^4x (x^0)^4 \sqrt{-g}\left[\frac{i}{2}\overline{\psi}\gamma^\mu\left(\frac{1}{x^0}\frac{\partial\psi}{\partial x^\mu} + S_\mu\psi\right) - \frac{i}{2}\left(\frac{1}{x^0}\frac{\partial\overline{\psi}}{\partial x^\mu} - \overline{\psi}S_\mu\right)\gamma^\mu\ \psi - m\overline{\psi}\psi\right] \tag{951}$$

To derive the equations of motion of the massless spinor field within the framework of Lyra Geometry, we take the total variation of the action integral (917) as follows:

$$\delta S = \int_\Omega d^4x 4(x^0)^3\ e\ L\ \delta x^0 - \int_\Omega d^4x(x^0)^4\ e\left[-\frac{i}{2(x^0)^2}\ \overline{\psi}\ \gamma^\mu\ \partial_\mu\psi + \frac{i}{2(x^0)^2}\ \psi\ \gamma^\mu\ \partial_\mu\overline{\psi}\right]\delta x^0$$
$$+ \int_\Omega d^4x(x^0)^4\ (\delta e)L + \int_\Omega d^4x(x^0)^4 e\left[\frac{i}{2}\ \overline{\psi}\ \ \delta\ \gamma^\mu\ \nabla_\mu\psi - \frac{i}{2}\ \nabla_\mu\overline{\psi}\ \delta\ \gamma^\mu\psi\right]$$
$$+ \int_\Omega d^4x(x^0)^4 e\left[\frac{i}{2}\ \overline{\psi}\gamma^\mu\delta S_\mu\psi + \frac{i}{2}\overline{\psi}\delta S_\mu\gamma^\mu\psi\right] + \int_\Omega d^4x(x^0)^4 e\delta\overline{\psi}\left[\frac{i}{2}\gamma^\mu\nabla_\mu\psi - m\psi + \frac{i}{2}S_\mu\gamma^\mu\psi\right]$$
$$- \int_\Omega d^4x(x^0)^4 e\left[\frac{i}{2}\nabla_\mu\overline{\psi}\gamma^\mu + m\overline{\psi} - \frac{i}{2}\overline{\psi}\gamma^\mu S_\mu\right]\delta\psi + \int_\Omega d^4x(x^0)^4 e\left[\frac{i}{2x^0}\overline{\psi}\gamma^\mu\delta\partial_\mu\psi - \frac{i}{2x^0}\delta\partial_\mu\overline{\psi}\gamma^\mu\ \psi\right] \tag{952}$$

If one assumes the functional variation only in the massless spinor field, then, $\delta x^0 = \delta e^\mu_b = \delta\omega_{\mu ab} = 0$. Note that $\delta e = -ee^b_\mu\delta e^\mu_b$. Considering $[\delta, \partial_\mu] = 0$, the equation (952) takes the form as

$$\delta S = \int_\Omega d^4x(x^0)^4 e\delta\overline{\psi}\left[\frac{i}{2}\gamma^\mu\nabla_\mu\psi - m\psi + \frac{i}{2}S_\mu\gamma^\mu\psi\right]$$
$$- \int_\Omega d^4x(x^0)^4 e\left[\frac{i}{2}\nabla_\mu\overline{\psi}\gamma^\mu + m\overline{\psi} - \frac{i}{2}\overline{\psi}\gamma^\mu S_\mu\right]\delta\psi$$
$$+ \int_\Omega d^4x\partial_\mu\left[\frac{i}{2}e(x^0)^3\left(\overline{\psi}\gamma^\mu\delta\psi - \delta\overline{\psi}\gamma^\mu\ \psi\right)\right] \tag{953}$$

Now following the action principle, we get the the generator of the variation of the spinor field,

$$G_{\partial\psi} = \int_\Omega d\sigma_\mu\left[\frac{i}{2}e(x^0)^3\left(\overline{\psi}\gamma^\mu\delta\psi - \delta\overline{\psi}\gamma^\mu\ \psi\right)\right] \tag{954}$$

and the Euler Lagrange equations for the spinor field in Lyra Manifold are

$$\frac{i}{2}\gamma^\mu\left[\nabla_\mu + S_\mu\right]\psi - m\psi = 0 \tag{955}$$

$$\frac{i}{2}\nabla_\mu\overline{\psi}\gamma^\mu - \frac{i}{2}\gamma^\mu S_\mu\overline{\psi} + m\overline{\psi} = 0 \tag{956}$$

7.8: GRAVITATIONAL FIELD EQUATIONS IN LYRA GEOMETRY

To derive field equations of a gravitational theory based on Lyra Geometry, one considers a four dimensional Lyra manifold endowed with metric tensor

$$ds^2 = g_{\mu\nu}x^0dx^\mu x^0dx^\nu \tag{957}$$

Here the dimensionless Lyra volume element is defined by $(x^0)^4\sqrt{-g}d^4x$.
The field equations are to be derived from a variational principle

$$\delta I = \delta\int\left[L_L\sqrt{-g} + 2\kappa L_m\sqrt{-g}\ \right]x^0dx^1.....\ x^0dx^4 \tag{958}$$

where we choose $L_L = K - \frac{3}{x^0}\phi^\alpha_{;\alpha}$ as the Lagrangian for the gravitational field with K is the curvature scalar in Lyra geometry.

Here,

$$K - \frac{3}{x^0}\phi^{\alpha}_{;\alpha} = \frac{R}{x^{02}} + \frac{3}{2}\phi_\alpha\phi^\alpha + \frac{3}{2}\tilde{\phi}_\alpha\phi^\alpha \tag{959}$$

where R is the Riemannian curvature scalar and $\tilde{\phi} = (x^0)^{-1}[\ln(x^0)^2]_{,\alpha}$. L_m is the Lagrangian for all the other fields and $\kappa =$ Einstein gravitational constant $= \frac{8\pi G}{c^4}$. Now we substitute (959) in (958) and considering independent variation of $g_{\alpha\beta}$ and ϕ_α, we get

$$\int \left[(x^0)^2\delta(R\sqrt{-g}\,) + \frac{3}{2}(x^0)^4\delta(\phi_\alpha\phi^\alpha\sqrt{-g}\,) + \frac{3}{2}(x^0)^4\delta(\tilde{\phi}_\alpha\phi^\alpha\sqrt{-g}\,) + 2\kappa(x^0)^4\delta(L_m\sqrt{-g}\,)\right] dx^1 \ldots\ldots dx^4 = 0 \tag{960}$$

The first term yields

$$\int (x^0)^2\left[R_{\alpha\beta} - \frac{1}{2}g_{\alpha\beta}R\right]\sqrt{-g}\;\delta g^{\alpha\beta} dx^1 \ldots\ldots dx^4$$

The second term yields

$$\int \left[-\frac{3}{4}(x^0)^4 g_{\alpha\beta}\phi_\gamma\phi^\gamma\sqrt{-g}\;\delta g^{\alpha\beta} + \frac{3}{2}(x^0)^4\phi_\alpha\phi_\beta\sqrt{-g}\;\delta g^{\alpha\beta} + 3(x^0)^4\phi^\alpha\sqrt{-g}\delta\phi_\alpha\right] dx^1 \ldots\ldots dx^4$$

The third term yields

$$\int \left[-\frac{3}{4}(x^0)^4 g_{\alpha\beta}\tilde{\phi}_\gamma\phi^\gamma\sqrt{-g}\;\delta g^{\alpha\beta} + \frac{3}{2}(x^0)^4\tilde{\phi}_\alpha\phi_\beta\sqrt{-g}\;\delta g^{\alpha\beta} + \frac{3}{2}(x^0)^4\tilde{\phi}^\alpha\sqrt{-g}\delta\phi_\alpha\right] dx^1 \ldots\ldots dx^4$$

The last term yields

$$\int \kappa\left[(x^0)^4 T_{\alpha\beta}\sqrt{-g}\;\delta g^{\alpha\beta}\right] dx^1 \ldots\ldots dx^4$$

Here we assume L_m is independent of ϕ_α and $T^{\alpha\beta} = \frac{2}{\sqrt{-g}}\frac{\partial}{\partial g_{\alpha\beta}}[\sqrt{-g}L_m]$.
Variation with respect to $g^{\alpha\beta}$ gives

$$R_{\alpha\beta} - \frac{1}{2}g_{\alpha\beta}R + \frac{3}{2}(x^0)^2\phi_\alpha\phi_\beta - \frac{3}{4}(x^0)^2 g_{\alpha\beta}\phi_\gamma\phi^\gamma - \frac{3}{4}(x^0)^2 g_{\alpha\beta}\tilde{\phi}_\gamma\phi^\gamma + \frac{3}{2}(x^0)^2\tilde{\phi}_\alpha\phi_\beta = -\kappa(x^0)^2 T_{\alpha\beta} \tag{961}$$

Variation with respect to ϕ_α gives

$$3\phi^\alpha + \frac{3}{2}\tilde{\phi}^\alpha = 0 \tag{962}$$

The above two sets of equations can be combined into the following single set of equations:

$$R_{\alpha\beta} - \frac{1}{2}g_{\alpha\beta}R - \frac{3}{2}(x^0)^2\phi_\alpha\phi_\beta + \frac{3}{4}(x^0)^2 g_{\alpha\beta}\phi_\gamma\phi^\gamma = -\kappa(x^0)^2 T_{\alpha\beta} \tag{963}$$

Since $\tilde{\phi}_\alpha = (x^0)^{-1}\;[\;\ln(x^0)^2\;]_{\;,\;\alpha} \equiv -2\partial_\alpha\left(\frac{1}{x^0}\right)$, using (962), one gets

$$\phi_\alpha = -\frac{1}{(x^0)^2}\;x^0_{\;,\;\alpha} \tag{964}$$

Substituting (964) in (963) yields

$$R_{\alpha\beta} - \frac{1}{2}g_{\alpha\beta}R - \omega\;(x^0)^{-2}x^0_{\;,\;\alpha}x^0_{\;,\;\beta} + \frac{1}{2}\omega\;(x^0)^{-2}g_{\alpha\beta}\;x^0_{\;,\;\gamma}\;x^{0,\;\gamma} = -\kappa(x^0)^2 T_{\alpha\beta} \tag{965}$$

with $\omega = \frac{3}{2}$.
It is to be noted that these equations are identical with the Brans-Dicke equations viz.

$$R_{\alpha\beta} - \frac{1}{2}g_{\alpha\beta}R - \omega\;\phi^{-2}\phi_{,\;\alpha}\phi_{,\;\beta} + \frac{1}{2}\omega\;\phi^{-2}g_{\alpha\beta}\;\phi_{,\;\gamma}\;x^{0,\;\gamma} = -\kappa(x^0)^2 T_{\alpha\beta} + \phi^{-1}(\phi_{,\;\alpha\;;\beta} - g_{\alpha\beta}\,\Box\,\phi)$$

if the scalar field φ satisfied the condition

$$\phi_{,\ \alpha\ ;\beta} - g_{\alpha\beta} \Box \phi = 0 \qquad and\ \omega\ =\ \frac{3}{2}.$$

The field equations take the following form in the normal gauge $x^0 = 1$:

$$R_{\alpha\beta} - \frac{1}{2} g_{\alpha\beta} R - \frac{3}{2} \phi_\alpha \phi_\beta + \frac{3}{4} g_{\alpha\beta} \phi_\gamma \phi^\gamma = -\kappa T_{\alpha\beta} \tag{966}$$

We write the above equation (966) in the form

$$R_{\alpha\beta} - \frac{1}{2} g_{\alpha\beta} R = -\kappa \left[T_{\alpha\beta} - f \left(\phi_\alpha \phi_\beta - \frac{1}{2} g_{\alpha\beta} \phi_\gamma \phi^\gamma \right) \right] \tag{967}$$

where $f = \frac{3c^4}{16\pi G}$.

This equation is identical to Hoyle's creation field (Hoyle's C-field) equation where the displacement vector ϕ_α can be identified as Hoyle's creation field C_α. It is noticed that the energy conservation law does not hold good in Lyra geometry (see equation (966)). So it is expected that some energy to be filled up again or depleting. Hence one may imagine that ϕ_α plays the role of a source or sink vector field. It is important to note that Hoyle introduced creation field in an ad hoc fashion where as in the present theory ϕ_α arises naturally in the geometry.

Again, we consider the displacement vector to be time like as $\phi_\alpha = (\beta, 0, 0, 0)$ where β is either an imaginary or a real constant. Then, assuming the Lyra manifold with signature (-, +, +, +), one can write the field equation (966) as

$$R_{00} - \frac{1}{2} g_{00} R + \frac{3}{4} \beta^2 g_{00} = -\kappa T_{00} \tag{968}$$

$$R_{ij} - \frac{1}{2} g_{ij} R - \frac{3}{4} \beta^2 g_{ij} = -\kappa T_{ij} \tag{969}$$

$$[i, j = 1, 2, 3]$$

These two equations are very similar to the field equation of General Relativity with Λ term apart from the sign difference in front of the Cosmological term.

(a) Static Spherically Symmetric Field:

Consider now the general solution to the scalar field theory in Lyra geometry for the static spherically symmetric field about a point mass. The field equation in the matter free region surrounding a point mass are

$$R_{\alpha\beta} - \frac{1}{2} g_{\alpha\beta} R - \omega\ (x^0)^{-2} x^0_{,\ \alpha} x^0_{,\ \beta} + \frac{1}{2} \omega\ (x^0)^{-2} g_{\alpha\beta}\ x^0_{,\ \gamma}\ x^{0,\ \gamma} = 0 \tag{970}$$

Halford [1972] obtained the general vacuum solution in isotropic coordinate as

$$ds^2 = -e^{2p} dt^2 + e^{2q} [dr^2 + r^2 (d\theta^2 + \sin^2 \theta d\varphi^2)] \tag{971}$$

where

$$e^{2p} = e^{2p_0} \left[\frac{1 - \frac{B}{r}}{1 + \frac{B}{r}} \right]^{\frac{2}{\lambda}}$$

$$e^{2q} = e^{2q_0} \left(1 + \frac{B}{r} \right)^4 \left[\frac{1 - \frac{B}{r}}{1 + \frac{B}{r}} \right]^{\frac{2(\lambda - 1)}{\lambda}}$$

$$x^0 = x_0^0 \left[\frac{1-\frac{B}{r}}{1+\frac{B}{r}}\right]^{\frac{C}{\lambda}}$$

Here, p_0, q_0, x_0^0, B & C are arbitrary constants and $\lambda^2 = 2 - \omega C^2$ with $\omega = \frac{3}{2}$.

Alternatively, Sen and Dunn provided a series solution to the field equation of (970) in Schwarzschild background for the metric

$$ds^2 = -e^\nu dt^2 + e^\lambda dr^2 + r^2(d\theta^2 + \sin^2\theta d\varphi^2) \tag{972}$$

as

$$e^\nu = D + C\phi(r) \tag{973}$$

$$e^\lambda = \frac{Ar^4(\phi')^2}{D + C\phi(r)} \tag{974}$$

$$\phi = \Sigma_{r=0}^{\infty} a_n r^{-n} \tag{975}$$

D,C,A are arbitrary constants.

The coefficients a_n are given by a_0 , arbitrary, $Aa_1^2 = D + Ca_0$, $a_2 = 0$, a_3 arbitrary, a_n , $n > 3$, are determined by the following recurrence relation

$$\begin{aligned}
&a_{n-1}[(D + Ca_0)(n-1)(n-4)] - Aa_1\Sigma_{k=3}^{n-1}(k-1)(n-k+1)a_{k-1}a_{n-k+1} \\
&-A\Sigma_{l=3}^{n-1}[(l-1)a_{l-1}][\Sigma_{k=2}^{n-l+2}(k-1)(n-l-k+3)a_{k-1}a_{n-l-k+3}] \\
&-\Sigma_{l=2}^{n-1}a_{n-l}a_{l-1}(n-l)(2l-n-1) = 0
\end{aligned}$$

Also

$$x^0 = k.exp \int \left[-\left(\frac{8}{3r^2} + \frac{4}{3r^2}\frac{\phi''}{\phi'}\right)\right]^{\frac{1}{2}} dr \tag{976}$$

where k is a constant.
Retaining only a few terms, we write equations (973), (974) and (976) as

$$e^\nu = C\left(b_0 + \frac{a_1}{r} + + \frac{a_3}{r^3} + \frac{a_4}{r^4}\right) \tag{977}$$

$$e^\lambda = \frac{C\left(a_1^2 + \frac{6a_1a_3}{r^2} + + \frac{8a_1a_4}{r^3} + \frac{9a_3^2}{r^4}\right)}{e^\nu} \tag{978}$$

$$(x^0)^{-1}\frac{dx_0}{dr} = 2C_0\left[r^{-2} + \frac{a_4}{a_3r^3} - \frac{a_1a_4^2 + 3a_3^3}{2a_1a_3^2r^4}\right] \tag{979}$$

where $b_0C = D + Da_0$ and $C_0^2 = \frac{2a_3}{a_1}$.

If we impose the usual boundary conditions at infinity i.e. e^ν and e^λ tend to 1 as r $\to \infty$, then one gets,

$$D + Ca_0 = 1 \tag{980}$$

and

$$D + Ca_0 = Aa_1^2 \tag{981}$$

These imply

$$Cb_0 = 1 \tag{982}$$

and

$$a_1^2 = \frac{1}{A} \tag{983}$$

i.e.

$$a_1 = \pm\frac{1}{\sqrt{A}} \tag{984}$$

For vanishing scale function *i.e.* $x_0 = 0$, these solutions reduce exactly to the Schwarzschild solution. Thus when $a_3 = 0$ *i.e.* $a_n = 0$ for $n > 1$, one gets

$$e^{\nu} = b_0 C \pm \frac{C}{\sqrt{A}r} = 1 \pm \frac{C}{\sqrt{A}r} \tag{985}$$

$$e^{\lambda} = \frac{Aa_1^2}{b_0 C + \frac{a_1 C}{r}} = \frac{1}{1 \pm \frac{C}{\sqrt{A}r}} \tag{986}$$

Since equations (985) and (986) represent Schwarzschild black hole solution, one should take negative sign and $\frac{C}{\sqrt{A}} = 2M' = M$ (say), (M' , mass of the black hole).
Thus one can write the solution (973) and (974) as

$$e^{\nu} = 1 - \frac{M}{r} + \frac{M\sqrt{A}a_3}{r^3} + \frac{M^2\sqrt{A}a_3}{r^4} \tag{987}$$

$$e^{\lambda} = \frac{\left(1 - \frac{6\sqrt{A}a_3}{r^2} - \frac{8M\sqrt{A}a_3}{r^3} + \frac{9a_3^2 A}{r^4}\right)}{1 - \frac{M}{r} + \frac{M\sqrt{A}a_3}{r^3} + \frac{M^2\sqrt{A}a_3}{r^4}} \tag{988}$$

These solutions represent black holes and we call it, Lyra black holes.

(b) Motion of test particles:

Recall the autoparallels of the affine connection in Lyra geometry as (equation (859))

$$\ddot{x}^{\mu} + \{ {}^{\mu}_{\alpha \;\; \beta} \}\dot{x}^{\alpha}\dot{x}^{\beta} + \frac{1}{2}x^0 \left(\delta^{\mu}_{\alpha}\, \phi_{\beta} + \delta^{\mu}_{\beta}\, \phi_{\alpha} - g_{\alpha\beta}\, \phi^{\mu}\right) \dot{x}^{\alpha}\dot{x}^{\beta} = \frac{1}{2}x^0(\phi_{\alpha} - \tilde{\phi}_{\alpha})\dot{x}^{\alpha}\dot{x}^{\mu} \tag{989}$$

where $\dot{x}^{\mu} = \frac{dx^{\mu}}{ds}$, $\ddot{x}^{\mu} = \frac{d^2x^{\mu}}{ds^2}$ and $\tilde{\phi}_{\alpha} = 2(x^0)^{-2}x^0_{,\alpha}$.

The relation between ϕ_{α} and $\tilde{\phi}_{\alpha}$ has been given in equation (962) as $2\phi_{\alpha} + \tilde{\phi}_{\alpha} = 0$. Using this relation, equation (989) yields

$$x^0\ddot{x}^{\mu} + x^0\{ {}^{\mu}_{\alpha \;\; \beta} \}\dot{x}^{\alpha}\dot{x}^{\beta} + \frac{1}{2}x^0_{,\alpha}\dot{x}^{\mu}\dot{x}^{\alpha} + \frac{1}{2}g_{\alpha\beta}g^{\mu\nu}x^0_{,\nu}\dot{x}^{\alpha}\dot{x}^{\beta} = 0 \tag{990}$$

where $g_{\alpha\beta}$ are the metric of the spherically symmetric spacetime given by

$$ds^2 = -A(r)dt^2 + B(r)dr^2 + r^2(d\theta^2 + \sin^2\theta d\varphi^2) \tag{991}$$

Assuming x^0 is a function r alone, i.e. $x^0 = x^0(r)$, one obtains (990) explicitly as

$$\ddot{r} + \left[\frac{1}{2}\frac{B'}{B} + (x^0)^{-1}\left(\frac{dx^0}{dr}\right)\right]\dot{r}^{\ 2} + \frac{1}{B}\left[-r + \frac{1}{2}r^2(x^0)^{-1}\left(\frac{dx^0}{dr}\right)\right]\dot{\theta}^{\ 2} + \frac{1}{B}\left[-r + \frac{1}{2}r^2(x^0)^{-1}\left(\frac{dx^0}{dr}\right)\right]\sin^2\theta\dot{\varphi}^{\ 2}$$
$$+ \frac{A}{B}\left[\frac{1}{2}\frac{A'}{B} - \frac{1}{2}(x^0)^{-1}\left(\frac{dx^0}{dr}\right)\right]\dot{t}^{\ 2} = 0 \tag{992}$$

$$\ddot{\theta} + \left[\frac{2}{r} + \frac{1}{2}(x^0)^{-1}\left(\frac{dx^0}{dr}\right)\right]\dot{\theta}\ \dot{r} - \sin\theta\cos\theta\dot{\varphi}^{\ 2} = 0 \tag{993}$$

$$\ddot{\varphi} + \left[\frac{2}{r} + \frac{1}{2}(x^0)^{-1}\left(\frac{dx^0}{dr}\right)\right]\dot{\varphi}\ \dot{r} + 2\cot\theta\ \dot{\theta}\dot{\varphi} = 0 \tag{994}$$

$$\ddot{t} + \left[\frac{A'}{A} + \frac{1}{2}(x^0)^{-1}\left(\frac{dx^0}{dr}\right)\right]\dot{t}\ \dot{r} = 0 \tag{995}$$

Taking into account the motion in the $\theta = \frac{\pi}{2}$ plane, the first integrals (994) and (995) yield

$$\dot{t}A = k_1(x^0)^{-\frac{1}{2}} \tag{996}$$

$$r^2\dot{\varphi} = k_2(x^0)^{-\frac{1}{2}} \tag{997}$$

where k_1 and k_2 are constants.
Instead of using (992), a third equation of motion can be found as

$$B\dot{r}^{\ 2} + r^2\dot{\varphi} - A\dot{t}^{\ 2} = -L \tag{998}$$

where L = 0, for photon
= 1, for massive particle. Putting (996) and (997) in (998), one obtains,

$$\dot{r}^{\ 2} + \frac{k_2^2}{x^0r^2B} - \frac{k_1^2}{x^0AB} + \frac{L}{B} = 0 \tag{999}$$

Assuming $r = \frac{1}{U}$, equation (999) yields

$$\left(\frac{dU}{d\varphi}\right)^2 = \frac{k_1^2}{k_2^2AB} - \frac{Lx^0}{k_2^2B} - \frac{U^2}{B} \tag{1000}$$

Equations (996) , (997) and (999) provide enough information to predict perihelion shift, bending of light etc.

(c) Linearized field equations:

In the weak field approximation of field equations in Lyra geometry, we assume that the metric components are given by

$$g_{\alpha\beta} = \eta_{\alpha\beta} + h_{\alpha\beta} \tag{1001}$$

with $|h_{\alpha\beta}| << 1$ and $\eta_{\alpha\beta} = diag.(-1, 1, 1, 1)$.
We also choose the gauge function

$$x^0 = x_0^0 + \xi\ , \quad with\ |\xi| << 1 \tag{1002}$$

For the sake of simplicity, we define

$$\gamma_{\alpha\beta} = h_{\alpha\beta} - \frac{1}{2}\eta_{\alpha\beta}h \tag{1003}$$

where , $h = \eta^{\alpha\beta}h_{\alpha\beta}$ to the first order.
Now, $R_{\alpha\beta} \cong \frac{1}{2}\Box h_{\alpha\beta}$, $R =\cong \frac{1}{2}\Box h$ and hence, $R_{\alpha\beta} - \frac{1}{2}g_{\alpha\beta}R \cong \frac{1}{2}\Box\gamma_{\alpha\beta}$.
The D-operator is the usual flat space wave operator. Using the field equations (965) based on Lyra geometry, we get

$$\Box\gamma_{\alpha\beta} = -2\kappa(x_0^0)^2T_{\alpha\beta} \tag{1004}$$

This is identical to the linearized form of Einstein field equations provided the gravitational constant G^* of the present theory is related to Newtonian constant G by $G(x_0^0)^2$.

REFERENCES

1. Lyra, G (1951), Math, Z **54**,52
2. Brans C and Dicke R.H(1961),Phys.Rev. **124**, 925
3. Sen D. K (1957), Phys. Z **149**, 311
4. Sen D. K and Dunn K. A (1971), J. Math. Phys **12**, 578
5. H H Soleng (1987) Gen.Rel.Grav. **19**,1213
6. Bharma K. S (1974), Aust. J. Phys. **27**, 541
7. J S Jeavons et al (1975), J. Math. Phys. **16**, 320
8. H Weyl , Sitzber.Preuss.Akad. Wiss. 465 (1918). Reprinted (English version) in L O'Raifeartaigh, The Dawning of Gauge Theory, Princeton Series in Physics, Princeton (1997).
9. R Casana, C A M de Melo and B Pimentel (2005) Braz.J.Phys.**35**, 1151
10. R Casana, C A M de Melo and B Pimentel (2007) Class.Quant.Grav.**24**, 723
11. R Casana, C A M de Melo and B Pimentel (2006) Astrophys.Space Sci. **305**, 125
12. W.D. Halfold (1972), J.Math. Phys, **13** , 1399
13. Sen D. K (1960), Can. Math. Bull. **3**, 255.
14. H H Soleng (1987) Gen. Rel.Grav. **19**, 1213
15. F. Rahaman, A.Ghosh and M.Kalam (2006), Nuovo. Cim. **121B**, 649
16. Hoyle. F and Narlikar. J. V.(1966), Proc.Roy.Soc. A **290** , 162

CHAPTER EIGHT:

TOPOLOGICAL DEFECTS IN LYRA GEOMETRY

8.1: INTRODUCTION

Over the last 30 years there has been a remarkable interaction between the physics of very small elementary physics (also in condensed matter physics) and the very large cosmology. Some characteristic parts of the early Universe have analogous in condensed matter physics. The Particle Physicists are trying to verify their theories at the early universe, while the Cosmologists are taking ideas from particle physics to understand the large-scale structure of the Universe. Interestingly the topological defects are the common idea in cosmology and in condensed matter physics. The topological defects might have been formed at phase transitions in the early Universe, similar to those formed in some phase transitions in condensed matter, and such defects might have left traces still visible today [Kibble, 1976]. It is believed that our Universe in the past was denser and hotter. So the early stage of the Universe was in a systematic phase and there were no topological defects. As the Universe expands, it cools down from its hot initial state. So the early Universe had undergone a number of phase transitions. Phase transitions in the early Universe can give rise to various forms of topological defects and their existence could have important implications in Cosmology. They can be monopoles, domain walls, cosmic strings or textures. A systematic exposition of the potential role of topological defects in our Universe, is provided by Vilenkin and Shellard [1994].

Global monopoles, created due to phase transitions when a global gauge symmetry was broken, may have been important for cosmology and astrophysics. A global monopole is a heavy object that forms in the phase transition of a system composed of a self coupling scalar field triplet ϕ^a whose original global 0(3) symmetry is spontaneously broken to U(1). From the topological point of view they are formed in the vacuum manifold M when the latter contains surfaces which cannot be shrunk continuously to a point i.e. when $\pi_2(M) \neq I$. Depending on the nature of the scalar field it can be shown that spontaneously symmetry breaking can give rise to such objects which are nothing but the topological knots in the vacuum expectation value of the scalar field and most of their energy is concentrated in a small region near monopole core. These monopoles have a Goldstone field with their energy density decreasing with the distance as r^{-2}. In their pioneering work Barriola and Vilenkin (BV) [1989] showed the existence of such a monopole solution resulting from the breaking of the global S0(3) symmetry of a triplet scalar field in a Schwarzschild background and the Lagrangian for the simplest model that gives rise to a global monopole is

$$L = \frac{1}{2}\partial_\mu \phi^a \partial^\mu \phi^a - \frac{\lambda}{4}(\phi^a\phi^a - \eta^2)^2 \tag{1005}$$

where η is the scale of symmetry breaking and λ is a constant.
The field configuration describing a monopole is

$$\phi^a = \eta f(r)\frac{x^a}{r} \tag{1006}$$

where $x^a x^a = r^2$.

They found a peculiar result: The space-time produced by a global monopole has no Newtonian gravitational potential in spite of the geometry produced by this heavy object having a non-vanishing curvature.

The appearance of domain walls is generally associated with the breaking of a discrete symmetry i.e. the vacuum manifold M consists of several disconnected components. So the homotopy group $\pi_0(M)$ is non trivial ($\pi_0(M) \neq I$). Recently the study of domain walls with finite thickness has renewed interest because of the proposal for a new scenario of galaxy formation due to Hill, Schramm and Fry [1989]. According to them, the formation of galaxies are due to domain walls produced during a phase transition after the time of recombination of matter and radiation. In general relativity domain walls are of special interest of research due to their peculiar and interesting gravitational effects. The domain wall models generally involve a set of real scalar field ϕ with the Lagrangian of the form,

$$L = \frac{1}{2}(\partial_\mu \phi)^2 - V(\phi) \tag{1007}$$

with

$$V(\phi) = \frac{\lambda}{2}(\phi^2 - \eta^2)^2 \tag{1008}$$

Stable domain walls occur when $V(\phi)$ has at least two degenerate minima.

Strings are line like defects. Strings arise in models when the vacuum manifold contains enclosed holes about which loops can be trapped. This corresponds to the vacuum manifold M having a non trivial homotopy group i.e. $\pi_1(M) \neq I$. Cosmic strings have received particular attention, mainly because of their cosmological implications. The double quasar problem can be well explained by strings, and galaxy formation might also be generated by density fluctuation in the early Universe due to the strings. Strings are of two general types: local and global. These are said to be local or global depending on their origin from the breakdown of local or global U(1) symmetry. Global strings are such that their energy extends to regions far beyond the central core. For a global string energy momentum tensor are calculated from the action density for a complex scalar field ψ along with a Mexican hat potential:

$$L = \frac{1}{2}g^{\mu\nu}\psi^*_{,\mu}\psi_{,\nu} - \frac{\lambda}{4}(\psi^*\psi - \eta^2)^2 \tag{1009}$$

where λ and η are constants and $\delta = (\eta\sqrt{\lambda})^{-1}$ is a measure of the core radius of the string.
Here the field configuration can be chosen as

$$\psi(r) = \eta f(r) e^{i\theta} \tag{1010}$$

in cylindrical coordinates.

The appearance of global textures in the early Universe during a phase transition is predicted by Grand Unified Theories. It is characterized by third order homotopy group (in fact $\pi_3(M) \neq I$, then texture will appear , M is the vacuum manifold). Textures are stable , non localized solutions to the classical equations of a spatial manifold with compact dimension. These structures collapse as soon as they come within the horizon i.e. when they become causally connected. It has been argued that this kind of topological defects might have been responsible for the formation of large scale structure. The entire energy of the texture comes from the gradient energy of the field. The simplest model that gives rise to texture is that of a four component real scalar fields Φ^a (a= 1,2,3,4) with the Lagrangian

$$L = \frac{1}{2}\partial_\mu \Phi^a \partial^\mu \Phi^a - \frac{\lambda}{4}(\Phi^a\Phi^a - \eta^2)^2 \tag{1011}$$

where η^2 is the vacuum expectation value (typically of the order 10^{16} GeV) and λ is a constant.
This model has a global 0(4) symmetry and the vacuum manifold is the three sphere S^3. The symmetry is spontaneously broken to 0(3) when the field Φ^a acquires a vacuum expectation value. It is to be noted that unlike other topological defects, the texture can remain in the vacuum manifold throughout the space.

Here Φ^a , a four component scalar field has the form:

$$\Phi^a = [\cos\chi, \sin\chi\sin\theta\cos\varphi, \sin\chi\sin\theta\sin\varphi, \sin\chi\cos\theta] \tag{1012}$$

for a spherically symmetric configuration of the texture , where θ and $\varphi$are the usual spherical angular co ordinates and χ depends on radial and time coordinates with $\chi \longrightarrow 0$ as $r \longrightarrow 0$ and $\chi \longrightarrow \pi$ as $r \longrightarrow \infty$. This model has an 0(4) which is spontaneously broken to 0(3) due to phase transition. In fact the vacuum manifold is characterized by the 3-sphere $\Phi^a\Phi_a = \eta^2$. When Φ^a acquires a vacuum expectation value , the residual symmetry is 0(3).

It is interesting to note that the defects formed during phase transitions in condensed matter systems are similar as in early Universe. For examples, cosmic strings are analogous to vortex lines in super fluid helium or to magnetic flux tubes in type two II super conductors. In a condensed matter context, textures appear in super fluid ^3He. Also monopoles and domain walls are found in liquid crystals.

Now in the following sections, we shall discuss different topological defects and their interesting features in alternative theory of gravity based on Lyra Geometry. In consecutive investigations, we use the field equations in normal gauge as

$$R_{\alpha\beta} - \frac{1}{2}g_{\alpha\beta}R - \frac{3}{2}\phi_\alpha\phi_\beta + \frac{3}{4}g_{\alpha\beta}\phi_\gamma\phi^\gamma = -8\pi G T_{\alpha\beta} \tag{1013}$$

where the displacement vector ϕ_α is chosen to be time like as $\phi_\alpha = (\beta, 0, 0, 0)$ where β is either a constant or a variable quantity (i.e. either space or time dependent or both).

8.2: GLOBAL MONOPOLES

In recent past, Barriolla and Vilenkin (1989) considered the monopole associated with a triplet of scalar field ϕ^a given in equation (1006) with the static spherically symmetric line element

$$ds^2 = e^{\nu(r)}dt^2 - e^{\mu(r)}dr^2 - r^2(d\theta^2 + sin^2\theta d\phi^2) \tag{1014}$$

The field equations for ϕ^a in the metric (1014) reduce to a single equation for f(r):

$$e^{-\mu}f'' + e^{-\mu}\left[\frac{2}{r} + \frac{\nu'}{2} - \frac{\mu'}{2}\right]f' - \frac{2f}{r^2} - \lambda\eta^2 f(f^2-1) = 0 \tag{1015}$$

$[{}'\prime'$ stand for $\frac{d}{dr}]$

The non vanishing components of the energy stress tensors are

$$T_t^t = \frac{1}{2}e^{-\mu}\eta^2(f')^2 + \frac{\eta^2 f^2}{r^2} + \frac{1}{4}\lambda\eta^4(f^2-1)^2 \tag{1016}$$

$$T_r^r = -\frac{1}{2}e^{-\mu}\eta^2(f')^2 + \frac{\eta^2 f^2}{r^2} + \frac{1}{4}\lambda\eta^4(f^2-1)^2 \tag{1017}$$

$$T_\theta^\theta = T_\phi^\phi = \frac{1}{2}e^{-\mu}\eta^2(f')^2 + \frac{1}{4}\lambda\eta^4(f^2-1)^2 \tag{1018}$$

In the flat space the core of the monopole has the size $\delta = \lambda^{-\frac{1}{2}}\eta^{-1}$ and its mass $M_{core} \sim \lambda\eta^4\delta^3 \sim \lambda^{-\frac{1}{2}}\eta$. Thus if $\eta << m_p$ where m_p is the plank mass, it is evident that we can still apply the flat space approximation of M_{core}. This follows from the fact that in this case the gravity would not much influence on monopole structure. Outside the core, however, one can assume f = 1.

Thus the energy momentum tensor of a static global monopole can be approximated (outside the core) as

$$T_t^t = T_r^r = \frac{\eta^2}{r^2} \quad ; \quad T_\theta^\theta = T_\phi^\phi = 0 \tag{1019}$$

For the metric (1014), the field equations (1013) reduce to

$$\frac{1}{2}e^{-\mu}\left[\nu'' + \frac{1}{2}(\nu')^2 - \frac{1}{2}\nu'\mu' + \frac{\nu' - \mu'}{r}\right] + \frac{3}{4}\beta^2 e^{-\nu} = 0 \tag{1020}$$

$$e^{-\mu}\left[\frac{1}{r^2} + \frac{\nu'}{r}\right] - \frac{1}{r^2} + \frac{3}{4}\beta^2 e^{-\nu} = -\frac{8\pi G\eta^2}{r^2} \tag{1021}$$

$$e^{-\mu}\left[\frac{1}{r^2} - \frac{\mu'}{r}\right] - \frac{1}{r^2} - \frac{3}{4}\beta^2 e^{-\nu} = -\frac{8\pi G\eta^2}{r^2} \tag{1022}$$

[here we assume β is a constant]

From the above two equations (1020) and (1021), one gets

$$\frac{\nu'}{4}\left[e^{-\mu}\left(\frac{\nu' + \mu'}{r}\right) + \frac{3}{4}\beta^2 e^{-\nu}\right] = 0 \tag{1023}$$

This implies either $\nu' = 0$ or

$$e^{-\mu}\left(\frac{\nu' + \mu'}{r}\right) + \frac{3}{4}\beta^2 e^{-\nu} = 0$$

But subtracting (1022) from (1021) yields

$$e^{-\mu}\left(\frac{\nu' + \mu'}{r}\right) + \frac{3}{4}\beta^2 e^{-\nu} = 0$$

So we have

$$\nu' \neq 0 \tag{1024}$$

This indicates the existence of gravitational force. Also, it is important to note that $\mu \neq \nu$ i.e. we we never get the Barriolla and Vilenkin like solutions for the monopole. At this stage, let us consider the weak field approximation and assume that

$$e^\nu = 1 + g(r) \quad ; \quad e^\mu = 1 + f(r) \tag{1025}$$

Here the functions f, g should be computed to the first order in η^2 and β^2. With these approximations equations (1020) - (1022) yield the following solutions

$$f = -\frac{1}{4}\beta^2 r^2 + 8\pi G\eta^2 + \frac{M}{r} \tag{1026}$$

$$g = -\frac{1}{2}\beta^2 r^2 - 8\pi G\eta^2 - \frac{M}{r} \tag{1027}$$

where M is an integration constant and can be considered as the mass of the monopole core.
Thus in the weak field approximation, the monopole metric in Lyra geometry takes the following form

$$ds^2 = \left(1 - \frac{1}{2}\beta^2 r^2 - 8\pi G\eta^2 - \frac{M}{r}\right)dt^2 - \left(1 - \frac{1}{4}\beta^2 r^2 + 8\pi G\eta^2 + \frac{M}{r}\right)dr^2 - r^2(d\theta^2 + sin^2\theta d\phi^2) \tag{1028}$$

From this solution (in the weak field approximation) one could come back to Barriolla and Vilenkin (i.e. in general relativity) monopole solution when $\beta \longrightarrow 0$; the contribution from the displacement becomes increasingly insignificant. Following Barriola and Vilenkin's reasoning, we drop the mass term in equation (24) as it is negligible on the astrophysical scale. Thus we have

$$e^\nu = 1 - \frac{1}{2}\beta^2 r^2 - 8\pi G\eta^2 \quad , \quad e^\mu = 1 - \frac{1}{4}\beta^2 r^2 + 8\pi G\eta^2 \tag{1029}$$

It is not difficult to show that the line element defined by the functions e^ν and e^μ above is conformally related to the Barriola-Vilenkin monopole solution.

To do so, let us consider the coordinate transformation given by the equations

$$e^{\nu} = (1 - 8\pi G\eta^2)p(R) \tag{1030}$$

$$e^{\mu}dr^2 = p(R)\left(1 - \frac{5}{4}\beta^2 r^2 + 8\pi G\eta^2\right)dR^2 \tag{1031}$$

$$r^2 = p(R)R^2 \tag{1032}$$

where p(R) is to be calculated and $p(R) = 1 + q(R)$, with $q(R) << 1$.
Differentiating (1032), one obtains

$$dr^2 = (1 + q + \dot{q}R)dR^2 \tag{1033}$$

where the overdot stands for a derivative with respect to R. Substituting equation (1033) into equation (1031) yields

$$q(R) = -\frac{1}{2}\beta^2 R^2 \tag{1034}$$

hence

$$p(R) = 1 - \frac{1}{2}\beta^2 R^2 \tag{1035}$$

In order to verify the consistency of this result with equation(1029), let us calculate e^{ν} directly from equations(1030) and (1035) (keeping the first order in η^2 and β^2)

$$e^{\nu} = (1 - 8\pi G\eta^2)(1 - \frac{1}{2}\beta^2 R^2) = 1 - \frac{1}{2}\beta^2 r^2 - 8\pi G\eta^2 \tag{1036}$$

Therefore, the line element (1014) which represents the space-time generated by the monopole may be written in terms of the new coordinate R as

$$ds^2 = \left(1 - \frac{1}{2}\beta^2 R^2\right)\left[\left(1 - 8\pi G\eta^2\right)dt^2 - \left(1 - \frac{5}{4}\beta^2 R^2 + 8\pi G\eta^2\right)dR^2 - R^2(d\theta^2 + sin^2\theta d\phi^2)\right] \tag{1037}$$

Rescaling the time and defining a new radial coordinate ρ as

$$\left(1 - \frac{5}{4}\beta^2 R^2 + 8\pi G\eta^2\right)^{\frac{1}{2}} dR = d\rho$$

i.e. approximately,

$$\left(1 - \frac{5}{8}\beta^2 R^2 + 4\pi G\eta^2\right)dR = d\rho \tag{1038}$$

From this one gets,

$$\left(1 + 4\pi G\eta^2\right)R - \frac{5}{24}\beta^2 R^3 = \rho \tag{1039}$$

This gives

$$R = S + F \tag{1040}$$

where

$$S = (P + \sqrt{Q^3 + P^2})^{\frac{1}{3}} \quad , \quad F = (P - \sqrt{Q^3 + P^2})^{\frac{1}{3}} \tag{1041}$$

$$Q = -\frac{8(1 + 4\pi G\eta^2)}{15\beta^2} \quad , \quad P = -\frac{4\rho}{5\beta^2} \tag{1042}$$

Thus the final form (1042) reads

$$ds^2 = \left(1 - \frac{1}{2}\beta^2(S + F)^2\right)[dT^2 - d\rho^2 - (S + F)^2(d\theta^2 + sin^2\theta d\phi^2)] \tag{1043}$$

Thus we have shown that in the weak field approximation equation (1043) represents the space-time generated by global monopole in Lyra geometry.

Analogously to the general relativity case this curved space-time also presents a deficit solid angle in the hypersurfaces t = constant. The area of a sphere of radius ρ in this space would different from $4\pi\rho^2$. Also a simple comparison of equation (1043) with Barriola and Vilenkin's solution shows that for small values of 'β^2 ' both space-times are related by a conformal transformation. In this case the motion of light rays is the same in the two space-times.
We would also like to point out an important property the gravitational force acts on surrounding the monopole. The radial component of the acceleration (A^r) acting on a test particle in the gravitational field of the monopole is given by $A^r = V^r_{;t}V^t$. For a co-moving particle $V^a = [\frac{1}{\sqrt{g_{tt}}}]\delta^a_t$. Hence using the line element (1014) and solution (1028) (neglecting the mass term), one can calculate A^r, which becomes

$$A^r = -\beta^2 r\left(1 - \frac{1}{2}\beta^2 r^2 - 8\pi G\eta^2\right)^{-2} \tag{1044}$$

Here one can see that the gravitational force varies with the radial distance and the particle accelerates towards in the radial direction in order to keep it at rest, which implies that a monopole has a repulsive influence on the test particle. This repulsive force is due to the presence of the displacement vector. From our results, the corresponding solutions (in the weak field approximations) in general relativity were re-obtained in the absence of the displacement vector. However our analysis shows that the monopole in presence of the displacement vector, does exert a gravitational repulsive force on a test particle. This observation is in striking contrast with the analogue of Einsteins theory.

8.3: DOMAIN WALLS

A domain wall can occur during a phase transition into a broken symmetry state where the vacuum manifold consists of several disconnected regions. The wall becomes boundary between regions lying in different vacua. Here we consider domain walls that arise in models with spontaneously broken discrete symmetries. The study of domain walls usually falls into two categories: Domain walls consist of a set of a real Higgs scalar field φ with a Lagrangian of the form

$$L = \frac{1}{2}g^{\alpha\beta}\varphi_{,\alpha}\varphi_{,\beta} - V(\varphi) \tag{1045}$$

and the classical field equation for the scalar field φ is

$$g^{\alpha\beta}\varphi_{,\alpha;\beta} - \frac{\partial V(\varphi)}{\partial\varphi} = 0 \tag{1046}$$

which contains a $\varphi \longrightarrow -\varphi$ symmetry. $V(\varphi)$ is assumed to have two or more minima at non zero values of φ. The second is to assume a certain form of energy stress tensors for a domain wall. And then solve the corresponding gravitational field equations.
The general metric for a plane symmetric spacetime can be parametrized as follows: The metric for a plane symmetric space time is taken as

$$ds^2 = e^A(dt^2 - dz^2) - e^C(dx^2 + dy^2) \tag{1047}$$

where $A = A(z,t); C = C(z,t)$.
The energy stress components in co-moving coordinates for the domain wall under consideration here are given by

$$T^t_t = \rho,\ T^x_x = T^y_y = p_1,\ T^z_z = p_2,\ T^z_t = 0 \tag{1048}$$

where ρ is the energy density of the wall , p_1 is the tension along X and Y directions in the plane of the wall and pressure in the perpendicular direction to the wall is p_2 .

In view of the above forms of energy stress tensors and using field equations (1013), we find the following specific solutions

$$e^A = [\sinh(DBz)]^{\frac{m}{B}}[\cosh(DA)]^{1-m} \tag{1049}$$

$$e^C = [\sinh(DBz)]^{\frac{1}{B}}[\cosh(DA)] \tag{1050}$$

where $D = \frac{N}{2(m+2)}; B = \frac{m+2}{m-1}$ and m & N are separation constants.
The stress energy components are

$$p_2 = 0 \tag{1051}$$

$$8\pi\rho = 8\pi p_1 = \frac{D^2}{[\sinh(DBz)]^{\frac{m}{B}}[\cosh(Dt)]^{1-m}}[1 + \frac{B}{\sinh(DBz)^2} - \coth(DBz)^2] \tag{1052}$$

Here $\beta^2(z,t)$ takes the following form,

$$\beta^2(z,t) = \frac{1}{3}[D^2[\coth(DBz)]^2 - \frac{(2m+2)D^2B}{\sinh(DBz)^2} - D^2\tanh(Dt)^2 - \frac{2(2-m)D^2}{\cosh(Dt)^2}] \tag{1053}$$

This model characterized of a thin domain wall. We see that the model does not exist for $m = 1$ and $m = -2$. One can also find the thick domain wall assuming $p_2 \neq 0$.
From the results given above, it is evident that at any instant the domain wall density, ρ decreases with the increase of the distance from the symmetry plane (both sides of the symmetry plane) and ρ vanishes as $z \to \pm\infty$.
The general expression for the three space volume is given by

$$\sqrt{|g_3|} = [\sinh(DBz)]^{\frac{m+2}{2B}}[\cosh(Dt)]^{\frac{3-m}{2}} \tag{1054}$$

Thus the temporal behaviour would be

$$\sqrt{|g_3|} \sim [\cosh(Dt)]^{\frac{3-m}{2}} \tag{1055}$$

If $m > 3$, then the three space collapses . On the other hand when $m < 3$, there are expansion along Z-direction . The proper distance , S_H(between $z = 0$, the centre of the wall and $z \to \infty$) measured along a space like curve running perpendicular to the wall$(t, x, y = const.)$ as

$$S_H = \int \exp\left(\frac{A}{2}\right) dz = [\cosh(Dt)]^{\frac{1-m}{2}} \int [\sinh(DBz)]^{\frac{m}{2B}} dz \tag{1056}$$

This distance diverges.
Hence there is no horizon in the Z-direction i.e. perpendicular to the wall.This is very similar to the result obtained in general relativity. The repulsive and attractive character of the wall can be discussed by either studying the time like geodesic in the space time or analyzing the acceleration of an observer who at rest relative to the wall.
Let us consider an observer with four velocity given by

$$V_i = [\sinh(DBz)]^{\frac{m}{2B}}[\cosh(Dt)]^{\frac{1-m}{2}}\delta_i^t \tag{1057}$$

Then we obtain the Acceleration Vector

$$A^i = V^i_{;k}V^k = \frac{Dm}{2}\frac{\coth(DBz)}{[\sinh(DBz)]^{\frac{m}{B}}[\cosh(Dt)]^{1-m}} \tag{1058}$$

It is evident that A^i is positive and it follows that an observer who wishes to remain stationary with respect to the wall must accelerate away from the wall. In other words,the wall exhibits an attractive nature to the observer. This result is in agreement with general relativity case. It is interesting to note that the displacement vector still exist after infinite time. So, concept of Lyra geometry is still exist even after the infinite times.

8.4: GLOBAL STRING

Cosmic strings are high energy relics which could be formed at symmetry breaking phase transition in the early Universe. The string is associated with the symmetry breaking of an Abelian group $G = U(1)$. The Lagrangian of this Abelian Higgs model is

$$L = \frac{1}{2}\partial_\mu \varphi^* \partial^\mu \varphi - V(\varphi) \tag{1059}$$

where φ is the complex scalar field and the potential $V(\varphi)$has a minimum at a non zero value of φ. The symmetry breaking potential has the form

$$V(\varphi) = \frac{\lambda}{4}(\varphi\varphi^* - \eta^2)^2 \tag{1060}$$

where λ is the self coupling constant of the Higgs field and η is the value of the symmetry breaking Higgs field.
The spontaneous symmetry breaking is obtained by introducing the Higgs field. The symmetry of the system after such a breaking is then determined by the degeneracy of the vacuum expectation value of the scalar field

$$\varphi = \eta f(r) e^{i\theta} \tag{1061}$$

To describe the spacetime geometry due to an infinitely long static cosmic string, the line element is taken to be the static cylindrically symmetric one given by

$$ds^2 = A(r)(-dt^2 + dr^2 + dz^2) + r^2 B(r) d\theta^2 \tag{1062}$$

The non vanishing components energy momentum tensor for the Lagrangian (1059) is given by

$$T_t^t = T_z^z = \frac{1}{2A}\left[\eta^2 (f')^2 + \frac{\eta^2 A f^2}{r^2 B} - \frac{\lambda\eta^4}{2}(f^2-1)^2\right] \tag{1063}$$

$$T_r^r = \frac{1}{2A}\left[\eta^2 (f')^2 - \frac{\eta^2 A f^2}{r^2 B} + \frac{\lambda\eta^4}{2}(f^2-1)^2\right] \tag{1064}$$

$$T_\theta^\theta = \frac{1}{2A}\left[-\eta^2 (f')^2 + \frac{\eta^2 A f^2}{r^2 B} + \frac{\lambda\eta^4}{2}(f^2-1)^2\right] \tag{1065}$$

The field equations for φ in the metric (1062) reduce to a single equation for f(r):

$$\frac{f''}{A} + \frac{f'}{A}\left[\frac{1}{r} + \frac{A'}{2A} + \frac{B'}{2B}\right] - \frac{f}{r^2 B} - \frac{f}{\delta^2}(f^2-1) = 0 \tag{1066}$$

['$\prime$' stand for $\frac{d}{dr}$]
Here $\delta = (\eta\sqrt{\lambda})^{-1}$ is the core radius of the string. The function $f(r)$ grows linearly when $r < \delta$ and approaches unity as soon as $r \geq \delta$. Thus taking $f = 1$ out side the core is a very good approximation to the exact solution. For our purpose, then, it is a good approximation to take

$$f(r) = 1\ , \quad f'(r) = 0 \tag{1067}$$

The nonzero components of the energy momentum tensors now become

$$T_r^r = T_z^z = T_r^r = -T_\theta^\theta = \frac{\eta^2}{r^2 B} \tag{1068}$$

The complete solutions of the line element (1062) under weak field approximation of the field equation (1013) with constant displacement vector are given by

$$A(r) = 1 - \frac{3}{8}\beta^2 r^2 - 4\pi G\eta^2 \ln\frac{r}{\delta} \tag{1069}$$

$$B(r) = 1 + \frac{3}{8}\beta^2 r^2 - 12\pi G\eta^2 \ln\frac{r}{\delta} \tag{1070}$$

The repulsive and attractive character of global string can be discussed either studying the time like geodesics in the space-time or analyzing the acceleration of an observer who is rest relative to the string. Let us consider an observer with four velocity $V_i = \sqrt{(g_{tt})}\delta_i^t$.

Now $A^r = V^r_{;t} = \Gamma^r_{tt} V^t V^t$.

Then we obtain the radial acceleration vector

$$A^r = -\left(\frac{3}{8}\beta^2 r^2 + \frac{2\pi G\eta^2}{r}\right)\left[1 - \frac{3}{8}\beta^2 r^2 - 4\pi G\eta^2 \ln\frac{r}{\delta}\right]^{-2} \tag{1071}$$

The above expression, being negative - definite, follows that an observer who wishes to remain stationary with respect to the string must accelerate towards the string core. This means that the string exhibits a repulsive nature to the observe. This implies that the gravitational force due to the string itself is repulsive. This result is in agreement with general relativity case.

Now, it is shown that local string is inconsistent in Lyra geometry. The general static cylindrically symmetric metric

$$ds^2 = e^{2(K-U)}(-dt^2 + dr^2) + e^{2U} dz^2 + W^2 e^{-2U} d\theta^2 \tag{1072}$$

is taken to describe the space-time given by an infinitely long static local string with the axis of symmetry being z axis . K, U and W are functions of the radial coordinate 'r' alone. The local string is characterized by an energy density and a stress along the symmetry axis given by

$$T_t^t = T_z^z = -\sigma \tag{1073}$$

and all other components are zero.
In view of equation (1073), the general solutions of the line element (1072) of the field equation (1013) with constant displacement vector are given by

$$W = \frac{1}{2}a(r+b) \tag{1074}$$

$$U = \frac{2s}{a}\ln(r+b) \tag{1075}$$

$$K = \frac{3}{8}\beta^2 r^2 + \frac{12s^2}{a^3}\ln(r+b) \tag{1076}$$

[s, a, b are integration constants]
Here the energy momentum tensor components are found as

$$T_t^t = T_z^z = 0\ ,\ T_r^r = T_\theta^\theta = 0 \tag{1077}$$

Thus, $\sigma = 0$ i.e. the string energy vanishes. Hence, it is evident that a straight local string, studied previously in general relativity, is incompatible with Lyra geometry.

8.5: GLOBAL TEXTURE

A texture configuration can be described by the action integral

$$I = \int d^4x\sqrt{-g}\left[\frac{1}{2}\partial_\mu \Phi^a \partial^\mu \Phi^a - \frac{\lambda}{4}(\Phi^a\Phi^a - \eta^2)^2 - \frac{1}{16\pi G}R\right] \tag{1078}$$

Here the index a on the real scalar field Φ^a runs from 1 to 4 (cf. equation (1012)). Here the Lagrangian in equation (1078) has an 0(4) symmetry. Due to phase transition the symmetry group 0(4) breaks down to 0(3) so that the relevant homotopy group that tells us the non trivial texture configuration is $\pi_3\,(0(4)/0(3)) = Z$, the integers of this group corresponding to winding number. In this paper we shall consider a texture configuration with winding number unity .

We consider a global texture configuration with the metric ansatz

$$ds^2 = A(r,t)dt^2 - B(r,t)dr^2) - r^2H(t)(d\theta^2 + \sin^2\theta d\varphi^2) \tag{1079}$$

The energy momentum tensor of the texture configuration is

$$T_{\mu\nu} = \nabla_\mu \Phi^a \nabla_\nu \Phi^a - \frac{1}{2} g_{\mu\nu} \nabla_\alpha \Phi^a \nabla^\alpha \Phi^a \tag{1080}$$

where ∇^μ is the covariant derivative operator.
For the texture configuration of equation (1080), this gives

$$T^t_t = \frac{(\chi')^2}{2B} + \frac{(\dot\chi)^2}{2A} + \frac{\sin^2\chi}{r^2H} \tag{1081}$$

$$T^r_r = -\frac{(\chi')^2}{2B} + \frac{(\dot\chi)^2}{2A} + \frac{\sin^2\chi}{r^2H} \tag{1082}$$

$$T^\theta_\theta = T^\varphi_\varphi = \frac{(\chi')^2}{2B} - \frac{(\dot\chi)^2}{2A} \tag{1083}$$

$$T^t_r = \chi'\dot\chi \tag{1084}$$

The equation of motion for Φ^a

$$\nabla^\mu \nabla_\mu \Phi^a = -\left[\frac{\nabla_\mu \Phi^b\ \nabla^\mu \Phi^b}{\eta^2}\right]\Phi^a$$

becomes for the above metric

$$\frac{\chi''}{2B} - \frac{\ddot\chi}{2A} + \frac{\chi'}{2B}\left[\frac{2}{r} - \frac{B'}{2B} + \frac{A'}{2A}\right] + \frac{\dot\chi}{A}\left[\frac{\dot A}{2A} + \frac{\dot H}{H} - \frac{\dot B}{2B}\right] = \frac{\sin 2\chi}{r^2H} \tag{1085}$$

Now if we ignore gravity , the equation for Φ^a results

$$-\chi'' + \ddot\chi - \frac{2\chi'}{r} = \frac{\sin 2\chi}{r^2} \tag{1086}$$

If one assumes, $U = \frac{r}{t}$,$V = t$ and χ is a function of U only. Then equation (1086) transforms to

$$\chi'' + \frac{2\chi'}{U} = \frac{\sin 2\chi}{U^2(1-U^2)} \tag{1087}$$

[Here '$\prime$' indicates differentiation with respect to U]

The solution of equation (1087) is

$$\begin{aligned}\chi(r,t) &= 2\tan^{-1}(-\tfrac{r}{t}) && t < 0\\ &= 2\tan^{-1}(\tfrac{r}{t}) && t > r > 0\\ &= 2\tan^{-1}(\tfrac{t}{r}) && r > t > 0\end{aligned} \tag{1088}$$

In view of the above forms of energy stress tensors and using field equations (9), we find the following specific solutions

$$B(r,t) = B_0[lr^{2-d} + gr^{-c}]^{-1}\ H^m \tag{1089}$$

$$A(r,t) = A_0 r^d \left[\frac{\dot{H}^2}{C_2 H^{-h} - a_2 H^{2-m}}\right] \tag{1090}$$

$$\chi(r,t) = cd\ \ln\frac{r}{r_0} + a_0 \ln\frac{H}{H_0} \tag{1091}$$

$$\beta^2(t) = \frac{2n\dot{H}^2(\frac{1}{m} - l)}{3B_0C_2H^{m-h} - 3B_0a_2H^2} - \frac{(m + a_1 + l)\dot{H}^2}{3H^2} \tag{1092}$$

where $A_0, B_0, r_0, H_0, C_2, g$ are integration constants, a_0, c, h, l, n, a_2 are arbitrary constants, m is the separation constant and $d = \frac{1-m}{\frac{1}{2}+2a_0cK}$ & $K = 8\pi G\eta^2$.
Thus time part of the metric coefficients and the displacement vector β^2 can be expressed in terms of H(t). So for arbitrary H(t), one can find the exact solutions of the field equations.

Weak field Approximation :

In this section we are discussing a case by taking a particular value of H(t) , $H(t) = e^{2t}$. Further we assume χ is a function of r only , say $\chi = \chi(r)$ and displacement vector β is a constant. Since χ is a function of r only, then the field equation $G_t^r = 0$ yields

$$B_2(t) = B_{00}H^q \quad , \quad A_1(r) = A_{00}r^{2(1-q)} \tag{1093}$$

[Assuming separable forms of the metric coefficients as $A = A_1(r)A_2(t)$ and $B = B_1(r)B_2(t)$].
Here A_{00} and B_{00} are integration constants and q is a separation constant.
If we choose the separation constant $q = 1$, then

$$B_2(t) = H(t) = e^{2t} \quad ; \quad A_1(r) = 1 \tag{1094}$$

[taking integration constants to be unity]

Now assuming

$$A_2(t) = H(t) = e^{2t} \tag{1095}$$

we get from the field equations (1013)

$$B_1(r) = \frac{1 + \frac{1}{2}Kr^2\chi'^2}{1 + ar^2 + K\sin^2\chi} \tag{1096}$$

where $a = 1 - \frac{3}{4}\beta^2$. So using the above solutions , the equation (1085) for χ results

$$r^2\chi''(1+ar^2+K\sin^2\chi)+Kr^2\chi'^3(1+3ar^2+\sin^2\chi)-\frac{1}{2}Kr^2\chi'^2\sin 2\chi+r\chi'(2+3ar^2+K\sin^2\chi) = \sin 2\chi \tag{1097}$$

Let us take the following coordinates transformations

$$T = e^t(1+r^2)^{\frac{1}{2}} \quad ; \quad R = e^t r \tag{1098}$$

Then the differential equation (1097) becomes

$$\chi''[1+(a-1)U^2+K(1-U^2)\sin^2\chi]+KU(1-U^2)\chi'^3[1+(3a-1)U^2+(1-U^2)\sin^2\chi]$$
$$-\tfrac{1}{2}K(1-U^2)\chi'^2\sin 2\chi+\tfrac{\chi'}{U}[2+3(a-1)U+K(1-3U^2)\sin^2\chi] = \tfrac{\sin 2\chi}{U^2(1-U^2)} \tag{1099}$$

where $U = \frac{R}{T}$ and χ is a function of U only & '$\prime$' indicates differentiation with respect to U. One can note that if we set K = 0 , then equation (1099) coincides with equation (1086) (K= 0 , i.e. in the absence of texture we neglect the displacement vector).

Thus we may conclude that the flat space solution for texture is self similar. Further in the region $1 << U << \frac{1}{\sqrt{K}}$, the equation (1099) can be written effectively as

$$(a-1)U^2\chi'' - K(3a-1)U^5\chi'^3 + (a-1)\frac{\chi'}{U} = 0 \tag{1100}$$

which has a first integral

$$\chi' = \left(\frac{e^F}{M}\right)^{\frac{1}{2}} \tag{1101}$$

where $F = \frac{6}{U}$ and

$$M = 6^4 S\left[-\frac{e^F}{4F^4} - \frac{e^F}{12F^3} - \frac{e^F}{24F^2} - \frac{e^F}{24F} + \left(\frac{1}{24}\right)\left\{\ln F + \frac{F}{1.1!} + \frac{F^2}{2.2!} + \frac{F^3}{3.3!} + ...\right\}\right] + D$$

[D is an integration constant and $S = \frac{2K(3a-1)}{(a-1)}$]

Hence the implicit expression of χ is

$$\chi = \int \left(\frac{e^F}{M}\right)^{\frac{1}{2}} dU + E \tag{1102}$$

Though we can not get the exact analytical form of χ but one can see that the solution is self similar form . We see that in the weak field approximation ($K << 1$) , the metric coefficients $B_1(r)$ takes the form (from equation(1096))

$$B_1(U) = \frac{(1-U^2)}{1+(a-1)U^2} \tag{1103}$$

which for $U >> 1$ and fixed t gives the line element

$$ds^2 = -\left(\frac{4}{3\beta^2}\right)dr^2 - r^2 d\Omega_2^2 \tag{1104}$$

This is just a conical metric with a deficit angle $2\pi\left[1-\left(\frac{4}{3\beta^2}\right)\right]$.
Thus deficit angle for the texture in the weak field depends on the displacement vector. For null radial geodesic, we set $ds^2 = d\Omega_2^2 = 0$ in the line element with metric coefficient (1103). This gives

$$\frac{dr}{dt} = \left[\frac{1-\left(\frac{3\beta^2}{4}\right)}{1-U^2}\right]^{\frac{1}{2}} \tag{1105}$$

This equation can be written in terms of a quardrature by writing it as a differential equation for U. That gives

$$|\,t\,|=|\,t_0\,|\;exp\left[\int_{U_0}^{U} dU \left\{\frac{1-\left(\frac{3\beta^2 U^2}{4}\right)}{1-U^2}\right\}^{\frac{1}{2}} - U\right]^{-1} \tag{1106}$$

where t_0 and U_0 refer to some initial time and position respectively . In the weak field case, the quardrature can be done explicitly to give

$$|\,t\,|=|\,t_0\,|\;\left[\frac{\frac{\sqrt{3}}{2}\beta - U_0}{\frac{\sqrt{3}}{2}\beta - U}\right] \tag{1107}$$

[in the region $U^2 >> 1$].

REFERENCES

1. A.Vilenkin and E.P.S. Shellard (1994), Cosmic String and other Topological Defects (Camb. Univ. Press) Cambridge.
2. Farook Rahaman, (2003), Nuovo Cim.**B118**, 17
3. F. Rahaman, (2000) Int.J.Mod.Phys.D**9**, 775
4. F. Rahaman, (2001) Int.J.Mod.Phys.D**10**, 579
5. F. Rahaman, M. Kalam, R. Mondal (2006) Astrophys.Space Sci.**305**, 337
6. F. Rahaman, S. Mal (2006) Astrophys.Space Sci.**302**, 3
7. M. Barriola and A. Vilenkin (1989) Phys. Rev. Lett. **63**, 341
8. T. W. B. Kibble (1976) J. Phys. A **9**, 1387
9. Hill C.T., D.N. Schram and J.N. Fry (1989) Nucl. Part. Phys. **19**,25 .

INDEX

C stands for chapter and P stands for page

◎ 编辑手记

著名数学家 F. J. Dyson 指出:

> 不能一劳永逸地定义数学在物理科学中的位置. 数学和科学的相互关系就像科学本身的纹理那样丰富和多样.

本书是一部英文版的微分几何专著,中文书名可译为《强子的芬斯勒几何和吕拉几何(宇宙学方面):强子结构的芬斯勒几何和吕拉几何(拓扑缺陷)》.

本书的第一作者为萨提亚·桑卡尔·德(Satya Sankar De),印度数学家,世界科学家联合会的会员,加尔各答大学的前任应用数学教授.

还有一位作者是法鲁克·拉哈曼(Farook Rahaman),印度贾达普大学数学系助教.

正如作者在前言中所述:

在过去的几年中,有关芬斯勒空间的一些书籍和专著对黎曼空间进行了概括,它们阐述了黎曼空间几何的各个方面,及其在理论物理学和生物系统的不同领域中的应用.本书的目的不是让这个领域的书籍变多.在这里我们要介绍一个称为强子的亚原子粒子扩展结构的芬斯勒空间,以此来发展其内部对称性.芬斯勒几何学的这一新的应用领域,及其在微局部域中对该时空的量化,实际上可以生成量子场方程.作为宇宙论的结果,人们已经详细描述了宇宙的非奇异起源,以及早期进化和标准宇宙论所遇到问题的解决.实际上,现在的应用展示了微局部时空中芬斯勒几何的各向异性特征.

在这里,我们还介绍了吕拉几何的解释性说明,它是对黎曼几何的一种修正,将度规函数引入到少结构流形中.此类说明无法以任何其他形式使用.在与外尔几何非常相似的吕拉几何标架中,我们讨论了拓扑缺陷的某些物理特征,例如畴壁、宇宙弦、单极、由假定的宇宙大爆炸附近的早期宇宙中的相变产生的纹理.我们要衷心感谢我们的妻子 Krishna De 和 Pakizah Yasmin 夫人在手稿酝酿期间的耐心和支持.非常感谢 M. Kalam 博士、A. Bhattacharya 博士和 A. Ghosh 博士为本书的编写提供的技术帮助.

数学工作室版权部主任李丹女士为了使读者能快速了解本书的

基本内容,特翻译了本书的目录如下:

第一章　强子物质扩展的芬斯勒空间

1.1　度量函数:一个第 n 阶形式的第 n 个正根

1.2　芬斯勒空间

1.3　芬斯勒空间的大地线

1.4　标形

1.5　曲率张量

1.6　作为芬斯勒空间的闵可夫斯基时空

1.7　芬斯勒的微局部时空

1.8　宏空间

1.9　微空间的大地线

第二章　芬斯勒空间中的波方程

2.1　经典波函数

2.2　量子化时空

2.3　作为量子化芬斯勒空间的微畴

2.4　芬斯勒空间中的场方程

2.5　外部电磁场

2.6　芬斯勒空间中的电磁场

2.7　电磁的相互作用

2.8　一般坐标变换

第三章　相伴黎曼空间的场方程

3.1　相伴黎曼空间

3.2　Robertson-Walker 时空中的狄拉克方程

3.3　$\phi(\tilde{\nu})$和 $\phi^{C}(\tilde{\nu})$的齐次解

芬斯勒几何是微分几何的一个重要的分支,包含比黎曼几何更为丰富的内容.芬斯勒几何是其度量没有二次型限制的黎曼几何,起源于黎曼的 1854 年著名的演说"论作为几何学基础的假设".在几何学家芬斯勒 1918 年的一篇博士论文中,芬斯勒较系统地讨论了基于弧长元素 $ds = F(x^1, \cdots, x^n, dx^1, \cdots, dx^n)$ 的度量空间中曲线与曲面的几何.关于空间中弧长形式 $ds = F(x^1, \cdots, x^n, dx^1, \cdots, dx^n)$ 的研究,特别是相关变分问题的研究早已为大数学家希尔伯特所关注,如他将其列为其在 1900 年巴黎国际数学家大会上提出的 23 个著名数学问题之最后一个,而其提出的第 4 个问题归结为在正则度量的情形,寻找所有欧氏空间中开集上的射影平坦的芬斯勒度量.

著名数学家陈省身先生对芬斯勒几何的研究非常重视,曾分别于 1995 年、2004 年在南开大学专门举行了"国际微分几何会议"以及"2004 国际芬斯勒几何会议".在陈省身先生的倡导和带领下,芬斯勒几何在国内外受到越来越广泛的重视,在遍及芬斯勒几何的多个重要方面都产生了一大批丰富的研究成果.芬斯勒几何的许多观点和方法不仅与数学的其他分支有紧密的联系,如李群、微分方程、拓扑空间理论等,而且在其他领域,如生物学、心理学、控制论、理论物理、数学物理、信息科学、图像处理等方面亦得到了广泛的应用,呈现出了蓬勃发展的势头.

关于黎曼流形上共形向量场与Killing向量场的研究以及它们的某些特殊性质与流形的曲率之间的关系已有大量的文献,积累了很丰富的结果.

我们知道在芬斯勒流形的射影球丛上有一个由芬斯勒度量自然诱导的黎曼度量,即Sasaki型度量,从而射影球丛关于此Sasaki型度量成为一个自然的黎曼流形.芬斯勒流形的几何与射影球丛的黎曼几何关系密切.另外,定义在射影球丛上的希尔伯特形式关于Sasaki型度量的对偶,给出了射影球丛上一个处处非零的向量场,即所谓的Reeb向量场.

莫小欢教授证明了芬斯勒流形具有常数旗曲率1当且仅当射影球丛上的Ricci曲率沿Reeb向量场方向为$(n-1)/2$,其中n为流形的维数.同时他还讨论了Reeb向量场的其他一些性质.Bejancu和Farran证明了芬斯勒流形具有常数旗曲率1当且仅当Reeb向量场为Killing向量场.另外,射影球丛各纤维的体积是否为常数也与芬斯勒流形的曲率紧密联系.Bao和Shen证明了若芬斯勒流形为Landsberg空间,则回答是肯定的.上述相关结论指出了射影球丛的黎曼几何与底流形的芬斯勒几何之间的某些联系,有关这方面的问题仍然值得进一步研究.

共形几何是许多领域的交叉领域,植根于纯数学领域,比如微分几何、黎曼面理论、代数拓扑,等等.它已经有了很长的历史,而且在现代几何与现代物理乃至工程领域中仍然十分活跃.在黎曼几何中,针对共形几何的相关研究已经取得许多丰富的成果. M. S. Knebelman 基于黎曼几何中的共形变换,首次提出了芬斯勒几何中类似的共形理论,在此基础上,Pandey 等几何学家们深入探讨了芬斯勒空间上与共形相关的共形变换群. 随后,Hashiguchi 研究了芬斯勒度量的共形变化理论. 基于上述的研究结果,S. Bacso 和程新跃考虑了芬斯勒几何中共形变换下的一些重要几何量的变化规律,从而给出了在共形变换下这些几何量保持不变的充要条件. 以上取得的研究成果不仅奠定了芬斯勒几何中共形理论的基础,而且还激励了现代几何学家们对这一领域的研究和思考. 这为促进和完善芬斯勒几何中共形几何的进展起了不容忽视的作用.

共形向量场一直是研究和探索共形几何的核心问题之一. 一方面,在黎曼几何中,Darboux,Yano 等人认为作为一维外球面的圆是测地圆. 也就是说,测地圆是一条具有常数曲率且挠率消失的光滑曲线. 保持测地圆不变的共形变换称为保圆变换,而由这个保圆变换生成的向量场称为保圆向量场. 在黎曼几何中,Tanno,Yaanno 等人基于保圆向量场的定义得到了一系列的重要结果. 近年来,Bidabad 和沈忠民把圆的概念推广到了芬斯勒几何中,进而探讨了芬斯勒空间中的保圆变换. 随后,Ishihara 通过对黎曼几何中保圆向量场的研究,以一种自然的方式,考虑了芬斯勒几何中保持测地圆不变的向量场的局部流. 直观地说,沿这些流的李导数应该保持测地圆的微分方程不变,此时,我们把保圆向量场称为保持测地圆的共形向量场. 换句话说,如果向量场保持测地圆不变,那么这个向量场是共形的.

另一方面,我们知道每一个单参数变换群$\{\varphi_t\}$可以诱导一个向量场,反过来,每一个向量场可以刻画成某一个单参数变换群所诱导的向量场. 特别地,当这个单参数变换群是一个共形变换群时,那么它所诱导的向量场即是共形向量场. 也就是说,芬斯勒流形 M 上的共形向量场可视为是由流形 M 上的单参数共形变换群所诱导的向量场. 研究和刻画芬斯勒流形上的共形向量场一直是芬斯勒几何中的热点问题之一. 目前,针对芬斯勒流形上共形向量场的探索和研究已获得一系列重要成果,2007 年,莫小欢、余昌涛考虑了 n 维黎曼流形上由导航术确定的 Randers 度量 F,他们得到了 V 是关于 h 的共形场的充要条件是 F 具有迷向 S - 曲率. 进一步地,2012 年,沈忠民和夏巧玲在探讨 Randers 流形上的共形向量场时,首先确定了导航术版的 Randers 度量的共形向量场的等价刻画,并完全分类了具有弱迷向旗曲率的 Randers 度量的共形向量. 更进一步地, 2013 年,夏巧玲和沈忠民在考察一类具有标量旗曲率的 Randers 度量时,证明了具有射影 R - 迷向的 Randers 流形上的共形向量场,并使用位似的方法构建具有标量旗曲率的新 Randers 度量. 同年,莫小欢和黄利兵考虑了具有迷向 S - 曲率的 Randers 流形上的共形向量场,证明了此流形上不存在非位似的共形向量场. 也就是说,具有迷向 S - 曲率的 Randers 度量的共形向量场一定是一个位似. 在刻画了一系列 (α,β) - 度量的共形向量场后,人们自然地考虑了广义 (α,β) - 度量的共形向量场. 2015 年,沈忠民和袁敏高在 *SCIENCECHINA Mathematics* 发表的文章中,探索了一定条件下广义 (α,β) - 度量的共形向量场,并得到在不同条件下球对称度量的共形向量场的具体表达. 最近,程新跃、李婷婷和殷丽给出了 Kropina 流形上的共形向量场的等价条件,并分别利用 α,β 和导航术刻画和分类了具有弱迷向旗曲率的 Kropina 度量的共形向量场. 从以上的一系列的研究结果可以看出,芬斯勒几何上共形向量场的相关研究成果从始至终都在鼓励和鞭策着人们对这一领域进行深入的探索和讨论,从而体现了共形向量场对芬斯勒几何的兴盛和发展起着不容忽视的推进作用.

所谓广义 (α,β) - 度量是比 (α,β) - 度量更一般的芬斯勒度量,

它是2011年由余昌涛和朱红梅首次提出的一类新的度量. 广义(α,β)-度量是形如$F=\alpha\phi(b^2,s)\left(s=\dfrac{\beta}{\alpha}\right)$的芬斯勒度量,其中$\alpha=\sqrt{a_{ij}(x)y^iy^j}$是一个黎曼度量,$\beta=b_i(x)y^i$是一个1-形式,$\phi(s)$是一个光滑非负函数且$b^2:=\|\beta\|_\alpha^2$. 特别地,如果$\phi$独立于$b^2$,那么广义$(\alpha,\beta)$-度量即为通常的$(\alpha,\beta)$-度量. 如果$\alpha=|y|,\beta=\langle x,y\rangle$,那么

$$F=|y|\phi\left(x,\frac{\langle x,y\rangle}{|y|}\right)$$

我们称之为球对称度量. 此外,广义(α,β)-度量还包括了Bryant度量的一部分和第四次根度量的一部分. 也就是说,广义(α,β)-度量是由一大类芬斯勒度量组成的,这使得我们对发现更多的芬斯勒度量成为可能. 比如,在(α,β)-度量中,我们无法找到任何非Ric平坦的爱因斯坦度量,除非它是Randers类型,主要原因是(α,β)-度量的范畴有点小. 然而,如果我们在广义(α,β)-度量中讨论爱因斯坦度量,那么就不难找到具有正负Ric常数的度量. 近年来,针对广义(α,β)-度量的研究和探索已经逐渐受到几何学家们的关注,而且还获得了一些有意义的研究成果. 2015年,余昌涛和朱红梅在一定条件下,完全分类了局部射影平坦且具有旗曲率的广义(α,β)-度量. 此外,在相同条件下,进一步地构造了在旗曲率$K=0,1,-1$的前提下,局部射影平坦的广义(α,β)-度量的例子. 随后,2017年,夏巧玲首先在α和β满足一定条件下,给出了具有标量旗曲率的广义(α,β)-度量的刻画,其次,在$\phi(b^2,s)$满足一定条件时,证明了具有标量旗曲率的广义(α,β)-度量等价于它是射影平坦的,而且还完全确定了这类具有标量旗曲率的广义(α,β)-度量的局部构造. 最后,在维数大于2的流形上,刻画了满足某些条件的广义(α,β)-度量的黎曼曲率表达式. 基于上述的研究结果,2019年,程新跃、吴莎莎和黄勤荣又进一步的探究和刻

画了一定条件下，广义(α,β)-度量是强爱因斯坦度量的等价结果，并且获得了广义(α,β)-度量是Ric齐次芬斯勒度量的一个充分必要条件. 然而，由于计算的复杂性，针对广义(α,β)-度量的曲率性质和几何结构的探索仍不够深入和完善，依然存在大量的研究工作值得我们做更进一步的探讨和考察. 因此，为了深入的研究和了解广义(α,β)-度量，我们自然的考虑了广义(α,β)-度量的共形向量场的等价刻画及其分类.

Zermelo导航术问题是在外力场的影响下，具有恒定内力的物体从一点运动到另一点的最短时间路程. 2003年，沈忠民考虑了一般情形下的导航术问题. 他证明了在芬斯勒流形(M,F)上，受外力场V的影响且满足$F(x,-V_x)\leq 1$的最短时间路程是通过以下方程定义的新度量$\widetilde{F}=\widetilde{F}(x,y)$的测地线

$$F\left(x,\frac{y}{\widetilde{F}}-V_x\right)=1 \tag{1}$$

其中$y\in T_xM$. 特别地，当M是一个黎曼流形，$h=\sqrt{h_{ij}(x)y^iy^j}$是其流形M上的黎曼度量，$W=W^i\dfrac{\partial}{\partial x^i}$是流形$M$上的向量场且$\|W\|_h<1$时，则我们可以通过解导航术问题(1)来构造一个Randers度量. 此时，(h,W)称为Randers度量的导航数据. 导航术一直是研究芬斯勒度量的曲率性质和几何结构的重要方法之一. 例如，2003年，D. Bao, C. Robles和沈忠民通过导航术获得了具有常旗曲率的Randers度量. 进一步地，D. Bao和G. Civi Yildirim分类了具有弱迷向旗曲率的Randers度量. 随后，在2012年，夏巧玲和沈忠民在探讨Randers流形上的共形向量场时，建立了在V是Randers流形(M,F)上的共形向量场的条件下，由解导航术问题(1)决定的新Randers度量$\widetilde{F}$与所给的Randers度量

F 的旗曲率、Ric 曲率及 S－曲率性质之间的关系. 基于上述结果,2013 年,夏巧玲考虑了 Randers 流形上的导航术问题,并确定了在 V 是流形 (M, F) 上的共形向量场的条件下,所给的 Randers 度量 F 的旗曲率、由解导航术(1)确定的新 Randers 度量 $\widetilde{F}$ 的旗曲率和共形向量场 V 三者之间的关系.

除 Randers 度量之外,在芬斯勒流形上还存在一类特殊的芬斯勒度量:Kropina 度量. 这类度量是通过俄罗斯物理学家 V. K. Kropina 引入且由 Matsumoto 等人在 1978 年定义的一类非黎曼－芬斯勒度量. 其形式是 $F=\dfrac{\alpha^2}{\beta}$,其中 $\alpha=\sqrt{a_{ij}(x)y^iy^j}$ 是一个黎曼度量,$\beta=b_i(x)y^i$ 是一个 1－形式. Kropina 度量和 Randers 度量都属于 (α,β)－度量且都是 C－可约的芬斯勒度量. 特别地,当 $\phi=1+s$ 时,(α,β)－度量就成为 Randers 度量:当 $\phi=\dfrac{1}{s}$ 时,(α,β)－度量就变为 Kropina 度量. Kropina 度量在物理学、磁场学、力学等方面都有着重要的作用. 但值得注意的是,从 ϕ 的表达式可以看出 Kropina 度量是含有奇异点的非正则的芬斯勒度量,而 Randers 度量是正则的芬斯勒度量. 因此,为了更好地谈论和探究 Kropina 度量的几何结构和曲率性质,我们考虑把它限制在切丛上的锥形领域中,使它成为一个正则的芬斯勒度量. 类似地,Kropina 度量同样可以通过黎曼流形 M 上的一个黎曼度量 h 和流形 (M,h) 上的一个向量场 W 在 $\|W\|_h=1$ 时通过解导航术问题(1)来表达. 此时,(h, W) 是与之对应的导航数据. 近年来,针对导航术版 Kropina 度量的研究已经获得了一系列的重要结果. 比如,2014 年,R. Yoshikawa 和 S. V. Sabau 通过导航术版 Kropina 度量表达,确定了这类 Kropina 度量具有常旗曲率的充要条件是:h 具有常数截面曲率,W 是一个具有单位长度的 Killing 向量场. 随后,2013 年,张晓玲和沈一兵

完全分类和刻画了 α,β 版和导航术版的 Kropina 度量是一个爱因斯坦度量的等价条件,而且还证明了每一个爱因斯坦 Kropina 度量具有消失的迷向 S-曲率. 此外,他们还获得两个爱因斯坦 Kropina 度量的共形映射是一个位似. 在此基础上,夏巧玲研究和探索了维数大于或等于 3 的流形上导航术(h,W)版的 Kropina 度量具有弱迷向旗曲率的等价条件. 基于上述结果,2018 年,程新跃、李婷婷和殷丽完全确定了导航术版本的 Kropina 度量的共形向量场的等价条件,并进一步给出了维数大于 2 的流形上具有弱迷向旗曲率的 Kropina 度量的共形向量场的具体分类. 经过上述丰富研究成果的启发,我们考虑了 Kropina 流形 M 上的导航术问题及其若干曲率性质,我们先证明了 Kropina 流形上导航术问题的解要么是一个 Randers 度量,要么是一个 Kropina 度量. 进一步,我们考虑了在 V 是流形(M,F)上的共形向量场的前提下,建立所给的 Kropina 度量 F 和由解导航术问题确定的 Randers 度量 $\widetilde{F}$ 或 Kropina 度量 $\widetilde{F}$ 的某些曲率性质之间的关系.

事实上,根据上述的研究动态可以看出,芬斯勒几何中的共形向量场是一个内容丰富且应用十分广泛的领域,它为完善和推动芬斯勒几何的发展奠定了一定的基础.

近二十年来,宇宙学获得蓬勃的发展,成为物理学最活跃的前沿之一. 宇宙学在量子场论、粒子物理理论等相互交叉,发展中又与其他学科如天文学直接联系,其意义之重大的确令人振奋,而我们关注的课题相信在一段时间内都将属于前沿问题.

首先在早期宇宙形成的过程中,由于局域或整体的对称性破缺,产生了许多拓扑缺陷,包括宇宙弦、磁壁、磁单极,等等. 其中整体磁单极作为一种球对称的拓扑缺陷出现在系统相变时,而这种系统是由标量场的一个自耦合三重态构成的,标量场原始的整体 $O(3)$ 对称破缺

成 $U(1)$ 对称. 一个磁单极外时空度规具有一个缺陷角. 整体磁单极作为一种拓扑缺陷带有早期宇宙的结构信息,对其研究是非常有必要且重要的. 其次,我们考虑了在 $f(R)$ 引力修正下带有整体磁单极的 Schwarzschild 黑洞. 现行的宇宙正处于加速膨胀的状态,这一点已经得到包括宇宙微波背景实验等一系列独立的实验所验证. 然而,这一现实与广义相对论的有些结果不符,于是许多新理论被提出来解释这一现象. 其中,最流行的理论之一就是对爱因斯坦方程进行推广,即在不引入其他物质的前提下修正原始的广义相对论. $f(R)$ 引力理论在解释宇宙加速膨胀问题的同时,更重要的是从本质上解释暗能量和暗物质的作用. 我们采用 $f(R)$ 理论修正后的度规,显然这样的时空背景是更完整且接近实际情况的. 也就是说,我们讨论的引力源或黑洞既包含整体磁单极,同时又考虑到 $f(R)$ 理论的引力修正.

华东理工大学的一篇博士学位论文研究了其他类型的黑洞,有 Bardeen 黑洞和带有共形反常的黑洞. 作为一种不存在奇异点但拥有视界的黑洞,Bardeen 黑洞是非线性的电动力学耦合爱因斯坦引力模型的一个特解. 由于这种特殊性,Bardeen 黑洞的时空结构值得仔细研究. 对于自引力吸引的非线性电动力学中的黑洞解,C. Moreno, O. Sarbach 给出了此类黑洞引力和电动力学方面的稳定性条件. 而满静赟从统计力学和热力学的角度讨论了在黑洞解存在的情况下,它是以何种状态稳定存在,又是经历怎样的相变演化而来. 满静赟介绍了带有整体磁单极的 Schwarzschild 黑洞,受 $f(R)$ 引力修正作用下包含整体磁单极的黑洞,Bardeen 黑洞,包括求解场方程、推导度规以及其时空特性等. 而带有共形反常的黑洞是包含 Weyl 反常的爱因斯坦半经典方程的静态球对称黑洞解. 这样的黑洞度规与 Gauss-Bonnet 黑洞解十分相似,或者说更接近在具有负常值曲率的空间内 Gauss-Bonnet 引力下的黑洞解. 当忽略反常时,度规变化为 Reissner-Nordstrom 黑洞的时空度规.由于共形反常在量子场论以及在宇宙学和统计力学上的广泛应用,我们很有必要更进一步研究共形反常对相变的作用,以显示其在黑洞热力学方面的表现.对于受 Gauss-Bonnet 引力影响的五维时空,强场引力透镜效应在光线偏折角方面的表现已经得到,并对可观测量进行了估值.然而引力透镜效应还包含相对论性像之间的时间延迟.

在黑洞时空背景下出射和吸收辐射方面的课题已经有大量的工作正在开展中并取得了相当多的成果.其中灰体因子是一个直接与吸收截面有关的物理量,其定义为一支给定的由无穷远处而来的波被黑洞吸收的概率.通过对吸收概率的讨论,为研究引力修正理论提供了一种全新的方法.满静赟给出了在带有 $f(R)$ 整体磁单极的黑洞背景下,辐射的吸收概率即灰体因子的解析表达式,能量出射率和广义吸收截面,从而研究了引力修正对时空结构的影响以及如何反映在可观测量上.

另一种用来解释暗物质的理论最近被提出,被称为 BSW 机制. Banados 等认为中等质量的黑洞是超大质量黑洞的早期形态,其周围若伴随有冷暗物质,则此黑洞可以作为一个粒子加速器.特别在 Kerr 黑洞的情况下,粒子的碰撞能量可以达到任意的无穷大.其任意性取决于碰撞粒子自身带有的角动量.粒子在黑洞周围的碰撞产生超高能量是一种可以解释暗物质的崭新理论.除了旋转的 Kerr 黑洞,还有各种黑洞模型也能够作为粒子碰撞出高能量的背景.满静赟考虑了这样一种情况,粒子碰撞后产生的能量可以作为一种可观测量,反映了不同黑洞作为粒子加速器的能力,不同的碰撞能量能够在一定程度上反映黑洞的特性,比如整体磁单极模型参量对碰撞能量的影响.

不同种类的拓扑缺陷已得到普遍的认识,比如畴壁、宇宙弦、磁单极,它们是在早期宇宙膨胀过程中形成的.所谓拓扑缺陷,其最基本的意义是发生局域或者整体的规范对称性破坏,即对称性破缺引发拓扑

缺陷. 假定一个 Higgs 场 ϕ^a,其中 $a = 1,2,\cdots,n$,它具有的势为

$$V(\phi) = \frac{\lambda}{4}(\phi^2 - \varepsilon^2)^2$$

其中 ε 是真空期望值,λ 是一个耦合常数. 此模型可以对应出许多不同类型的拓扑缺陷,例如,$n = 1$ 时代表畴壁,$n = 2$ 时代表宇宙弦,$n = 3$ 则是磁单极.

整体磁单极是在整体的规范球对称性遭到破坏时形成的,对于宇宙学和高能物理学,整体磁单极都具有十分重要的意义. Barriola 和 Vilenkin 在 1989 年给出了一个对于整体磁单极度规下的爱因斯坦方程的近似解,这个整体磁单极来自整体 SO(3) 对称破裂. 而后,Banerji 等在前者的基础上将方程推广到高维,得到高维的带有磁单极的球对称度规. 整体磁单极模型的拉格朗日量写作

$$L = \frac{1}{2}(\partial_\mu \phi^a)(\partial^\mu \phi^a) - \frac{1}{4}\lambda(\phi^a \phi^a - \eta^2)^2 \tag{2}$$

其中描述磁单极的场构型是 $\phi^a = \eta\alpha(r)\frac{x^a}{r}$. 一般球对称时空度规

$$ds^2 = A(r)dt^2 - B(r)dr^2 - r^2(d\theta^2 + \sin^2\theta d\phi^2) \tag{3}$$

在此度规描述的时空背景下,关于场 $\alpha(r)$ 的欧拉 - 拉格朗日方程为

$$\frac{(r^2\alpha')'}{Ar^2} + \frac{1}{2B}\left(\frac{B}{A}\right)'\alpha' - \frac{2\alpha}{r^2} - \lambda\eta^2\alpha(\alpha^2 - 1) = 0 \tag{4}$$

利用拉格朗日量(2)和度规(3),得到能量 - 动量的张量

$$\begin{aligned} T_t^t &= \eta^2\left[\frac{\alpha^2}{r^2} + \frac{\alpha'^2}{2A} + \frac{\lambda}{4}\eta^2(\alpha^2 - 1)^2\right] \\ T_r^r &= \eta^2\left[\frac{\alpha^2}{r^2} - \frac{\alpha'^2}{2A} + \frac{\lambda}{4}\eta^2(\alpha^2 - 1)^2\right] \\ T_\theta^\theta &= T_\varphi^\varphi = \eta^2\left[\frac{\alpha'^2}{2A} + \frac{\lambda}{4}\eta^2(\alpha^2 - 1)^2\right] \end{aligned} \tag{5}$$

平坦空间中磁单极的尺度为 $\delta \sim \frac{1}{\sqrt{\lambda\eta^2}}$，质量 $M \sim \frac{\eta}{\sqrt{\lambda}}$. 核外可做近似 $\alpha(r) \sim 1$，能量－动量的张量可近似为 $T_t^t \sim T_r^r \sim -\frac{\eta^2}{r^2}$，$T_\theta^\theta = T_\phi^\phi \sim 0$. 所以，在磁单极核外的点，与静态整体磁单极有关的能动张量近似表示为

$$T_\nu^\mu = \mathrm{diag}\left(\frac{\eta^2}{r^2}, \frac{\eta^2}{r^2}, 0, 0\right) \tag{6}$$

在早期宇宙的演化过程中产生具有拓扑结构的真空流形，伴随这种相变，生成一系列拓扑缺陷，如畴壁、宇宙弦和磁单极子，其中，整体磁单极是一种系统相变时出现的球对称引力拓扑缺陷，具有一个自耦合三重态，自发地从整体 $O(3)$ 对称破缺到 $U(1)$. 整体磁单极的引力效应具有这样一种特性，即其周围时空的立体角欠缺，使得所有光线都以相同的角度弯曲. 由于此特性，我们可以从天体物理学的角度来研究整体磁单极.

近年来，带有整体磁单极的黑洞引起了人们广泛的关注，同时，人们对带有整体磁单极的黑洞的相结构也有了深入的研究. 其中，两点关联函数就是研究黑洞相变结构的有利的工具之一. 沈阳师范大学物理科学与技术学院的李慧玲、陈阳、林榕三位教授 2021 年不仅利用黑洞熵，还利用了两点关联函数来探讨带有整体磁单极电荷的黑洞的相变行为，而且在对应的平面内可以观察到和普通热力学液－气流体完全相类似的范德瓦尔斯相变. 在 $T-\delta L$ 平面中，他们也考虑了磁单极参数对黑洞相变行为的影响，在对应的平面中显示的结果仍然呈现与普通热力学液－气流体相类似的范德瓦尔斯相变. 研究结果表明，在全息框架下，带有磁单极电荷的 AdS 时空中黑洞展示出与黑洞热力学熵相类似的范德瓦尔斯相变.

霍金发现黑洞具有热辐射，人们由此把黑洞作为热力学系统，那么相应的，黑洞也应该具有相变行为. 因为 AdS 时空的相变结构更加丰富，人们大量研究的是 AdS 时空的相变行为. 基于 AdS/CFT 对偶，从全息的角度来研究黑洞的相变行为，这能帮助人们更深入的了解黑洞. 1999 年，Chamblin 等[①]首次在温度 - 熵平面发现了类范德瓦尔斯相变的存在. 在 2017 年，曾晓雄等[②]在全息框架下用两点相关函数研究了典型的 R - N AdS 黑洞的相结构.

对 Schwarzschild 黑洞施加整体电荷等效于整体单极子破坏 Schwarzschild 时空的真空和渐近平直性，由于出现整体单极子，该黑洞具有了实心赤角[③]. 因此，具有整体单极子的黑洞的某些性质值得进一步研究. 1994 年，余洪伟[④]通过表面重力和欧氏路径积分这两种方法研究了静态球对称系统——整体单极黑洞的热力学. Dadhich 等[⑤]研究了整体单极电荷对粒子轨道和霍金辐射的影响，并在相关文献[⑥]中校

---

① CHAMBLIN A, EMPARAN R, JOHNSON C V, et al. Holography, thermodynamics, and fluctuations of charged AdS black holes[J]. Phys. Rev. D, 1999, 60(10): 104026.

CHAMBLIN A, EMPARAN R, JOHNSON C V, et al. Charged AdS black holes and catastrophic holography[J]. Phys. Rev. D, 1999, 60(6): 064018.

NIU C, TIAN Y, WU X N. Critical phenomena and thermodynamic geometry of Reissner-Nordström-anti-de Sitter black holes[J]. Phys. Rev. D, 2012, 85(2): 24017.

② ZENG X X, LI L F. Van der Waals phase transition in the framework of holography[J]. Phys. Lett. B, 2017, 764(C): 100-108.

③ YU H W. Black hole thermodynamics and global monopoles[J]. Nucl. Phys. B, 1994, 430(2): 427-440.

BARRIOLA M, VILENKIN A. Gravitational field of a global monopole[J]. Phys. Rev. Lett., 1989, 63(4): 341-343.

④ YU H W. Black hole thermodynamics and global monopoles[J]. Nucl. Phys. B, 1994, 430(2): 427-440.

⑤ DADHICH N, NARAYAN K, YAJNIK U A. Schwarzschild black hole with global monopole charge[J]. Pramana, 1998, 50(4): 307-314.

⑥ LI H L, YANG S Z. Correction to Hawking tunneling radiation from global monopole charged black hole[J]. Commun. Theor. Phys., 2009, 51(1): 190-192.

正了整体单极黑洞的霍金辐射. 2016 年,Ahmed 等[①]研究了单极参数的质量分数,发现它可以抑制最大吸积率. 彭俊金和吴双清[②]通过采用不同的过程并执行各种坐标转换,通过要求消除地平线上的引力反常,应用 RWs 方法从具有整体单极子的 Schwarzschild 型黑洞中推导出 Hawking 通量.

在全息框架下,两点关联函数是一个研究黑洞相结构非常有用的工具,在以前的工作中,两点关联函数已经被广泛用来研究全息热化[③]、全息奇点[④]等.

基于上述动机,李慧玲等在全息框架下,用两点关联函数来研究带有整体磁单极电荷的 AdS 黑洞是否具有类范德瓦尔斯相变.

带有整体磁单极电荷的 AdS 弯曲时空度规为

$$\mathrm{d}s^2 = -f(r)\mathrm{d}t^2 + \frac{1}{f(r)}\mathrm{d}r^2 + r^2(\mathrm{d}\theta^2 + \sin^2\theta\mathrm{d}\varphi^2) \tag{7}$$

其中

$$f(r) = 1 - \frac{2M}{r} + \frac{Q^2}{r^2} + \eta^2 + \frac{r^2}{L^2} \tag{8}$$

M 为质量参数,Q 为 AdS 黑洞的电荷,η 为磁单极参数,L 为 AdS 半径

---

① AHMED A K, CAMCI U. Accretion on Reissner-Nordström-(anti)-de sitter black hole with global monopole[J]. Class Quantum Gravity, 2016, 33(21): 215012.

② PENG J J, WU S Q. Hawking radiation from the Schwarzschild black hole with a global monopole via gravitational anomaly[J]. Chin. Phys. B, 2008, 17(3): 825-828.

③ ZENG X X, LIU W B. Holographic thermalization in Gauss-Bonnet gravity[J]. Phys. Lett. B, 2013, 726(1/2/3): 481-487.

ZENG X X, LIU X M, LIU W B. Holographic thermalization with a chemical potential in Gauss-Bonnet gravity[J]. J. High Energy Phys., 2014, 2014(3): 280-283.

④ ENGELHARDT N, HERTOG T, HOROWITZ G T. Further holographic investigations of big bang singularities[J]. J. High Energy Phys., 2015, 2015(7): 1-20.

ENGELHARDT N, HERTOG T, HOROWITZ G T. Holographic signatures of cosmological singularities[J]. Phys. Rev. Lett., 2014, 113(12): 121602.

且$L=(-\Lambda/3)^{-1/2}$,带有整体磁单极的黑洞的温度为

$$T=\frac{3r_H^4+L^2(-Q^2+r_H^2(1+\eta^2))}{4L^2\pi r_H^3} \tag{9}$$

当$f(r_H)=0$时,所得最大根为事件视界半径r_H,且熵S与r_H之间满足$S=\pi r_H^2$,可以得到温度T关于熵S的关系式

$$T(S,Q,\eta)=\frac{\sqrt{\pi}\left[\frac{3S^2}{\pi^2}+L^2\left(-Q^2+\frac{S(1+\eta^2)}{\pi}\right)\right]}{4L^2S^{3/2}} \tag{10}$$

由上式可知相变结构不只取决于电荷Q,还取决于磁单极参数η. 为了研究相变结构,可以通过上面公式求解相变结构的临界值

$$\left(\frac{\partial T}{\partial S}\right)_Q=\left(\frac{\partial^2 T}{\partial S^2}\right)_Q=0 \tag{11}$$

可以得到临界熵

$$S_{\mathrm{Cr}}=\frac{1}{6}\pi(L^2+L^2\eta^2),Q_{\mathrm{Cr}}=\frac{1}{6}L(1+\eta^2) \tag{12}$$

此时的临界温度为

$$T_{\mathrm{Cr}}=\frac{3\sqrt{\frac{3}{2}}\left[\frac{1}{12}(L^2+L^2\eta^2)^2+L^2\left(-\frac{1}{36}L^2(1+\eta^2)^2+\frac{1}{6}(1+\eta^2)(L^2+L^2\eta^2)\right)\right]}{L^2\pi(L^2+L^2\eta^2)^{3/2}} \tag{13}$$

由此可以得到温度T与熵S之间的关系,如图1所示. 在图1中,令$L=1,\eta=0.1$,对应曲线从上到下电荷的取值依次为$Q<Q_{\mathrm{Cr}},Q=Q_{\mathrm{Cr}},Q>Q_{\mathrm{Cr}}$.

从图1中很明显可以看到,电荷的数值对相变有很大的影响. 当$Q=Q_{\mathrm{Cr}}$时,图像存在拐点,说明此时存在二阶相变. 当$Q<Q_{\mathrm{Cr}}$时,随着熵的增加,温度随之先增大再减小,对应一阶相变. 当$Q>Q_{\mathrm{Cr}}$时,温度随着熵的增加单调增加,无相变的产生. 在温度-熵平面,有着与普通热力学中液-气体系类似的范德瓦尔斯相变的产生.

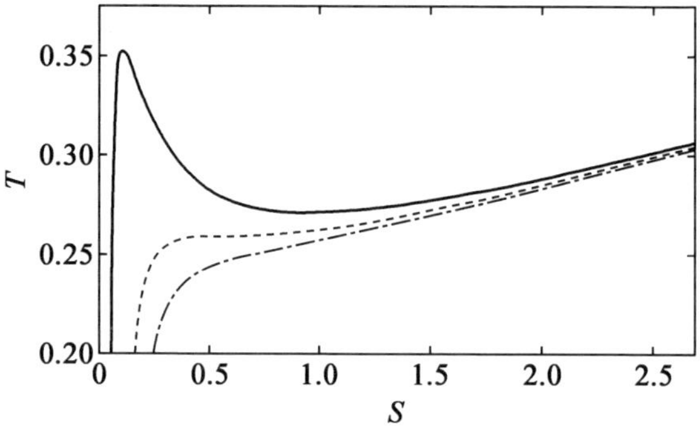

图1　温度 T 与熵 S 之间的关系图像

本书既可以当作专著来研读,也可以当作大学生的课外读物.因为大学生是求知欲最旺盛的群体,他们将来是最有希望去从事创造性劳动的群体,勤奋学习是这个群体的特性之一.我们相信,凡创造者必定都是热爱工作、养成了工作习惯的人,工作是他自己选定的,是由他的精神欲望发动的,所以他乐此不疲,欲罢不能.那些无此体验的人从外面看他,觉得不可理解,便勉强给了一个解释,叫作勤奋.世上许多人是在外在动机的推动下做工作的,他们的确无法理解为自己工作是怎么一回事,一旦没有了外来的推动,他们就不知道自己该做什么了.还有一些聪明人或有才华的人,也总是不能养成工作的习惯,终于一事无成.他们往往有怀才不遇之感,可是,在我们看来,一个人不能养成工作的习惯,这本身即已是才华不足的证明,因为创造欲正是才华最重要的组成部分.

刘培杰

2022.8.27

于哈工大

刘培杰数学工作室
已出版(即将出版)图书目录——原版影印

书　　名	出版时间	定　价	编号
数学物理大百科全书.第1卷(英文)	2016—01	418.00	508
数学物理大百科全书.第2卷(英文)	2016—01	408.00	509
数学物理大百科全书.第3卷(英文)	2016—01	396.00	510
数学物理大百科全书.第4卷(英文)	2016—01	408.00	511
数学物理大百科全书.第5卷(英文)	2016—01	368.00	512
zeta函数,q-zeta函数,相伴级数与积分(英文)	2015—08	88.00	513
微分形式:理论与练习(英文)	2015—08	58.00	514
离散与微分包含的逼近和优化(英文)	2015—08	58.00	515
艾伦·图灵:他的工作与影响(英文)	2016—01	98.00	560
测度理论概率导论,第2版(英文)	2016—01	88.00	561
带有潜在故障恢复系统的半马尔柯夫模型控制(英文)	2016—01	98.00	562
数学分析原理(英文)	2016—01	88.00	563
随机偏微分方程的有效动力学(英文)	2016—01	88.00	564
图的谱半径(英文)	2016—01	58.00	565
量子机器学习中数据挖掘的量子计算方法(英文)	2016—01	98.00	566
量子物理的非常规方法(英文)	2016—01	118.00	567
运输过程的统一非局部理论:广义波尔兹曼物理动力学,第2版(英文)	2016—01	198.00	568
量子力学与经典力学之间的联系在原子、分子及电动力学系统建模中的应用(英文)	2016—01	58.00	569
算术域(英文)	2018—01	158.00	821
高等数学竞赛:1962—1991年的米洛克斯·史怀哲竞赛(英文)	2018—01	128.00	822
用数学奥林匹克精神解决数论问题(英文)	2018—01	108.00	823
代数几何(德文)	2018—04	68.00	824
丢番图逼近论(英文)	2018—01	78.00	825
代数几何学基础教程(英文)	2018—01	98.00	826
解析数论入门课程(英文)	2018—01	78.00	827
数论中的丢番图问题(英文)	2018—01	78.00	829
数论(梦幻之旅):第五届中日数论研讨会演讲集(英文)	2018—01	68.00	830
数论新应用(英文)	2018—01	68.00	831
数论(英文)	2018—01	78.00	832

刘培杰数学工作室
已出版(即将出版)图书目录——原版影印

书　名	出版时间	定　价	编号
湍流十讲(英文)	2018—04	108.00	886
无穷维李代数:第3版(英文)	2018—04	98.00	887
等值、不变量和对称性(英文)	2018—04	78.00	888
解析数论(英文)	2018—09	78.00	889
《数学原理》的演化:伯特兰·罗素撰写第二版时的手稿与笔记(英文)	2018—04	108.00	890
哈密尔顿数学论文集(第4卷):几何学、分析学、天文学、概率和有限差分等(英文)	2019—05	108.00	891
偏微分方程全局吸引子的特性(英文)	2018—09	108.00	979
整函数与下调和函数(英文)	2018—09	118.00	980
幂等分析(英文)	2018—09	118.00	981
李群,离散子群与不变量理论(英文)	2018—09	108.00	982
动力系统与统计力学(英文)	2018—09	118.00	983
表示论与动力系统(英文)	2018—09	118.00	984
分析学练习.第1部分(英文)	2021—01	88.00	1247
分析学练习.第2部分,非线性分析(英文)	2021—01	88.00	1248
初级统计学:循序渐进的方法:第10版(英文)	2019—05	68.00	1067
工程师与科学家微分方程用书:第4版(英文)	2019—07	58.00	1068
大学代数与三角学(英文)	2019—06	78.00	1069
培养数学能力的途径(英文)	2019—07	38.00	1070
工程师与科学家统计学:第4版(英文)	2019—06	58.00	1071
贸易与经济中的应用统计学:第6版(英文)	2019—06	58.00	1072
傅立叶级数和边值问题:第8版(英文)	2019—05	48.00	1073
通往天文学的途径:第5版(英文)	2019—05	58.00	1074
拉马努金笔记.第1卷(英文)	2019—06	165.00	1078
拉马努金笔记.第2卷(英文)	2019—06	165.00	1079
拉马努金笔记.第3卷(英文)	2019—06	165.00	1080
拉马努金笔记.第4卷(英文)	2019—06	165.00	1081
拉马努金笔记.第5卷(英文)	2019—06	165.00	1082
拉马努金遗失笔记.第1卷(英文)	2019—06	109.00	1083
拉马努金遗失笔记.第2卷(英文)	2019—06	109.00	1084
拉马努金遗失笔记.第3卷(英文)	2019—06	109.00	1085
拉马努金遗失笔记.第4卷(英文)	2019—06	109.00	1086
数论:1976年纽约洛克菲勒大学数论会议记录(英文)	2020—06	68.00	1145
数论:卡本代尔 1979:1979年在南伊利诺伊卡本代尔大学举行的数论会议记录(英文)	2020—06	78.00	1146
数论:诺德韦克豪特 1983:1983年在诺德韦克豪特举行的Journees Arithmetiques 数论大会会议记录(英文)	2020—06	68.00	1147
数论:1985—1988年在纽约城市大学研究生院和大学中心举办的研讨会(英文)	2020—06	68.00	1148

刘培杰数学工作室
已出版(即将出版)图书目录——原版影印

书　　名	出版时间	定　价	编号
数论:1987 年在乌尔姆举行的 Journees Arithmetiques 数论大会会议记录(英文)	2020—06	68.00	1149
数论:马德拉斯 1987:1987 年在马德拉斯安娜大学举行的国际拉马努金百年纪念大会会议记录(英文)	2020—06	68.00	1150
解析数论:1988 年在东京举行的日法研讨会会议记录(英文)	2020—06	68.00	1151
解析数论:2002 年在意大利切特拉罗举行的 C. I. M. E. 暑期班演讲集(英文)	2020—06	68.00	1152
量子世界中的蝴蝶:最迷人的量子分形故事(英文)	2020—06	118.00	1157
走进量子力学(英文)	2020—06	118.00	1158
计算物理学概论(英文)	2020—06	48.00	1159
物质,空间和时间的理论:量子理论(英文)	2020—10	48.00	1160
物质,空间和时间的理论:经典理论(英文)	2020—10	48.00	1161
量子场理论:解释世界的神秘背景(英文)	2020—07	38.00	1162
计算物理学概论(英文)	2020—06	48.00	1163
行星状星云(英文)	2020—10	38.00	1164
基本宇宙学:从亚里士多德的宇宙到大爆炸(英文)	2020—08	58.00	1165
数学磁流体力学(英文)	2020—07	58.00	1166
计算科学:第 1 卷,计算的科学(日文)	2020—07	88.00	1167
计算科学:第 2 卷,计算与宇宙(日文)	2020—07	88.00	1168
计算科学:第 3 卷,计算与物质(日文)	2020—07	88.00	1169
计算科学:第 4 卷,计算与生命(日文)	2020—07	88.00	1170
计算科学:第 5 卷,计算与地球环境(日文)	2020—07	88.00	1171
计算科学:第 6 卷,计算与社会(日文)	2020—07	88.00	1172
计算科学. 别卷,超级计算机(日文)	2020—07	88.00	1173
多复变函数论(日文)	2022—06	78.00	1518
复变函数入门(日文)	2022—06	78.00	1523
代数与数论:综合方法(英文)	2020—10	78.00	1185
复分析:现代函数理论第一课(英文)	2020—07	58.00	1186
斐波那契数列和卡特兰数:导论(英文)	2020—10	68.00	1187
组合推理:计数艺术介绍(英文)	2020—07	88.00	1188
二次互反律的傅里叶分析证明(英文)	2020—07	48.00	1189
旋瓦兹分布的希尔伯特变换与应用(英文)	2020—07	58.00	1190
泛函分析:巴拿赫空间理论入门(英文)	2020—07	48.00	1191
卡塔兰数入门(英文)	2019—05	68.00	1060
测度与积分(英文)	2019—04	68.00	1059
组合学手册. 第一卷(英文)	2020—06	128.00	1153
*一代数、局部紧群和巴拿赫*一代数丛的表示. 第一卷,群和代数的基本表示理论(英文)	2020—05	148.00	1154
电磁理论(英文)	2020—08	48.00	1193
连续介质力学中的非线性问题(英文)	2020—09	78.00	1195
多变量数学入门(英文)	2021—05	68.00	1317
偏微分方程入门(英文)	2021—05	88.00	1318
若尔当典范性:理论与实践(英文)	2021—07	68.00	1366
伽罗瓦理论. 第 4 版(英文)	2021—08	88.00	1408

刘培杰数学工作室
已出版(即将出版)图书目录——原版影印

书　名	出版时间	定　价	编号
典型群,错排与素数(英文)	2020－11	58.00	1204
李代数的表示:通过 gln 进行介绍(英文)	2020－10	38.00	1205
实分析演讲集(英文)	2020－10	38.00	1206
现代分析及其应用的课程(英文)	2020－10	58.00	1207
运动中的抛射物数学(英文)	2020－10	38.00	1208
2－纽结与它们的群(英文)	2020－10	38.00	1209
概率,策略和选择:博弈与选举中的数学(英文)	2020－11	58.00	1210
分析学引论(英文)	2020－11	58.00	1211
量子群:通往流代数的路径(英文)	2020－11	38.00	1212
集合论入门(英文)	2020－10	48.00	1213
酉反射群(英文)	2020－11	58.00	1214
探索数学:吸引人的证明方式(英文)	2020－11	58.00	1215
微分拓扑短期课程(英文)	2020－10	48.00	1216
抽象凸分析(英文)	2020－11	68.00	1222
费马大定理笔记(英文)	2021－03	48.00	1223
高斯与雅可比和(英文)	2021－03	78.00	1224
π与算术几何平均:关于解析数论和计算复杂性的研究(英文)	2021－01	58.00	1225
复分析入门(英文)	2021－03	48.00	1226
爱德华・卢卡斯与素性测定(英文)	2021－03	78.00	1227
通往凸分析及其应用的简单路径(英文)	2021－01	68.00	1229
微分几何的各个方面.第一卷(英文)	2021－01	58.00	1230
微分几何的各个方面.第二卷(英文)	2020－12	58.00	1231
微分几何的各个方面.第三卷(英文)	2020－12	58.00	1232
沃克流形几何学(英文)	2020－11	58.00	1233
仿射和韦尔几何应用(英文)	2020－12	58.00	1234
双曲几何学的旋转向量空间方法(英文)	2021－02	58.00	1235
积分:分析学的关键(英文)	2020－12	48.00	1236
为有天分的新生准备的分析学基础教材(英文)	2020－11	48.00	1237
数学不等式.第一卷.对称多项式不等式(英文)	2021－03	108.00	1273
数学不等式.第二卷.对称有理不等式与对称无理不等式(英文)	2021－03	108.00	1274
数学不等式.第三卷.循环不等式与非循环不等式(英文)	2021－03	108.00	1275
数学不等式.第四卷.Jensen 不等式的扩展与加细(英文)	2021－03	108.00	1276
数学不等式.第五卷.创建不等式与解不等式的其他方法(英文)	2021－04	108.00	1277

刘培杰数学工作室
已出版(即将出版)图书目录——原版影印

书　　名	出版时间	定　价	编号
冯・诺依曼代数中的谱位移函数:半有限冯・诺依曼代数中的谱位移函数与谱流(英文)	2021－06	98.00	1308
链接结构:关于嵌入完全图的直线中链接单形的组合结构(英文)	2021－05	58.00	1309
代数几何方法.第1卷(英文)	2021－06	68.00	1310
代数几何方法.第2卷(英文)	2021－06	68.00	1311
代数几何方法.第3卷(英文)	2021－06	58.00	1312

书　　名	出版时间	定　价	编号
代数、生物信息和机器人技术的算法问题.第四卷,独立恒等式系统(俄文)	2020－08	118.00	1199
代数、生物信息和机器人技术的算法问题.第五卷,相对覆盖性和独立可拆分恒等式系统(俄文)	2020－08	118.00	1200
代数、生物信息和机器人技术的算法问题.第六卷,恒等式和准恒等式的相等 问题、可推导性和可实现性(俄文)	2020－08	128.00	1201
分数阶微积分的应用:非局部动态过程,分数阶导热系数(俄文)	2021－01	68.00	1241
泛函分析问题与练习:第2版(俄文)	2021－01	98.00	1242
集合论、数学逻辑和算法论问题:第5版(俄文)	2021－01	98.00	1243
微分几何和拓扑短期课程(俄文)	2021－01	98.00	1244
素数规律(俄文)	2021－01	88.00	1245
无穷边值问题解的递减:无界域中的拟线性椭圆和抛物方程(俄文)	2021－01	48.00	1246
微分几何讲义(俄文)	2020－12	98.00	1253
二次型和矩阵(俄文)	2021－01	98.00	1255
积分和级数.第2卷,特殊函数(俄文)	2021－01	168.00	1258
积分和级数.第3卷,特殊函数补充:第2版(俄文)	2021－01	178.00	1264
几何图上的微分方程(俄文)	2021－01	138.00	1259
数论教程:第2版(俄文)	2021－01	98.00	1260
非阿基米德分析及其应用(俄文)	2021－03	98.00	1261
古典群和量子群的压缩(俄文)	2021－03	98.00	1263
数学分析习题集.第3卷,多元函数:第3版(俄文)	2021－03	98.00	1266
数学习题:乌拉尔国立大学数学力学系大学生奥林匹克(俄文)	2021－03	98.00	1267
柯西定理和微分方程的特解(俄文)	2021－03	98.00	1268
组合极值问题及其应用:第3版(俄文)	2021－03	98.00	1269
数学词典(俄文)	2021－01	98.00	1271
确定性混沌分析模型(俄文)	2021－06	168.00	1307
精选初等数学习题和定理.立体几何.第3版(俄文)	2021－03	68.00	1316
微分几何习题:第3版(俄文)	2021－05	98.00	1336
精选初等数学习题和定理.平面几何.第4版(俄文)	2021－05	68.00	1335
曲面理论在欧氏空间 En 中的直接表示(俄文)	2022－01	68.00	1444
维纳—霍普夫离散算子和托普利兹算子:某些可数赋范空间中的诺特性和可逆性(俄文)	2022－03	108.00	1496
Maple中的数论:数论中的计算机计算(俄文)	2022－03	88.00	1497
贝尔曼和克努特问题及其概括:加法运算的复杂性(俄文)	2022－03	138.00	1498

刘培杰数学工作室
已出版(即将出版)图书目录——原版影印

书　　名	出版时间	定　价	编号
复分析:共形映射(俄文)	2022—07	48.00	1542
微积分代数样条和多项式及其在数值方法中的应用(俄文)	2022—08	128.00	1543
蒙特卡罗方法中的随机过程和场模型:算法和应用(俄文)	2022—08	88.00	1544
线性椭圆型方程组:论二阶椭圆型方程的迪利克雷问题(俄文)	2022—08	98.00	1561
动态系统解的增长特性:估值、稳定性、应用(俄文)	2022—08	118.00	1565
群的自由积分解:建立和应用(俄文)	2022—08	78.00	1570

狭义相对论与广义相对论:时空与引力导论(英文)	2021—07	88.00	1319
束流物理学和粒子加速器的实践介绍:第 2 版(英文)	2021—07	88.00	1320
凝聚态物理中的拓扑和微分几何简介(英文)	2021—05	88.00	1321
混沌映射:动力学、分形学和快速涨落(英文)	2021—05	128.00	1322
广义相对论:黑洞、引力波和宇宙学介绍(英文)	2021—06	68.00	1323
现代分析电磁均质化(英文)	2021—06	68.00	1324
为科学家提供的基本流体动力学(英文)	2021—06	88.00	1325
视觉天文学:理解夜空的指南(英文)	2021—06	68.00	1326
物理学中的计算方法(英文)	2021—06	68.00	1327
单星的结构与演化:导论(英文)	2021—06	108.00	1328
超越居里:1903 年至 1963 年物理界四位女性及其著名发现(英文)	2021—06	68.00	1329
范德瓦尔斯流体热力学的进展(英文)	2021—06	68.00	1330
先进的托卡马克稳定性理论(英文)	2021—06	88.00	1331
经典场论导论:基本相互作用的过程(英文)	2021—07	88.00	1332
光致电离量子动力学方法原理(英文)	2021—07	108.00	1333
经典域论和应力:能量张量(英文)	2021—05	88.00	1334
非线性太赫兹光谱的概念与应用(英文)	2021—06	68.00	1337
电磁学中的无穷空间并矢格林函数(英文)	2021—06	88.00	1338
物理科学基础数学. 第 1 卷,齐次边值问题、傅里叶方法和特殊函数(英文)	2021—07	108.00	1339
离散量子力学(英文)	2021—07	68.00	1340
核磁共振的物理学和数学(英文)	2021—07	108.00	1341
分子水平的静电学(英文)	2021—08	68.00	1342
非线性波:理论、计算机模拟、实验(英文)	2021—06	108.00	1343
石墨烯光学:经典问题的电解解决方案(英文)	2021—06	68.00	1344
超材料多元宇宙(英文)	2021—07	68.00	1345
银河系外的天体物理学(英文)	2021—07	68.00	1346
原子物理学(英文)	2021—07	68.00	1347
将光打结:将拓扑学应用于光学(英文)	2021—07	68.00	1348
电磁学:问题与解法(英文)	2021—07	88.00	1364
海浪的原理:介绍量子力学的技巧与应用(英文)	2021—07	108.00	1365
多孔介质中的流体:输运与相变(英文)	2021—07	68.00	1372
洛伦兹群的物理学(英文)	2021—08	68.00	1373
物理导论的数学方法和解决方法手册(英文)	2021—08	68.00	1374
非线性波数学物理学入门(英文)	2021—08	88.00	1376
波:基本原理和动力学(英文)	2021—07	68.00	1377
光电子量子计量学. 第 1 卷,基础(英文)	2021—07	88.00	1383
光电子量子计量学. 第 2 卷,应用与进展(英文)	2021—07	68.00	1384
复杂流的格子玻尔兹曼建模的工程应用(英文)	2021—08	68.00	1393

刘培杰数学工作室

已出版(即将出版)图书目录——原版影印

书　　名	出版时间	定　价	编号
电偶极矩挑战(英文)	2021—08	108.00	1394
电动力学:问题与解法(英文)	2021—09	68.00	1395
自由电子激光的经典理论(英文)	2021—08	68.00	1397
曼哈顿计划——核武器物理学简介(英文)	2021—09	68.00	1401
粒子物理学(英文)	2021—09	68.00	1402
引力场中的量子信息(英文)	2021—09	128.00	1403
器件物理学的基本经典力学(英文)	2021—09	68.00	1404
等离子体物理及其空间应用导论.第1卷,基本原理和初步过程(英文)	2021—09	68.00	1405
拓扑与超弦理论焦点问题(英文)	2021—07	58.00	1349
应用数学:理论、方法与实践(英文)	2021—07	78.00	1350
非线性特征值问题:牛顿型方法与非线性瑞利函数(英文)	2021—07	58.00	1351
广义膨胀和齐性:利用齐性构造齐次系统的李雅普诺夫函数和控制律(英文)	2021—06	48.00	1352
解析数论焦点问题(英文)	2021—07	58.00	1353
随机微分方程:动态系统方法(英文)	2021—07	58.00	1354
经典力学与微分几何(英文)	2021—07	58.00	1355
负定相交形式流形上的瞬子模空间几何(英文)	2021—07	68.00	1356
广义卡塔兰轨道分析:广义卡塔兰轨道计算数字的方法(英文)	2021—07	48.00	1367
洛伦兹方法的变分:二维与三维洛伦兹方法(英文)	2021—08	38.00	1378
几何、分析和数论精编(英文)	2021—08	68.00	1380
从一个新角度看数论:通过遗传方法引入现实的概念(英文)	2021—07	58.00	1387
动力系统:短期课程(英文)	2021—08	68.00	1382
几何路径:理论与实践(英文)	2021—08	48.00	1385
论天体力学中某些问题的不可积性(英文)	2021—07	88.00	1396
广义斐波那契数列及其性质(英文)	2021—08	38.00	1386
对称函数和麦克唐纳多项式:余代数结构与 Kawanaka 恒等式(英文)	2021—09	38.00	1400
杰弗里·英格拉姆·泰勒科学论文集:第1卷.固体力学(英文)	2021—05	78.00	1360
杰弗里·英格拉姆·泰勒科学论文集:第2卷.气象学、海洋学和湍流(英文)	2021—05	68.00	1361
杰弗里·英格拉姆·泰勒科学论文集:第3卷.空气动力学以及落弹数和爆炸的力学(英文)	2021—05	68.00	1362
杰弗里·英格拉姆·泰勒科学论文集:第4卷.有关流体力学(英文)	2021—05	58.00	1363

刘培杰数学工作室
已出版(即将出版)图书目录——原版影印

书　　名	出版时间	定　价	编号
非局域泛函演化方程:积分与分数阶(英文)	2021—08	48.00	1390
理论工作者的高等微分几何:纤维丛、射流流形和拉格朗日理论(英文)	2021—08	68.00	1391
半线性退化椭圆微分方程:局部定理与整体定理(英文)	2021—07	48.00	1392
非交换几何、规范理论和重整化:一般简介与非交换量子场论的重整化(英文)	2021—09	78.00	1406
数论论文集:拉普拉斯变换和带有数论系数的幂级数(俄文)	2021—09	48.00	1407
挠理论专题:相对极大值,单射与扩充模(英文)	2021—09	88.00	1410
强正则图与欧几里得若尔当代数:非通常关系中的启示(英文)	2021—10	48.00	1411
拉格朗日几何和哈密顿几何:力学的应用(英文)	2021—10	48.00	1412
时滞微分方程与差分方程的振动理论:二阶与三阶(英文)	2021—10	98.00	1417
卷积结构与几何函数理论:用以研究特定几何函数理论方向的分数阶微积分算子与卷积结构(英文)	2021—10	48.00	1418
经典数学物理的历史发展(英文)	2021—10	78.00	1419
扩展线性丢番图问题(英文)	2021—10	38.00	1420
一类混沌动力系统的分歧分析与控制:分歧分析与控制(英文)	2021—11	38.00	1421
伽利略空间和伪伽利略空间中一些特殊曲线的几何性质(英文)	2022—01	68.00	1422
一阶偏微分方程:哈密尔顿—雅可比理论(英文)	2021—11	48.00	1424
各向异性黎曼多面体的反问题:分段光滑的各向异性黎曼多面体反边界谱问题:唯一性(英文)	2021—11	38.00	1425
项目反应理论手册.第一卷,模型(英文)	2021—11	138.00	1431
项目反应理论手册.第二卷,统计工具(英文)	2021—11	118.00	1432
项目反应理论手册.第三卷,应用(英文)	2021—11	138.00	1433
二次无理数:经典数论入门(英文)	2022—05	138.00	1434
数,形与对称性:数论,几何和群论导论(英文)	2022—05	128.00	1435
有限域手册(英文)	2021—11	178.00	1436
计算数论(英文)	2021—11	148.00	1437
拟群与其表示简介(英文)	2021—11	88.00	1438
数论与密码学导论:第二版(英文)	2022—01	148.00	1423

刘培杰数学工作室

已出版(即将出版)图书目录——原版影印

书　　名	出版时间	定　价	编号
几何分析中的柯西变换与黎兹变换:解析调和容量和李普希兹调和容量、变化和振荡以及一致可求长性(英文)	2021－12	38.00	1465
近似不动点定理及其应用(英文)	2022－05	28.00	1466
局部域的相关内容解析:对局部域的扩展及其伽罗瓦群的研究(英文)	2022－01	38.00	1467
反问题的二进制恢复方法(英文)	2022－03	28.00	1468
对几何函数中某些类的各个方面的研究:复变量理论(英文)	2022－01	38.00	1469
覆盖、对应和非交换几何(英文)	2022－01	28.00	1470
最优控制理论中的随机线性调节器问题:随机最优线性调节器问题(英文)	2022－01	38.00	1473
正交分解法:涡流流体动力学应用的正交分解法(英文)	2022－01	38.00	1475
芬斯勒几何的某些问题(英文)	2022－03	38.00	1476
受限三体问题(英文)	2022－05	38.00	1477
利用马利亚万微积分进行 Greeks 的计算:连续过程、跳跃过程中的马利亚万微积分和金融领域中的 Greeks(英文)	2022－05	48.00	1478
经典分析和泛函分析的应用:分析学的应用(英文)	2022－03	38.00	1479
特殊芬斯勒空间的探究(英文)	2022－03	48.00	1480
某些图形的施泰纳距离的细谷多项式:细谷多项式与图的维纳指数(英文)	2022－05	38.00	1481
图论问题的遗传算法:在新鲜与模糊的环境中(英文)	2022－05	48.00	1482
多项式映射的渐近簇(英文)	2022－05	38.00	1483
一维系统中的混沌:符号动力学,映射序列,一致收敛和沙可夫斯基定理(英文)	2022－05	38.00	1509
多维边界层流动与传热分析:粘性流体流动的数学建模与分析(英文)	2022－05	38.00	1510
演绎理论物理学的原理:一种基于量子力学波函数的逐次置信估计的一般理论的提议(英文)	2022－05	38.00	1511
R^2 和 R^3 中的仿射弹性曲线:概念和方法(英文)	2022－08	38.00	1512
算术数列中除数函数的分布:基本内容、调查、方法、第二矩、新结果(英文)	2022－05	28.00	1513
抛物型狄拉克算子和薛定谔方程:不定常薛定谔方程的抛物型狄拉克算子及其应用(英文)	2022－07	28.00	1514
黎曼-希尔伯特问题与量子场论:可积重正化、戴森-施温格方程(英文)	2022－08	38.00	1515
代数结构和几何结构的形变理论(英文)	2022－08	48.00	1516
概率结构和模糊结构上的不动点:概率结构和直觉模糊度量空间的不动点定理(英文)	2022－08	38.00	1517

刘培杰数学工作室
已出版(即将出版)图书目录——原版影印

书　名	出版时间	定　价	编号
反若尔当对:简单反若尔当对的自同构(英文)	2022-07	28.00	1533
对某些黎曼-芬斯勒空间变换的研究:芬斯勒几何中的某些变换(英文)	2022-07	38.00	1534
内诣零流形映射的尼尔森数的阿诺索夫关系(英文)	即将出版		1535
与广义积分变换有关的分数次演算:对分数次演算的研究(英文)	即将出版		1536
强子的芬斯勒几何和吕拉几何(宇宙学方面):强子结构的芬斯勒几何和吕拉几何(拓扑缺陷)(英文)	2022-08	38.00	1537
一种基于混沌的非线性最优化问题:作业调度问题(英文)	即将出版		1538
广义概率论发展前景:关于趣味数学与置信函数实际应用的一些原创观点(英文)	即将出版		1539
纽结与物理学:第二版(英文)	2022-09	118.00	1547
正交多项式和q-级数的前沿(英文)	2022-09	98.00	1548
算子理论问题集(英文)	2022-09	108.00	1549
抽象代数:群、环与域的应用导论:第二版(英文)	即将出版		1550
菲尔兹奖得主演讲集:第三版(英文)	即将出版		1551
多元实函数教程(英文)	2022-09	118.00	1552
球面空间形式群的几何学:第二版(英文)	2022-09	98.00	1566

联系地址:哈尔滨市南岗区复华四道街 10 号　哈尔滨工业大学出版社刘培杰数学工作室
网　　址:http://lpj.hit.edu.cn/
邮　　编:150006
联系电话:0451-86281378　　13904613167
E-mail:lpj1378@163.com